Biotechnology and Crop Improvement

The green revolution led to the development of improved varieties of crops, especially cereals, and since then classical or molecular breeding has resulted in the creation of economically valuable species. Thanks to recent developments in biotechnology, it has become possible to introduce genes from different sources, such as bacteria, fungi, viruses, mice and humans, to plants. This technology has made the scientific community aware of the critical role of transgenic, not only as a means of producing stress tolerant crops but also as a platform for the production of therapeutics through molecular farming. *Biotechnology and Crop Improvement: Tissue Culture and Transgenic Approaches* focuses on important field crops to highlight germplasm enhancement for developing resistance to newly emerging diseases, pests, nutrient- and water-use efficiency, root traits and improved tolerance to increasing temperature and introduces significant recent achievements in crop improvement using methods such as micropropagation, somaclonal variation, somatic embryogenesis, anther/pollen/embryo culture, and compressing the breeding cycle for accelerated breeding and early release of crop varieties.

Plant biotechnology has now become an integral part of tissue culture research. The tremendous impact generated by genetic engineering and consequently of transgenic now allows us to manipulate plant genomes at will. There has indeed been a rapid development in this area with major successes in both developed and developing countries. Development of transgenic crop plants, their utilization for improved agriculture, health, ecology and environment and their socio-political impacts are currently important fields in education, research, and industry and also of interest to policy makers, social activists and regulatory and funding agencies. This work prepared with a classroom approach on this multidisciplinary subject will fill an existing gap and meet the requirements of such a broad section of readers. It describes the recent biotechnological advancement and developments in plant tissue culture and transgenic. Plant tissue culture techniques such as such as micropropagation, regeneration, somaclonal variation, somatic embryogenesis, anther/pollen/embryo culture are discussed for genetic improvement of crop plants. Transgenic techniques are discussed for developing resistance to newly emerging diseases, pests, nutrient- and water-use efficiency, root traits, and improved tolerance to increasing temperature.

Key Features

- Shows the importance of plant tissue culture and transgenic technology on plant biology research and its application to agricultural production
- Provides insight into what may lie ahead in this rapidly expanding area of plant research and development
- Contains contributions from major leaders in the field of plant tissue culture and transgenic technology

This book is devoted to topics with references at both graduate and postgraduate levels. The book traces the roots of plant biotechnology from the basic sciences to current applications in the biological and agricultural sciences, industry, and medicine. The processes and methods used to genetically engineer plants for agricultural, environmental, and industrial purposes along with bioethical and biosafety issues of the technology are vividly described in the book.

Biotechnology and Crop Improvement

Tissue Culture and Transgenic Approaches

Edited by
Nitish Kumar
Central University of South Bihar, India

CRC Press
Taylor & Francis Group
Boca Raton London New York

CRC Press is an imprint of the
Taylor & Francis Group, an informa business

First edition published 2023
by CRC Press
6000 Broken Sound Parkway NW, Suite 300, Boca Raton, FL 33487-2742

and by CRC Press
2 Park Square, Milton Park, Abingdon, Oxon, OX14 4RN

CRC Press is an imprint of Taylor & Francis Group, LLC

Library of Congress Cataloging-in-Publication Data
Names: Kumar, Nitish, editor.
Title: Biotechnology and crop improvement : tissue culture and transgenic approaches / Nitish Kumar.
Description: First edition. | Boca Raton, FL : CRC Press, 2022. | Includes bibliographical references and index.
Identifiers: LCCN 2022006037 (print) | LCCN 2022006038 (ebook) | ISBN 9781032145594 (hardback) | ISBN 9781032145600 (paperback) | ISBN 9781003239932 (ebook)
Subjects: LCSH: Crop improvement. | Transgenic plants. | Plant genetics.
Classification: LCC SB123.57.B573 2022 (print) | LCC SB123.57 (ebook) | DDC 631.5/233—dc23/eng/20220218
LC record available at https://lccn.loc.gov/2022006037
LC ebook record available at https://lccn.loc.gov/2022006038

ISBN: 9781032145594 (hbk)
ISBN: 9781032145600 (pbk)
ISBN: 9781003239932 (ebk)

DOI: 10.1201/9781003239932

Typeset in Times
by Newgen Publishing UK

Contents

Preface

At present, biotechnology is characterized by the application of the discoveries made in biological science concerning the use of plants as suppliers of more and new useful products. The group of technologies that use biological matter or processes to generate new and useful products and processes are defined as biotechnology. Plant biotechnology is increasingly gaining importance because it is related to many facets of our lives, particularly in connection with global warming, alternative energy initiatives, food production, and medicine. This book, *Biotechnology and Crop Improvement: Tissue Culture and Transgenic Approaches*, is devoted to topics with relevance at both graduate and postgraduate levels. The book traces the roots of plant biotechnology from the basic sciences to current applications in the biological and agricultural sciences, industry, and medicine. The processes and methods used to genetically engineer plants for agricultural, environmental, and industrial purposes, along with bioethical and biosafety issues of the technology, are vividly described. It is also an ideal reference for teachers and researchers, filling the gap between fundamental and high-level approaches. Each chapter has been written by one or more eminent scientists in the field and then carefully edited to ensure thoroughness and consistency. The book will be valuable as a reference for undergraduate and postgraduate students and can also be used as a reference for plant biologists, biochemists, molecular biologists, plant breeders, and geneticists in academia and industry.

Dr. Nitish Kumar
Gaya, Bihar, India

Acknowledgments

Thanks to all the authors of the various chapters for their contributions. It was quite a long process, from the initial outlines to developing the full chapters and then revising them in the light of reviewers' comments. We sincerely acknowledge the authors' willingness to go through this process. I also acknowledge the work and knowledge of the members of our review panels, many of whom were called on at short notice. Thanks to all the people at CRC Press, India, especially Ms. Renu Upadhyay, Ms. Jyotsna Jangra and Ms. Mansi Kabra, with whom we corresponded for their advice and facilitation in the production of this book. I am grateful to my family members, Mrs. Kiran (my wife), Miss Kartika Sharma and Laavanya Sharma (my daughters), and my parents, for their unconditional, selfless and loving support all the time.

Dr. Nitish Kumar
Gaya, Bihar, India

Editor's Biography

Dr. Nitish Kumar is Senior Assistant Professor at the Department of Biotechnology, Central University of South Bihar, Gaya, Bihar, India. Dr. Kumar completed his doctoral research at the Council of Scientific & Industrial Research–Central Salt & Marine Chemicals Research Institute, Bhavnagar, Gujarat, India. He has published more than 60 research articles and book chapters in leading international and national journals and books. He has a wide area of research experience in the field of genetic improvement of crop plants and has received many awards/fellowships/projects from various organizations, including the CSIR, DBT, ICAR and SERB-DST, BRNS-BARC, among others. He is an active reviewer for journals, including *Biotechnology Reports*, *Aquatic Botany*, *Industrial Crops and Products*, *PLoS One*, *Plant Biochemistry and Biotechnology*, and *3Biotech*, to name a few. He also serves as an associate editor of the journal *Gene*.

Contributors

A.C. Anugraha
Department of Botany
St. Thomas College Palai
Kerala, India

Nandeibam Apana
Department of Biotechnology
Manipur University
Canchipur
Manipur, India

Abubakker Aslam
Plant Molecular Biology Laboratory
Department of Botany
Jamal Mohamed College
Tiruchirappalli
Tamil Nadu, India

A.V. Deepa
Department of Plant Science
Central University of Kerala
Kerala, India

Pallavi Billowria
Department of Biotechnology
University of Jammu
Jammu (J&K), India

Jasdeep Chatrath Padaria
National Institute for Plant Biotechnology
Pusa Campus
New Delhi, India

Priyanka Chaudhary
Department of Botany
DPG Degree College
Gurugram, India

Abhijit Dey
Department of Life Sciences
Presidency University
Kolkata, India

Kunnampalli Faizal
Plant Molecular Biology Laboratory
Department of Botany
Jamal Mohamed College
Tiruchirappalli
Tamil Nadu, India

Gourab Ghosh
National Institute for Plant Biotechnology
Pusa Campus
New Delhi, India

Ambika Gupta
Department of Biotechnology
Central University of South Bihar
Gaya
Bihar, India

Imlitoshi Jamir
Department of Biotechnology
Nagaland University
Dimapur, India

Nisha Kapoor
Department of Biotechnology
University of Jammu
Jammu (J&K), India

Nitish Kumar
Department of Biotechnology
Central University of South Bihar
Gaya
Bihar, India

Shweta Kumari
P. G Department of Botany
Patna University
Patna
Bihar, India

Varsha Kumari
Department of Plant Breeding and Genetics
SKN College of Agriculture
Sri Karan Narendra Agriculture University
Jobner-Jaipur
Rajasthan, India

Jitendra Kumar Sharma
Centre for Biotechnology
Maharshi Dayanand University
Rohtak
Haryana, India

Ritu Mahajan
Department of Biotechnology
University of Jammu
Jammu (J&K), India

S.R. Maloo
Pacific College of Agriculture
Pacific University
Udaipur
Rajasthan, India

Kizhakke Modongal Shamsudheen
Plant Molecular Biology Laboratory
Department of Botany
Jamal Mohamed College
Tiruchirappalli
Tamil Nadu, India

Valiyaparambath Musfir Mehaboob
Department of Botany
MES Ponnani College
Ponnani
Kerala, India

Potshangbam Nongdam
Department of Biotechnology
Manipur University
Canchipur
Manipur, India

Nater Pal Singh
Centre for Biotechnology
Maharshi Dayanand University
Rohtak
Haryana, India

Maheshwar Prasad Trivedi
P. G Department of Botany
Patna University
Patna
Bihar, India

Palusamy Raja
Plant Molecular Biology Laboratory
Department of Botany
Jamal Mohamed College
Tiruchirappalli
Tamil Nadu, India

Anita Rani Santal
Department of Microbiology
Maharshi Dayanand University
Rohtak
Haryana, India

Tania Sagar
Department of Biotechnology
University of Jammu
Jammu (J&K), India

Amar A. Sakure
Department of Agricultural Biotechnology
Anand Agricultural University
Anand
Gujarat, India

Appakan Shajahan
Plant Molecular Biology Laboratory
Department of Botany
Jamal Mohamed College
Tiruchirappalli
Tamil Nadu, India

Radheshyam Sharma
Biotechnology Centre
Jawaharlal Nehru Krishi Vishwa Vidhalya
Jabalpur
Madhya Pradesh, India

Shivika Sharma
Sardar Swaran Singh National Institute of Bio-Energy
Kapurthala Punjab, India

Vikas Sharma
Molecular Biology & Genetic Engineering
School of Bioengineering & Biosciences
Lovely Professional University
Phagwara-Jalandhar, India

Manish Shukla
Amity University Uttar Pradesh
Lucknow Campus
Lucknow, India

Vikram Singh Gaur
College of Agriculture
Waraseoni
Jawaharlal Nehru Krishi Vishwa Vidhalya
Jabalpur
Madhya Pradesh, India

Sumer Singh Punia
Department of Plant Breeding and Genetics
SKN College of Agriculture
Sri Karan Narendra Agriculture University
Jobner-Jaipur
Rajasthan, India

Paonam Sonia
Department of Biotechnology
Manipur University
Canchipur
Manipur, India

Ganesan Thiagu
Plant Molecular Biology Laboratory
Department of Botany
Jamal Mohamed College
Tiruchirappalli
Tamil Nadu, India

Toji Thomas
Department of Botany
St. Thomas College Palai
Kerala, India

Dennis T. Thomas
Department of Plant Science
Central University of Kerala
Kerala, India

Leimapokpam Tikendra
Department of Biotechnology
Manipur University
Canchipur
Manipur, India

Rajesh K. Tiwari
Amity University Uttar Pradesh
Lucknow Campus
Lucknow, India

Mala Trivedi
Amity University Uttar Pradesh
Lucknow Campus
Lucknow, India

Sharanabasappa B. Yeri
Plant Biotechnology
Zonal Agricultural Research Station
Kalaburgi
University of Agricultural Sciences
Raichur
Karnataka, India

1

Transgenic Technology in Crop Improvement

A.C Anugraha
St. Thomas College Palai
Kerala, India

Toji Thomas
St. Thomas College Palai
Kerala, India

Dennis T. Thomas
Central University of Kerala
Kerala, India

CONTENTS

DOI: 10.1201/9781003239932-1

1.1 Introduction

At the dawn of the 19th century, the world population reached a record level of one billion (Van Bavel, 2013). The trend has exponentially increased ever since, reaching almost 10 billion people by 2015, and is expected to surpass 10 billion by 2050 (Low et al. 2018). Such a rapid growth in population size can indeed cause poverty, famine and malnutrition (Van Bavel, 2013). Hence, in order to cope with the nutritional requirement of a growing population, it is imperative to enhance the yield and nutritional value of food crops (Passricha et al. 2020). In spite of the massive population growth, farming and food production have declined considerably due to unavailability of agricultural land, diseases of food crops and various biotic and abiotic stresses. According to Mesterházy et al. (2020), abiotic stresses, pests, weeds and diseases accounted for an annual pre-harvest loss of 1051.5 million tons of total grain production. Most of the traditional crop varieties are unable to withstand extreme climate changes and pathogen attack as compared with transgenic crops (Fang et al. 2016). Another important aspect associated with the population boom was the substantial loss of arable land and water resources for agriculture. Conventional seed stocks that lack efficient nutrient and water utilisation characteristics fail to withstand the conditions in less fertile areas, which often results in lower productivity (Morison et al. 2008).

In the past, classical breeding techniques were practised to develop desirable traits in crop plants. The technique simply relied on normal sexual recombination of selected parental genes for potentially useful novel combinations in subsequent generations (Manshardt, 2004). However, conventional breeding methods failed to meet the food production demand in terms of quality and quantity, and also, the techniques seem to be restricted to individuals belonging to the same species or closely related species (Low et al. 2018). Moreover, the normal sexual reproduction process allows segregation and recombination of all the traits between selected parental strains, which often results in unpredicted genetic mutations in offspring. In order to overcome this, several backcrosses for introgression of a desirable trait are required in conventional breeding practices, which are time consuming and can compromise the quality of the product (Tomar et al. 2019). With the advent of modern biotechnological approaches, conventional breeding practices were practically replaced, and a new era of agriculture has begun.

Plant genetic engineering has made a breakthrough in the agriculture sector (Basso et al. 2020). Advancements in plant molecular biology and transgenic technology enabled modification of gene function by insertion or deletion of specific gene sequences (Rani and Usha, 2013). The basic aspects of genetic engineering were borrowed from the studies of Armin Braun, a plant pathologist whose studies were centred on the tumour-inducing capability of *Agrobacterium tumefaciens* upon infecting plants (Grunewald et al. 2013). Later on, it was revealed that *A. tumefaciens* can deliver its DNA segment into host cell and can be integrated into the plant genome via the plasmid integration system, which explains the mechanism of natural DNA transfer (Somssich, 2019). An

TABLE 1.1
List of Selected Crops Developed by Different Transgenic Approaches

Crop	Transformation method	Improved characteristics	Reference
Apple	ZFNs	Stable transgenic expression	Peer et al. (2015)
Banana	Particle-bombardment method	Resistance to virus	Ismail et al. (2011)
Cabbage	Particle-bombardment method	Plastid transformation	Tseng et al. (2014)
Guava	Agrobacterium-mediated transformation	Disease resistance	Mishra et al. (2014)

antibiotic-resistant tobacco herb was introduced in 1982 as the foremost genetically modified (GM) crop plant (Fraley et al. 1983). In subsequent years, several GM crops were developed and by the mid-1990s, biotech crops had become more popularized. The Calgene company marketed the first genetically engineered food, the Flavr-Savr tomato, for human consumption in 1994 (Dunwell, 2000). In the same year, the European Union approved a bromoxynil-resistant tobacco plant, and this was marketed in Europe. In 1995, the US Environmental Protection Agency sanctioned the Bt potato as the first pesticide-producing crop in the US. Following the trend, several other GM crops, like canola with improved oil composition, virus-resistant squash, Bt cotton, Bt brinjal, Bt maize, etc., were licensed for commercialization (Brookes and Barfoot, 2014). According to the International Service for the Acquisition of Agri-biotech Applications (ISAAA), in 2018, 191 million hectares of GM crops were grown by 26 countries, among which the US, Canada, Australia, Spain and Portugal contribute 46% of total production (ISAAA, 2018, 2020; Turnbull et al. 2021).

Like other modern technologies, transgenic technology also imposes safety concerns (Bawa and Anilakumar, 2013). The presence of foreign genes in transgenic crops has limited their public acceptance due to potential allergic reactions or toxicity (Kumar et al. 2020). Controversies over GM crops have often resulted in trade disputes, international protest and litigation (Sheldon, 2002). In fact, each and every GM crop available on the market has passed safety assessments and poses no harmful effect to human health or the environment (Giraldo et al. 2019). Moreover, novel transgenic techniques like genome editing, cisgenesis and intragenesis have been developed as alternative techniques to modify the crop genome without transgenes, thereby eliminating the associated uncertainties (Kumar et al. 2020). Table 1.1 lists some of the selected GM crops developed.

The present chapter deals with the recent developments in plant transgenic techniques in crop improvement by highlighting some commercially successful transgenic crops. We also address the current status of GM crops and potential biosafety concerns regarding the use of transgenic crops. Their future prospects and current challenges are also discussed in this chapter.

1.2 Brief Account of Plant Transformation Methods

The introduction of a foreign gene into a host plant cell, followed by its subsequent integration into the plant genome, is generally regarded as plant transformation (Keshavareddy et al. 2018). Such genetic modification in agronomic traits has resulted in strategic improvements in the agriculture sector, in terms of both quality and quantity. In general, plant transformation involves two stages: the delivery of a desired gene into a host cell, and regeneration of the plant cell/explant into a whole plant. Consequently, the success of plant transformation is determined by the selection of a healthy and viable explant and the adoption of a suitable tissue culture technique. Common methods employed in plant transformation studies are *Agrobacterium*-mediated transformation and particle bombardment. However, because of the urge for increased transformation efficiency, existing methods have been modified and novel methods developed for a successful transformation.

1.2.1 *Agrobacterium*-mediated Transformation

Agrobacterium tumefaciens causes crown gall disease in several plants. *Agrobacterium* has an exceptional ability to transfer part of its tumour-inducing (Ti) plasmid to the plant cell upon infection. The T-DNA region of the Ti plasmid holds genes for phytohormone synthesis and is transferred to the plant nuclear DNA, leading to random integration of T-DNA into genomic DNA (Gelvin, 2010). The manifestation of oncogenes in the T-DNA region triggers the synthesis of phytohormones, ultimately leading to crown gall formation. This unique ability of *Agrobacterium* made it an excellent tool for plant transformation. It is possible to transfer any DNA sequence positioned between T-DNA borders into plant cells, and this can be incorporated into the host plant genome (Gelvin, 2012). The left and right borders of T-DNA contain a border sequence of 25 base pair repeat sequences vital for T-DNA transfer. A Vir region composed of a group of virulence genes such as Vir A, B, C, D, E, F and Vir G encodes vir proteins that mediate T-DNA processing and transfer (Passricha et al. 2020). The T-DNA can be engineered to substitute tumour-causing genes with a particular gene of interest along with promoter and transcription termination sequences. $GV3_{101}$, $C_{58}C1$, AGL1 and EHA_{105} are a few *Agrobacterium tumefaciens* strains with varied degrees of virulence that can be used for efficient plant transformation (Basso et al. 2020). Monocotyledonous plants and recalcitrant dicot plants can be efficiently transformed using hypervirulent EHA_{105}, AGL_1 and LBA_{404} strains of *A. tumefaciens*.

Agrobacterium-mediated wheat transformation was generally considered to be inefficient and challenging, with average transformation efficiencies of approximately 5% (Risacher et al. 2009). Hence, for a considerable period of time, the biolistic method remained a choice for wheat transformation. Hayta et al. (2019) reported a repeatable protocol for *Agrobacterium*-mediated wheat transformations with 25% transformation efficiency. Rashid et al. (2014) developed drought-resistant transgenic tobacco via *Agrobacterium*-mediated transformation. Wheat DREB2 gene cloned into pCAMBIA1304 under CaMV 35S promoter was transferred into the LBA4404 strain of *Agrobacterium tumefaciens*. The transformed *Agrobacterium* colonies were used to develop transgenic tobacco via leaflet transformation.

1.2.2 Protoplast Transformation

Protoplast-mediated plant transformation usually leads to transient or temporary expression of genes. The technique involves direct uptake of DNA by the protoplast via electroporation or using polyethylene glycol. High transformation frequency and non-necessity of binary vectors remain the major advantages of protoplast transformation. Transformed protoplast occasionally results in incorporation of useful agronomic traits in regenerated plants (Baltes et al. 2017). Protoplast transformation has now been successfully applied in several crop varieties, including rice, wheat and potato (Cocking, 1993). Nanjareddy et al. (2016) developed protoplast isolation and transient transformation protocols for gene functional analysis in *Phaseolus vulgaris*.

1.2.3 Biolistic Transformation

Biolistic-mediated DNA transfer was developed especially for plants that produce recalcitrant seeds as an alternative to protoplast-mediated transformation. The method allows direct penetration of transgenes via particle bombardment or the gene gun method (Tomar et al. 2019). The desired DNA construct is mixed with microcarriers like gold or tungsten particles having a size range 0.6–1 μm in diameter followed by high-velocity bombardment against plant cells (Gan, 1989). Even though the method is suitable for transformations of a wide range of tissues, including embryos, meristems, pollen, etc., it is quite expensive. Furthermore, optimum DNA concentration per shot is of utmost importance, as integration of multiple copies can reduce stability of transformation. Cry10Aa protein of *Bacillus thuringiensis* was identified as toxic to cotton boll weevil (Aguiar et al. 2012), which is a major devastating pest of cotton plants. Ribeiro et al. (2017) successfully employed the biolistic method to develop GM Bt cotton. Particle bombardment of cotton embryos with a Cry10Aa expression cassette led to the expression of Cry10Aa protein under the control of uceA1.7 promoter in transformed cotton plants.

1.2.4 Chloroplast Engineering

Chloroplast engineering has been considered as a valuable biotechnological tool in developing biopharmaceuticals; resistance to insects, herbicides, pests and diseases; and drought and salt tolerance (Jana, 2010). Chloroplast transformation is regarded as eco-friendly and safe because maternal inheritance of chloroplast DNA restricts it from being transferred to other plant species (Daniell et al. 2002). Polyploidy of the plastid genome facilitates the insertion of multiple copies of transgene per cell, allowing increased level of protein accumulation (Grevich and Daniell, 2005). The plastid genome of tobacco leaves engineered using unmodified Cry1A(c) coding sequence of *Bacillus thuringiensis* resulted in the accumulation of insecticidal cry protein (McBride et al. 1995). Glyphosate resistance is achieved in *Nicotiana tabacum* using plastid transformation with particle bombardment of mutated EPSP(5-enoylpyruvyl shikimate-3-phosphate) synthase gene (Roudsari et al. 2009). Although chloroplast engineering seems to be promising for successful incorporation of agronomic traits, still, there are challenges to be resolved. The development of a shoot regeneration protocol for recalcitrant cereal plants, phenotypic alterations in transplastomic plants due to massive accumulation of foreign proteins, and optimization of the transgene delivery protocol impose difficulties in chloroplast technology (Ahmad et al. 2012; Adem et al. 2017). In spite of these hurdles, chloroplast engineering has been successfully applied in *Arabidopsis*, cabbage, cotton, carrot, petunia, soybean, sugar cane, sugar beet, cauliflower, potato, tomato, poplar etc. (Ahmadabadi et al. 2007).

1.2.5 Microinjection

Microinjection is a host -independent, vector-less, direct physical approach to deliver DNA into target cell. The DNA construct is injected into the cytoplasm or nucleus using glass micropipettes or metal microinjection needles. The technique has been widely used to transform animal cells (Rakoczy-Trojanowska, 2002). Even though microinjection technology is not routinely used for plant transformation studies, it allows introduction of DNA plasmids as well as whole chromosomes into target cells.

1.2.6 Electroporation

Electroporation is a physical method of gene transfer in which plant cells are subjected to high-voltage electric pulses to make them permeable to DNA uptake, leading to transient gene expression as well as stable transformation (Bates, 1999). Transgene expression in electroporated plant protoplasts was first described by Fromm et al. (1985) in carrot, tobacco and maize protoplasts. The exogenous DNA enters through transient pores in the plasma membrane formed in response to heat shock, leading to transformed protoplast and the development of transgenic calli (Joersbo and Brunstedt, 1996).

1.2.7 Chemical Methods

1.2.7.1 Calcium Phosphate Co-precipitation

As DNA is mixed with calcium chloride solution and isotonic phosphate buffer, a DNA–calcium phosphate precipitate is formed. Actively dividing cells are then incubated with DNA–calcium phosphate precipitate for several hours. Concentration of DNA and stability of DNA–calcium phosphate precipitate affect transformation frequency (Passricha et al. 2020). Calcium phosphate nanoparticles (20–50 nm in diameter) have been developed as an efficient carrier of foreign genes into plant cells (González-Melendi et al. 2008). Naqvi et al. (2012) encapsulated transgenic pCambia1301 in calcium phosphate nanoparticles (CaP) to transform hypocotyl of *Brassica juncea*. Incubation of the hypocotyl explant with CaP nanoparticles has shown a transformation efficiency of 80.7%, suggesting the use of CaP nanoparticles to deliver scientifically and economically important traits into crop plants.

1.2.7.2 DEAE-dextran-mediated Transfer

Diethylaminoethyl (DEAE)-dextran-mediated DNA transfer was first reported by Vaheri and Pogano in 1965 (Lalani and Misra, 2011). Electrostatic interaction between DEAE-dextran and DNA forms a DEAE-dextran–DNA complex, which can bind to the plasma membrane of cells, and internalization can occur via endocytosis. However, this method can only be used for transient transfection and often causes cytotoxicity at higher concentrations (Passricha et al. 2020).

1.2.7.3 PEG-mediated DNA Delivery

PEG (polyethylene glycol)-mediated DNA delivery is a widely employed direct method of gene transfer suitable for different plant species without host specificity (Mathur and Koncz, 1998). Rasmussen and Rasmussen (1993) performed PEG-mediated DNA delivery into the protoplast of carrot, rapeseed and soybean. Transient gene expression was monitored under CaMV35S-GUS reporter gene. GUS (β-glucuronidase) activity was expressed in protoplast of rapeseed immediately after 1.5 hours of DNA uptake. Optimum GUS activity in transfected carrot and soybean protoplast was detected after 66–72 hours. The success of PEG-mediated DNA uptake relies on PEG concentration, order of PEG and DNA application, and concentration of protoplast (Rasmussen and Rasmussen, 1993). PEG-mediated protoplast transformation of rice plant produced transgenic Indica rice IR43. Hygromycinphosphotransferase (*hpt*) gene under the control of CaMV 35S promotor was introduced into the protoplast of rice plant in the presence of PEG. Protoplast cultured in maltose-containing medium produced transgenic plantlets (Biswas et al. 1994).

1.3 Modern Transgenic Approaches for Crop Improvement

Traditional transgenic approaches are found to be successful in the production of transgenic crop plants. Efforts have been made to modify existing methods, and novel techniques are being introduced over time (Basso et al. 2020). The modern era of biotechnology offers exciting tools suitable for *in vivo* modifications of genomes of plant cells, thereby eliminating the presence of transgenes in GM crops. Genome editing tools offer site-specific mutations, insertions and deletions at the target locus with high precision. RNA interference, cisgenesis, intragenesis, fine tuning of microRNA and genome editing strategies have now been successfully applied in crop plants to improve agronomic traits.

1.3.1 Cisgenesis and Intragenesis

In spite of the strategic progress in the agronomic field made by GM crops, their public acceptance is still a major concern. The presence of selectable markers and transgenes in engineered crop plants has greatly affected their reception by consumers (Purchase, 2005). To address safety concerns, alternative technologies called cisgenesis and intragenesis were introduced (Espinoza et al. 2013). Cisgenesis involves the production of GM crops with a natural copy of a gene having regulatory elements (cisgene) obtained from the same species or from a sexually compatible species (Schouten et al. 2006). A cisgenic plant contains only desired genetic elements from a source plant that can be crossed. Like cisgenic plants, introgenic plants also receive the gene from the same species or one that can be crossed, but are hybrid in nature, i.e., parts of different genes are recombined into a single construct. In most cases, the promoter is derived from one gene and the coding sequence is obtained from a different gene so as to construct new genetic combinations that give innovative properties. Several economically important crop plants, such as potato (de Vetten et al. 2003), alfalfa (Weeks et al. 2008), durum wheat (Gadaleta et al. 2008), strawberry (Schaart et al. 2004) and apple (Joshi et al. 2011), use intragenesis/cisgenesis technology.

1.3.2 RNA Interference

RNAi (RNA interference) was a breakthrough discovery in molecular biology that involved sequence-specific gene silencing at the post-transcriptional level (Yogindran and Rajam, 2015). Andrew Fire and Craig C Mello discovered RNAi in *Caenorhabditis elegans* and won the Nobel Prize in Physiology or Medicine in 2006. The process of RNAi is initiated by the presence of either endogenous or exogenous double-stranded RNA (dsRNA) molecules in the cytoplasm. The dsRNA triggers DICER, a ribonuclease protein, which cleaves dsRNA into short siRNAs (small interfering RNAs), which are then incorporated into RISC (RNA-Induced Silencing Complex). siRNAs unwind to form a sense and an antisense strand, of which the sense strand is discarded by RNA helicase activity. It is the retained antisense siRNA that pairs with complementary mRNA, which is then cleaved by ARGONAUTE 2 of RISC, inducing gene suppression (Kusaba, 2004; Vaucheret, 2008).

Research has shown the potential use of RNAi in crop improvement programmes to improve stress tolerance and modified agronomic traits by gene silencing (Kupferschmidt, 2013; Yogindran and Rajam, 2015). Transgenic banana lines with increased resistance against *Fusarium oxysporum* f. sp. *cubens* were developed using intron hairpin RNA (ihpRNA)-mediated gene silencing. ihpRNAs were prepared using a partial sequence of targeted fungal genes, velvet and fusarium transcription factor1, and introduced into embryonic banana cell suspension via *Agrobacterium*-mediated transformation. Transgenic banana lines derived from ihpRNA-VEL and ihpRNA-FTF1 exhibited enhanced resistance (Ghag et al. 2014). RNAi offers a promising way to alter the nutritional profile of agronomic crops. For instance, the RNAi approach can be used to improve the beta carotene content of potato tubers by silencing the beta carotene hydroxylase (bch) gene that converts beta carotene to zeaxanthin. Two RNAi constructs, one with tuber-specific granule bound starch synthase (GBSS) promoter and other with a strong and constitutive cauliflower mosaic virus 35 S (CaMV 35 S) promoter, were made to silence the beta carotene hydroxylase gene. *Agrobacterium*-mediated transfer of RNAi constructs into three *Solanum tuberosum* lines, Desiree, Yema de Huevo and a breeding line 91E22, produced transformants with altered carotenoid content. Most of the bch (beta-carotene hydroxylase gene)-silenced lines exhibited increased beta carotene and lutein content in the tubers. Beta carotene content was found to be higher in GBSS-derived transformants as compared with CaMV 35S transformants (Eck et al. 2007). Jørgensen et al. (2005) successfully employed RNAi technology to deplete the cyanogenic glycosides, such as linamarin and lotaustralin, in cassava tubers and leaves. Transgenic cassava plants blocked the expression of CYP79D1 and CYP79D2 genes, which encode enzymes responsible for the first committed step in linamarin and lotaustralin biosynthesis. Transgenic cassava plants developed through RNAi technology exhibited 99% and 92% reduction in cyanide potential in leaves and tubers, respectively (Jørgensen et al. 2005). RNAi is emerging as a powerful technology to develop novel crops incorporating desirable traits, such as decaffeinated coffee, Arctic apples without enzymatic browning, nicotine-devoid tobacco and hypoallergenic crops (Saurabh et al. 2014; Gavilano et al. 2006).

1.3.3 Fine Tuning of miRNAs in Crop Improvement

The introduction of one or more desirable genes into crop plants through recombinant DNA technology develops transgenic plants with enhanced agronomic performance. However, among thousands of agronomically important genes identified, only a few sets of genes are found to be suitable for successful transformations, as the introduction of a beneficial trait is always associated with secondary issues hampering other beneficial traits (Tang and Chu, 2017).

The utilization of genetic modulators can effectively and precisely regulate agronomic traits in crop plants. Plant microRNAs are regarded as master modulators of gene expression and precisely regulate spatiotemporal accumulation of target mRNA via translation inhibition or sequence-specific cleavage (Borges and Martienssen, 2015). Plant miRNAs are single-stranded microRNAs transcribed from MIR (microRNA) genes. Transcription of MIR genes by RNA polymerase II gives 5′ capped and 3′ polyadenylated pri-miRNA, which is further managed by DICER like 1 enzyme to yield pre-miRNA. Pre-miRNA is converted to miRNA-miRNA duplexes by DCL1 enzyme. 3′ methylated duplex miRNAs are shuttled to the cytoplasm by HST protein and disassemble. Mature miRNA strands bind with Argonaute to develop functional RISC (Lelandais-Brière et al. 2010). miRNA specifically targets complementary mRNA, and gene silencing is accomplished by mRNA degradation by Ago 2 or by preventing mRNA translation (Lim et al. 2005).

Recent studies revealed that MIR genes could be finely tuned to improve agronomic characteristics in crop plants (Zhang, 2015; Teotia et al. 2016). The emergence of Bt-resistant insects stimulated alternative approaches to effective pest control. Agrawal et al. (2015) developed transgenic tobacco plants showing increased resistance towards *Helicoverpa armingera*. Artificial miRNA was designed to specifically target larval chitinase gene and cloned into pUC57 vector. *Agrobacterium*-mediated transformation of tobacco yielded transgenic tobacco expressing amiR-24 (artificial microRNA). *H. armingera* larvae feeding on transgenic tobacco leaves showed downregulation of chitinase gene and increased mortality rate (Agrawal et al. 2015).

1.3.4 Genome Editing

Genome editing is a novel technique in biotechnology, which allows manipulation of the target genome in a highly efficient and proper manner (Chen and Gao, 2013). Targeted genome editing relies on sequence-specific, engineered endonucleases, which can induce single- or double-stranded breaks in particular DNA sequences (Zhang et al. 2018). These breaks are then fixed either by non-homologous end joining, leading to knockouts, insertions or alterations, or by homology-directed repair (Nadakuduti et al. 2018; Tomar et al. 2019). Genome editing in agricultural crops provides several transgene-free varieties with improved stress tolerance, nutritional yield and productivity (Zhang et al. 2018). Potato, a polyploid heterozygous crop, failed to develop novel agronomic traits via conventional breeding. The availability of genomic sequence information and regeneration procedures facilitated successful genome editing in potato cultivars with enhanced traits like self-incompatibility, processing efficiency, modified starch quality and development of cold-induced sweetening-resistant potato cultivars (Nadakuduti et al. 2018). During the past years, several approaches have been developed to edit the genome of plants. Zinc finger nucleases (ZFNs), transcription activator like effector nucleases (TALENs) and the clustered regularly interspaced short palindromic repeats/cas9 (CRISPR/Cas9) nuclease system have offered targeted genome editing in a variety of plants (Shah et al. 2018).

1.3.4.1 ZFNs

Zinc fingers are small protein domains typically found as a part of transcription factors. The stability of the domain is mainly attributed to zinc; it can bind to nucleic acids, proteins, etc. (Krishna et al. 2003). ZFNs are artificially made hybrid nucleases generated by the joining of a zinc finger DNA attaching domain and a DNA-cleavage domain (Tomar et al. 2019). The DNA-cleavage domain in ZFNs is usually derived from type II restriction endonuclease Fok1 (Kim et al. 1996). The Fok1 cleavage domain must dimerize to cleave DNA, and hence, ZFN-mediated DNA cleavage also requires dimerization of the cleavage domain (Bitinaite et al. 1998). Due to the weak dimer interface, a pair of fingers is used to achieve cleavage. ZFN genome modification has been reported in *Arabidopsis*, apple, fig, rapeseed, rice, maize, soybean, nicotiana, corn and petunia (Martinez-Fortun et al. 2017). Conventional transgenesis and random mutagenesis brought about inefficient genetic modifications at target loci.

Shukla et al. (2009) employed ZFN-driven gene addition in maize. ZFNs were directed against IPK1 gene, encoding inositol-1,3,4,5,6-penta-kisphosphate-2-kinase, responsible for phytate biosynthesis in seeds. Phytate reduction can enhance the nutritional profile of maize grains. Four pairs of ZFNs targeted IPK1 at two positions in exon 2. Disruption of the gene by insertion of PAT (phosphinothricinacetyltransferase) gene cassettes resulted in phytate reduction and increased herbicide tolerance (Shukla et al. 2009).

1.3.4.2 TALENs

TALENs are also joining products consisting of a DNA-binding domain and DNA-cutting nuclease domains. TALENs are engineered nucleases having a DNA-cleavage domain fused to a DNA-binding domain. TALEs are proteins from which the DNA-binding domain is taken to engineer TALENs. TALEs are secreted by plant pathogens like *Xanthomonas* bacteria (Boch and Bonas, 2010). The DNA-binding domain of TALE is characterized by a highly conserved 33–34-amino-acid sequence repeat, in which the 12th and 13th positions show variation and directly influence DNA-binding specificity. Engineering of TALEs can facilitate their binding to the desired site in the target genome. Hybridization of TALE with Fok1 cleavage domain gives TALENs. As a genome editing tool, TALENs have been used to create economically important rice, barley, potato, soybean, maize, tomato, wheat, flax, sugarcane, etc. (Ran et al. 2017). Shan et al. (2015) exemplified the use of TALEN technology in the production of fragrant rice. Fragrance in rice is imparted by 2-acetyl-1-pyrroline (2AP) synthesized from defective badh2 allele. Shan et al. designed TALENs that specifically target the fourth exon of BADH2 gene. Sanger sequencing in transgenic plants showed additions and deletions at the target site. Considerable 2AP content was noted in homozygous mutants generated by targeted knockouts of a non-fragrant variety exploiting TALEN technology (Shan et al. 2015). Conversion of sucrose into glucose and fructose in potato tubers is controlled by vacuolarinvertase gene (Vinv). Build-up of reducing sugars during low-temperature storage affects the quality of potato tubers by elevating acrylamide content. Targeted knockout of Vinv using TALENs improved the storage properties of knockout plants, with an undetectable level of reducing sugar in tubers (Clasen et al. 2016). Even though TALEN-mediated crop improvement programmes have been successfully implemented in several crops, the selection of appropriate TALE DNA-binding domain repeats is challenging.

1.3.4.3 CRISPR/Cas9 System

CRISPRs, otherwise known as clustered regularly interspaced short palindromic repeats, are found in the genome of bacteria and archaea (Barrangou, 2015). CRISPR sequences and associated proteins (Cas) are involved in the defence mechanism of prokaryotes in the form of acquired immunity (Redman et al. 2016). The spacers between DNA repeats of CRISPRs are valuable sequences derived from the genome of previously attacked pathogens. Subsequent infection of the pathogen is immediately destroyed by the genetic memory of spacers (Barrangou et al. 2007). Cas9 is a CRISPR-associated protein that can specifically cleave a DNA sequence complementary to the CRISPR sequence. The spacers are expressed as guide CRISPR RNAs (crRNAs). crRNAs along with Cas proteins consequently provide adaptive immunity. CRISPR/Cas systems, particularly the CRISPR/Cas9 system, have emerged as a genome-editing tool in plants during the last decade (Osakabe et al. 2016). crRNA-guided nucleases induce double-stranded breaks at a target locus. DNA repair mechanisms in plants cause insertion or deletion of nucleotides during DNA repair. The CRISPR/Cas9 non-homologous end joining approach can be used to induce indel mutations in the coding sequence of a desired gene, leading to frameshift and knockdown. CRISPR/Cas9 homology-directed repair and homology and recombination-directed repair using engineered donor DNA facilitate the achievement of sequence-specific editing in a gene or promoter sequence (Sun et al. 2016). As compared with ZFN- or TALEN-mediated genome editing, the CRISPR/Cas system offers an efficient and low-cost technology to edit the plant genome with an option to target multiple genes (Cong et al. 2013). CRISPR/Cas9-mediated gene editing for improved crop yield, better stress management and disease resistance has been done in several crops, including apple, cotton, maize, camelina, cucumber, soybean, barley rice, potato, tomato, etc. (Ricroch et al. 2017).

Li et al. (2016) demonstrated the use of CRISPR/Cas9 gene editing to modify yield-related genes in rice cultivar Zhonghua II. Number of panicles, number of grains in panicle and weight of grain are key determinants of rice yield. Mutations were induced in Gnla gene, controlling grain number; DEPl, relating to panicle architecture; GS3, regulating grain size; and IPA1, regulating plant architecture. Mutations in these four genes resulted in Gnla, DEP1 and GS3 mutants showing an increase in the number of grains, dense panicles and increased grain size (Li et al. 2016). CRISPR/Cas9 nuclease-edited FAD2 gene in *Camelia sativa* exhibited an increase in oleic acid content from 16% to 50% and a simultaneous decrease in unpleasant linoleic acid from 16% to less than 4% and in linolenic acid to less than 10%. CRISPR/Cas9 gene editing has significantly enhanced the fatty acid profile in camelina seeds by targeting FAD2 genes (Jiang et al. 2017). In addition to the crop yield and nutritional profile, the CRISPR/Cas9 system offers effective stress management in crops (Jaganathan et al. 2018). Non-transgenic cucumber with virus resistance has been developed by Cas9/sgRNA technology targeting the recessive eIF4 gene. Transformed cucumber plants exhibited deletions and single nucleotide polymorphisms (SNPs) in targeted sites. Backcrossing for three generations resulted in homozygous mutants exhibiting increased resistance against ipomovirus and polyviruses (Chandrasekaran et al. 2016).

1.4 Application of Transgenic Techniques in Crop Improvement

The history of genetic manipulation in crop plants has been traced back to the onset of agriculture. Advancements in the transgenic technologies have sped up the process of genetic manipulation to incorporate novel traits such as high yield, stress tolerance and nutritional fortification (Figure 1.1). Major applications of transgenic technologies are exemplified in this section.

1.4.1 Enhanced Pest Resistance and Disease Control

Transgenic technology in the production of herbicide-resistant, insect-resistant or disease-resistant crops has gained much attention, as it can effectively control insect pests and diseases and decrease the use of chemicals. It helps in reducing the associated pollution effect and increases crop yield (Raney, 2006). Herbicide-resistant crops, especially glyphosate-resistant corn, soybean and cotton, have been developed to manage weeds (Green and Owen, 2010).

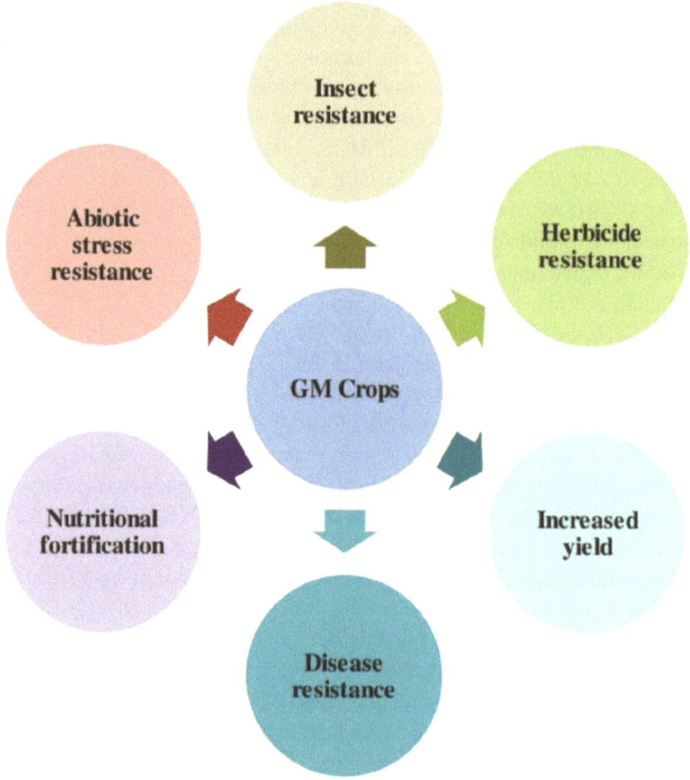

FIGURE 1.1 Genetic manipulations in crop plants for novel and commercially desirable traits.

Bacillus thuringiensis crops, or the well-known Bt crops, are developed by recombinant DNA technology; they express cry toxins of Bt strains (Abbas, 2018). Even though cry toxins are safe for humans and vertebrates, they show insecticidal activity against insect species belonging to Coleoptera, Diptera, Hymenoptera, Nematoda and lLepidoptera (Hofte and Whitely, 1989). Commercial production of Bt crops was approved by the Environmental Protection Agency (EPA) in the US in 1995. The most widely cultivated crops included Bt corn, Bt cotton, Bt tobacco, Bt potato, Bt brinjal, Bt soybean, Bt sweetcorn, etc. (Koch et al. 2015).

1.4.2 Enhanced Abiotic Stress Resistance

Many of the important crop species are vulnerable to various abiotic stresses like drought, salt and extreme temperature (Wang et al. 2016). Plant stress responses are obviously associated with changes in the morphology, physiology, phenology and biochemistry of plants. As the stress response mechanisms are controlled genetically, efforts are made to manipulate stress-induced genes in order to improve abiotic stress resistance in crop plants (Bhatnagar et al. 2008). Genes conferring abiotic stress resistance have been identified and manipulated to produce GM crops (Wang et al. 2016). Late embryogenesis abundant (LEA) proteins are expressed in seeds during late developmental stages under oxidative stress, draught, salinity, dehydration and elevated abscisic acid (ABA) concentration (Dos-Reis et al. 2018). The osLEA3-2 gene in *Oryza sativa* produces LEA proteins in the embryo. It was found that osLEA 3-2 gene expression in rice could be triggered by abiotic stresses. The osLEA3-2 gene introduced into Zhonghua 11 rice cultivar showed better growth under drought, salinity and osmotic stresses (Duan and Cai, 2012). Heat shock proteins (HSP) are potential candidates for genetic engineering, as HSPs are strongly involved in abiotic stress tolerance (Dos-Reis et al. 2018). Small heat shock protein, $ZmHSP_{16.9}$ from maize is overexpressed in GM tobacco, resulting in increased tolerance to heat and oxidative stress in terms of seed germination rate, antioxidant enzyme activity and root length (Sun et al. 2012).

1.4.3 Quality Improvement

Malnutrition and food insecurity are major concerns in many developing countries. In this context, transgenic techniques offer genetic modification of crop plants to ensure increase in yield along with quality improvement, including amino acid composition, protein content, starch composition, lipid content, etc. (Fang et al. 2016). Biofortification is the method of refining the nutritional profile of food crops via conventional breeding or genetic engineering. Biofortified crops are an effective strategy to deal with malnutrition and micronutrient deficiencies (Nestel et al. 2006).

Golden rice was developed with an intention to solve the vitamin A deficiency in many countries, where rice is a staple food (Tang et al. 2009). Golden rice contains 1.6–2.0 μg beta carotene per gram of dry rice. Genetic modification of rice with two beta carotene biosynthesis genes, such as *psy* gene from daffodil and *crtl* gene from a soil bacterium, has created golden rice, which expresses beta carotene in the rice endosperm (Schaub et al. 2005). Likewise, several crop plants have been developed by recombinant DNA technology to contain enhanced vitamin B, vitamin C, vitamin E and other micronutrients (Malik and Maqbool, 2020).

1.4.4 Enhanced Shelf-life

Transgenic approaches to improve shelf-life of horticultural crops have gained much interest, as they can reduce post-harvest losses by up to 50% (Khabbazi et al. 2020). The short shelf-life of tomatoes often causes problems during storage and transport. To address this problem, the Calgene company developed the GM Flavr-Savr tomato with increased shelf-life properties. The Flavr-Savr tomato was genetically engineered to contain antisense polygalacturonase gene encoding beta polygalacturonase enzyme. However, the inferior quality of Flavr-Savr tomatoes ultimately led to their withdrawal from the markets (Dias and Ortiz, 2014). Arctic apples resist enzymatic browning after being sliced, keeping the food appealing and nutritious. Silencing the polyphenol oxidase gene in Arctic apples blocked the production of polyphenol oxidase enzyme, thereby inhibiting the browning reaction (Armen, 2015).

1.5 Current Challenges and Future Prospects

The adoption of transgenic techniques in crop improvement programmes has brought about revolutionary changes in crop yield, farmers' income, and less dependency on chemical insecticides, pesticides and herbicides (Brookes and Barfoot, 2018). In spite of these beneficial aspects, GM crops invite several controversies regarding the biosafety of genetically engineered crops for human health and the environment (Kumar et al. 2020). Disputes over potential toxicity and allergy-related problems in the consumption of GM food have not yet been resolved. 'Starlink' maize expressing cry 9c has not been approved, as it is not fit for human consumption due to its potential interaction with the immune system. Moreover, the protein is stable enough to trigger allergic reactions in humans. However, 'Starlink maize' was licensed for industrial use and as an animal feed in the US in 1998 (Bucchini and Goldman, 2002). In addition to the human health hazards, the possibility of gene transfer from modified crops to sexually compatible species could result in unintentional creation of specific traits in weed plants; this also creates ethical and cultural issues and reduction in biodiversity (Uzogara, 2000). Introgression of transgenes can create herbicide-resistant weed plants termed 'superweeds'. *Amaranthus palmeri* and *Amaranthus tuberculatus* are glyphosate-resistant weed plants that cause huge economic losses across the globe (Heap and Duke, 2018). The development of resistance in crop species can cause co-evolution of more resistant insect pests and superweeds (Gilbert, 2013). Extensive cultivation of Bt cotton led to the development of resistance in pink bollworm, *Pectinophora gossypiella*, against the cry1Ac gene product expressed in engineered Bt cotton (Bagla, 2010). Hence, it is necessary to take safety measures for long-term cultivation of GM crops, as this could result in co-evolution of highly resistant insect pest and weed species.

Cultivation of GM crops poses risks to non-targeted organisms. Losey et al. (1999) reported adverse effects of GM crops on Monarch caterpillars. A hike in the mortality rate was observed for *Danus plexippus* larvae reared on milkweed leaves stored with Bt corn under laboratory conditions. These results suggested the possibility of dispersal of Bt corn pollen to other nearby plants, which can be consumed by non-targeted organisms (Losey et al. 1999).

The release of a GM crop requires the expenditure of millions of dollars and complex and lengthy regulatory approval procedures (Davison, 2010). It was roughly estimated that about 35.01 million US dollars are required for regulatory safety assessments (McDougall, 2011). Moreover, the commercial launch of transgenic products is a quite lengthy and time-consuming process. An average time period of 5 years is required for a GM crop to pass regulatory pipelines in the EU, and 7 years in the US (McDougall, 2011). Concerns associated with the use of biotech crops have been discussed by the scientific community and policy makers with the introduction of transgenic crops (Tsatsakis et al. 2017). As transgenic crops have been available for only a short time; research output regarding the long-term effects of genetically modified organisms (GMO) on the environment and human health is limited at this point (Prakash et al. 2011). It is crucial to follow systematic, well-planned research for the development and commercialization of engineered crops that strongly adhere to the regulatory guidelines and post-release monitoring to assess the long-term effects (Shukla et al. 2018). The safety of GMOs is ensured and regulated by different federal agencies with separate guidelines for the cultivation of GM crops for consumption as food and for animal feed; this is applicable to both import and export due to the difference in risk associated with cultivation, trade and consumption (Turnbull et al. 2021). In the US, three federal agencies, the US food and administration (FDA), the US department of agriculture (USDA) and the US EPA are responsible for regulatory assessment of GM products. In the EU, a case-by-case evaluation of GMOs is required for food or feed derived from GMOs to be marketed or imported. Risk assessments are performed by the European Food Safety Authority (EFSA) (Shukla et al. 2018). Even though the application of transgenic technology in crop improvement is still in its infancy, the future seems to be promising. Increasing population, dietary preferences, nutritional requirements, urbanization and economic status in developing countries demand novel methodologies and sustainable research to optimize crop productivity (Choudhary et al. 2014). Currently, about 525 transgenic events in 32 crop species have been cultivated globally (Kumar et al. 2020). Advancements in novel biotechnological tools enabled identification, isolation, cloning and transfer of the desirable genes associated with agronomic traits. Modern transgenic technologies now offer overexpression of exogenous genes or regulation of endogenous gene expression under tissue-specific promoters and stress-inducible promoters. However, insufficient expression of the desired phenotype under commonly used tissue-specific promoters and the time lag in achieving the desired resistance with the use of stress-inducible promoters have stimulated the search for new promoter sequences to achieve sufficient expression of the gene to confer the desired phenotype. Until recently, transgenic technologies like genome editing using the CRISPR/Cas 9 system allowed the creation of conventional crops without transgene or mutant forms of conventional crops (non-GMOs) (Basso et al. 2020). Transgene-free technologies have gained much attention in recent years in crop improvement programmes due to the reduction in regulatory cost, minimized impact on the environment and above all, the wide public acceptance of transgene-free products. However, more research efforts are still needed to develop new and improved crop plants in a short time to feed the rapidly increasing global population with minimum cost.

1.6 Conclusions

Both conventional and modern transgenic approaches have been successfully implemented in developing elite cultivars with economically and scientifically important traits incorporated. Conventional plant breeding techniques have been used initially to transfer simple traits to crops. Over time, new technologies have been discovered to open up new avenues in crop improvement research. Genetic engineering provides tools for horizontal gene transfer to introduce desirable traits in crop plants. Likewise, modern approaches and genome editing tools like ZFNs, TALENs and the CRISPR/Cas 9 system offer precise and sequence-specific editing of crop genomes for enhanced agronomic performance. However, GM crops have been a matter of controversy since their introduction. The safety of GM crops for human health and the environment is debated worldwide. Nevertheless, enhanced traits such as stress tolerance, nutritional profile and high yield of GM crops have made them an excellent option to achieve global food security.

Acknowledgment

Ms. Anugraha acknowledges financial assistance from Council of Scientific and Industrial Research (CSIR), Govt of India, in the form of a junior research fellowship (08/528(0010)/2019-EMR-I).

REFERENCES

Abbas, M.S.T., 2018. Genetically engineered (modified) crops (*Bacillus thuringiensis* crops) and the world controversy on their safety. *Egypt J Biol Pest Control.* 28(1)1–12. https://doi.org/10.1186/s41938-018-0051-2

Adem, M., Beyene, D., Feyissa, T., 2017. Recent achievements obtained by chloroplast transformation. *Plant Methods.* 13(1),1–11. https://doi.org/10.1186/s13007-017-0179-1

Agrawal, A., Rajamani, V., Reddy, V.S., Mukherjee, S.K., Bhatnagar, R.K., 2015. Transgenic plants over-expressing insect-specific microRNA acquire insecticidal activity against *Helicoverpa armigera:* An alternative to Bt-toxin technology. *Transgenic Res.* 24(5),791–801. https://doi.org/10.1007/s11248-015-9880-x

Aguiar, R.W.S., Martins, E.S., Ribeiro, B.M., Monnerat, R.G., 2012. Cry10Aa protein is highly toxic to *Anthonomus grandis* Boheman (Coleoptera: Curculionidae), an important insect pest in Brazilian cotton crop fields. *Bt Res.* 3(1),20–28.

Ahmad, N., Michoux, F., Nixon, P.J., 2012. Investigating the production of foreign membrane proteins in tobacco chloroplasts: Expression of an algal plastid terminal oxidase. *PLoS ONE.* 7(7),1–10.

Ahmadabadi, M., Ruf, S., Bock, R., 2007. A leaf-based regeneration and transformation system for maize (*Zea mays* L.). *Transgenic Res.* 16(4),437–448. https://doi.org/10.1007/s11248-006-9046-y

Armen, J., 2015. Arctic apples: Leading the 'next wave' of biotech foods with consumer benefits. *Australasian Biotechnology.* 25(2),1–50. https://search.informit.org/doi/10.3316/informit.296007511823496

Bagla, P., 2010. Hardy cotton-munching pests are latest blow to GM crops. *Science.* 327(5972),1439. https://doi.org/10.1126/science.327.5972.1439

Baltes, N.J., Gil-Humanes, J., Voytas, D.F., 2017. Genome engineering and agriculture: Opportunities and challenges. In: Donald P Weeks, Bing Yang (ed) *Progress in Molecular Biology and Translational Science.* Published by: Academic Press, USA. pp.1–26. https://doi.org/10.1016/bs.pmbts.2017.03.011

Barrangou, R., 2015. The roles of CRISPR-Cas systems in adaptive immunity and beyond. *Curr Opin Immunol.* 32(1),36–41. https://doi.org/10.1016/j.coi.2014.12.008

Barrangou, R., Fremaux, C., Deveau, H., Richards, M., Boyaval, P., Moineau, S., Romero, D., Horvath, P., 2007. CRISPR provides acquired resistance against viruses in prokaryote. *Science.* 315(5819),1709–1712. https://doi.org/10.1126/science.1138140

Basso, M.F., Arraes, F.B.M., Grossi-de-Sa, M., Moreira, V.J.V., Alves-Ferreira, M., Grossi-de-Sa, M.F., 2020. Insights into genetic and molecular elements for transgenic crop development. *Front Plant Sci.* 11(1),509–713. https://doi.org/10.3389/fpls.2020.00509

Bates, G.W., 1999. Plant transformation via protoplast electroporation. *Methods Mol Biol.* 111(1),359–366. https://doi.org/10.1385/1-59259-583-9:359

Bawa, A.S., Anilakumar, K.R., 2013. Genetically modified foods: Safety, risks and public concerns-a review. *J Food Sci Technol.* 50(6),1035–1046. https://doi.org/10.1007/s13197-012-0899-1

Bhatnagar- Mathur, P., Vadez, V., Sharma, K.K., 2008. Transgenic approaches for abiotic stress tolerance in plants: Retrospect and prospects. *Plant Cell Rep.* 27(3),411–424. https://doi.org/10.1007/s00299-007-0474-9

Biswas, G.C., Iglesias, V., Datta, S., Potrykus, I., 1994. Transgenic Indica rice (*Oryza sativa* L.) plants obtained by direct gene transfer to protoplasts. *J Biotechnol.* 32(1),1–10. https://doi.org/10.1016/0168-1656(94)90114-7

Bitinaite, J., Wah, D.A., Aggarwal, A.K., Schildkraut, I., 1998. FokI dimerization is required for DNA cleavage. *Proc Natl Acad Sci USA.* 95(18),10570–10575. https://doi.org/10.1073/pnas.95.18.10570

Boch, J., Bonas, U., 2010. *Xanthomonas* AvrBs3 family-type III effectors: Discovery and function. *Annu Rev Phytopathol.* 48(1),419–436. https://doi.org/10.1146/annurev-phyto-080508-081936

Borges, F., Martienssen, R.A., 2015.The expanding world of small RNAs in plants. *Nat Rev Mol Cell Biol.* 16(12),727–741. https://doi.org/ 10.1038/nrm4085

Brookes, G., Barfoot, P., 2014. Economic impact of GM crops: The global income and production effects 1996-2012. *GM Crops Food.* 5(1),65–75. https://doi.org/10.4161/gmcr.28098

Brookes, G., Barfoot, P., 2018. Farm income and production impacts of using GM crop technology 1996–2016. *GM Crops Food.* 9(2),59–89. https://doi.org/10.1080/21645698.2018.1464866

Bucchini, L., Goldman, L., 2002. Starlink Corn: A risk analysis. *Environ. Health Perspect.* 110(1),5–13. https://doi.org/10.1289/ehp.021105

Chandrasekaran, J., Brumin, M., Wolf, D., Leibman, D., Klap, C., Pearlsman, M., Sherman, A., Arazi, T., Gal-On, A., 2016. Development of broad virus resistance in non-transgenic cucumber using CRISPR/Cas9 technology. *Mol Plant Pathol.* 17(7),1140–1153. https://doi.org/10.1111/mpp.12375

Chen, K., Gao, C., 2013. Targeted genome modification technologies and their applications in crop improvements. *Plant Cell Rep.* 33(4), 575–583.

Choudhary, B., Gheysen, G., Buysse, J., van der Meer, P., Burssens, S., 2014. Regulatory options for genetically modified crops in India. *Plant Biotechnol J.* 12(2),135–146. https://doi.org/10.1111/pbi.12155

Clasen, B.M., Stoddard, T.J., Luo, S., Demorest, Z.L., Li, J., Cedrone, F., Tibebu, R., Davison, S., Ray, E.E., Daulhac, A., Coffman, A., Yabandith, A., Retterath, A., Haun, W., Baltes, N.J., Mathis, L., Voytas, D.F., Zhang, F., 2016. Improving cold storage and processing traits in potato through targeted gene knockout. *Plant Biotechnol J.* 14(1),169–176. https://doi.org/ 10.1111/pbi.12370

Cocking, E.C., 1993. The role of plant regeneration from protoplasts in agricultural biotechnology. In: Chongbiao You, Zhangliang Chen, Yong Ding (ed) *Biotechnology in Agriculture. Current Plant Science and Biotechnology in Agriculture*. Published by: Springer, Dordrecht. pp. 27–32. https://doi.org/10.1007/978-94-011-1779-1_5

Cong, L., Ran, F.A., Cox, D., Lin, S., Barretto, R., Habib, N., Hsu, P.D., Wu, X., Jiang, W., Marraffini, L.A., Zhang, F., 2013. Multiplex genome engineering using CRISPR/Cas systems. *Science*. 339(6121),819–823. https://doi.org/10.1126/science.1231143

Daniell, H., Khan, M.S., Allison, L., 2002. Milestones in chloroplast genetic engineering: An environmentally friendly era in biotechnology. *Trends Plant Sci*. 7(2),84–91. https://doi.org/10.1016/s1360-1385(01)02193-8

Davison, J., 2010. GM plants: Science, politics and EC regulations. *Plant Sci*. 178(2), 94–98.

de Vetten, N., Wolters, A.M., Raemakers, K., van der Meer, I., terStege, R., Heeres, E., Heeres, P., Visser, R., 2003. A transformation method for obtaining marker-free plants of a cross-pollinating and vegetatively propagated crop. *Nat Biotechnol*. 21(4),439–442. https://doi.org/10.1038/nbt801

Dias, J.S., Ortiz, R., 2014. Advances in transgenic vegetable and fruit breeding. *Agric. Sci*. 5(14),1448–1467. https://doi.org/10.4236/as.2014.514156

Dos Reis, S.P., Marques, D.N., Ferreira Barros, N.L., de NazaréMonteiro Costa, C., Batista de Souza, C.R., 2018. Genetically engineered food crops to abiotic stress tolerance. In: Alina Maria Holban, Alexandru Mihai Grumezescu (ed) *Genetically Engineered Foods*. Published by: Academic Press. pp. 247–279. https://doi.org/10.1016/b978-0-12-811519-0.00010-8

Duan, J., Cai, W., 2012. OsLEA3-2, an abiotic stress induced gene of rice plays a key role in salt and drought tolerance. *PLoS ONE*. 7(9), 1–11. https://doi.org/10.1371/journal.pone.0045117

Dunwell, J.M., 2000. Transgenic approaches to crop improvement. *J Exp Bot*. 51(1),487–496. https://doi.org/10.1093/jexbot/51.suppl_1.487

Eck, J.V., Conlin, B., Garvin, D.F., Mason, H., Navarre, D.A., Brown, C.R., 2007. Enhanced beta-carotene content in potato via RNAi silencing of the beta-carotene hydroxylase gene. *Amer J of Potato Res*. 84(4),331–342.

Espinoza, C., Schlechter, R., Herrera, D., Torres, E., Serrano, A., Medina, C., Arce-Johnson, P., 2013. Cisgenesis and intragenesis: New tools for improving crops. *Biol Res*. 46(4),323–331. https://doi.org/10.4067/S0716-97602013000400003

Fang, J., Zhu, X., Wang, C., Shangguan, L., 2016. Applications of DNA technologies in agriculture. *Curr Genomics*. 17(4),379–386. https://doi.org/10.2174/1389202917666160331203224

Fraley, R.T., Rogers, S.G., Horsch, R.B., Sanders, P.R., Flick, J.S., Adams, S.P., Bittner, M.L., Brand, L.A., Fink, C.L., Fry, J.S., Galluppi, G.R., Goldberg, S.B., Hoffmann, N.L., Woo, S.C., 1983. Expression of bacterial genes in plant cells. *Proc Natl Acad Sci USA*. 80(15),4803–4807. https://doi.org/10.1073/pnas.80.15.4803

Fromm, M., Taylor, L.P., Walbot, V., 1985. Expression of genes transferred into monocot and dicot plant cells by electroporation. *Proc. Natl. Acad. Sci. USA* 82(17), 5824–5828.

Gadaleta, A., Giancaspro, A., Blechl, A.E., Blanco, A., 2008. A transgenic durum wheat line that is free of marker genes and expresses 1DY10. *J Cereal Sci*. 48(2),439–445. https://doi.org/10.1016/j.jcs.2007.11.005

Gan, C., 1989. Gene gun accelerates DNA-coated particles to transform intact cells. *The Scientist*. 3(18),25–42.

Gavilano, L.B., Coleman, N.P., Burnley, L.E., Bowman, M.L., Kalengamaliro, N.E., Hayes, A., Bush, L., Siminszky, B., 2006. Genetic engineering of *Nicotiana tabacum* for reduced nornicotine content. *J Agric Food Chem.* 54(24),9071–9078. https://doi.org/10.1021/jf0610458

Gelvin, S.B., 2010. Plant proteins involved in Agrobacterium-mediated genetic transformation. *Annu Rev Phytopathol.* 48(1),45–68.

Gelvin, S.B., 2012. Traversing the cell: Agrobacterium T-DNA's journey to the host genome. *Front Plant Sci.* 3(52),1–11. https://doi.org/10.3389/fpls.2012.00052

Ghag, S.B., Shekhawat, U.K.S., Ganapathi, T.R., 2014. Host induced post-transcriptional hairpin RNA-mediated gene silencing of vital fungal genes confers efficient resistance against Fusarium wilt in banana. *Plant Biotechnol J.* 12(5),541–553. https://doi.org/10.1111/pbi.12158.

Gilbert, N., 2013. A hard look at GM crops. *Nature.* 497(7447),24–26. https:// doi.org/10.1038/497024a

Giraldo, P.A., Shinozuka, H., Spangenberg, G.C., Cogan, N., Smith, K.F., 2019. Safety assessment of genetically modified feed: Is there any difference from food? *Front Plant Sci.* 10(1),1592–1609. https://doi.org/10.3389/fpls.2019.01592

González-Melendi, P., Fernández-Pacheco, R., Coronado, M.J., Corredor, E., Testillano, P.S., Risueño, M.C., Marquina, C., Ibarra, M.R., Rubiales, D., Pérez-de-Luque, A., 2008. Nanoparticles as smart treatment-delivery systems in plants: assessment of different techniques of microscopy for their visualization in plant tissues. *Ann Bot.* 101(1),187–195. https://doi.org/ 10.1093/aob/mcm283

Green, J.M., Owen, M.D., 2011. Herbicide-resistant crops: Utilities and limitations for herbicide-resistant weed management. *J Agric Food Chem.* 59(11),5819–5829. https://doi.org/10.1021/jf101286h

Grevich, J. J., and Daniell, H., 2005. Chloroplast genetic engineering: recent advances and future perspectives. *Crit Rev Plant Sci.* 24(2),83–107. https://doi.org/10.1080/07352680590935387

Grunewald, W., Bury, J., Inzé, D., 2013. Thirty years of transgenic plants. *Nature.* 497(7447),40–40. https://doi.org/10.1038/497040a

Hayta, S., Smedley, M.A., Demir, S.U., Blundell, R., Hinchlife, A., Atkinson, N., Harwood, W.A., 2019. An efficient and reproducible *Agrobacterium*-mediated transformation method for hexaploid wheat (*Triticum aestivum* L.). *Plant Methods.* 15(1),121–136. https://doi.org/10.1186/s13007-019-0503-z

Heap, I., Duke, S.O., 2018. Overview of glyphosate-resistant weeds worldwide. *Pest Manag Sci.* 74(5),1040–1049. https://doi.org/10.1002/ps.4760

Hofte, H., Whitely, H.P., 1989. Insecticidal crystal proteins of *Bacillus thuringiensis. Microbiol Mol Biol Rev.* 53(2),242–255.

ISAAA, 2018. *Global Status of Commercialized Biotech/GM Crops in 2018: Executive Brief.* Published by: ISAAA, Ithaca, NY.

ISAAA, 2020. ISAAA Brief 55-2019: Executive Summary. Available Online at: www.isaaa.org/resources/publications/briefs/55/executivesummary/default.asp

Ismail, R.M., El-Domyati, F.M., Wagih, E.E., Sadik, A.S., Abdelsalam, A.Z.E., 2011. Construction of banana bunchy top nanovirus-DNA-3 encoding the coat protein gene and its introducing into banana plants cv. Williams. *J Genet Eng Biotechnol.* 9(1),35–41. https://doi.org/10.1016/j.jgeb.2011.05.012

Jaganathan, D., Ramasamy, K., Sellamuthu, G., Jayabalan, S., Venkataraman, G., 2018. CRISPR for crop improvement: An update review. *Front Plant Sci.* 9(1),985–1002. https://doi.org/10.3389/fpls.2018.00985

Jana, Ř., 2010. Potential of chloroplast genome in plant breeding. *Czech J Genet Plant Breed.* 46(3),103–113.

Jiang, W.Z., Henry, I.M., Lynagh, P.G., Comai, L., Cahoon, E.B., Weeks, D.P., 2017. Significant enhancement of fatty acid composition in seeds of the allohexaploid, *Camelina sativa*, using CRISPR/Cas9 gene editing. *Plant Biotechnol J.* 15(5),648–657. https://doi.org/10.1111/pbi.12663.

Joersbo, M., Brunstedt, J., 1996. Electroporation and transgenic plant production. In: Paul T. Lynch, Michael R. Davey (ed) *Electrical Manipulation of Cells.* Published by: Springer, Boston. pp. 201–222. https://doi.org/10.1007/978-1-4613-1159-1_10

Jørgensen, K., Bak, S., Busk, P.K., Sørensen, C., Olsen, C.E., Puonti-Kaerlas, J., Møller, B.L., 2005. Cassava plants with a depleted cyanogenic glucoside content in leaves and tubers. Distribution of cyanogenic glucosides, their site of synthesis and transport, and blockage of the biosynthesis by RNA interference technology. *Plant Physiol.* 139(1),363–374. https://doi.org/10.1104/pp.105.065904

Joshi, S.G., Schaart, J.G., Groenwold, R., Jacobsen, E., Schouten, H.J., Krens, F.A., 2011. Functional analysis and expression profiling of HcrVf1 and HcrVf2 for development of scab resistant cisgenic and intragenic apples. *Plant Mol Biol.* 75(6),579–591.

Keshavareddy, G., Kumar, A.R.V., Ramu, V.S., 2018. Methods of plant transformation: A review. *Int J Curr Microbiol App Sci.* 7(7),2656–2668. https://doi.org/10.20546/ijcmas.2018.707.312

Khabbazi, S.D., Khabbazi, A.D., Cevik, V., Ergül, A., 2020. Genetic engineering of horticultural crops contributes to the improvement of crop nutritional quality and shelf life. In: Usha Kiran, Malik Zainul Abdin, Kamaluddin (ed) *Transgenic Technology Based Value Addition in Plant Biotechnology.* Published by: Academic Press. pp. 247–272. https://doi.org/10.1016/b978-0-12-818632-9.00011-3

Kim, Y.G., Cha, J., Chandrasegaran, S., 1996. Hybrid restriction enzymes: Zinc finger fusions to Fok I cleavage domain. *Proc Natl Acad Sci USA.* 93(3),1156–1160. https://doi.org/10.1073/pnas.93.3.1156

Koch, M.S., Ward, J.M., Levine, S.L., Baum, J.A., Vicini, J.L., Hammond, B.G., 2015. The food and environmental safety of Bt crops. *Front Plant Sci.* 6(1),283–305. https://doi.org/10.3389/fpls.2015.00283

Krishna, S.S., Majumdar, I., Grishin, N.V., 2003. Structural classification of zinc fingers: Survey and summary. *Nucleic Acids Res.* 31(2),532–550. https://doi.org/10.1093/nar/gkg161

Kumar, K., Gambhir, G., Dass, A., Tripathi, A.K., Singh, A., Jha, A.K., Yadava, P., Choudhary, M., Rakshit, S., 2020. Genetically modified crops: Current status and future prospects. *Planta.* 251(4),1–27. https://doi.org/10.1007/s00425-020-03372-8

Kupferschmidt, K., 2013. A lethal dose of RNA. *Science.* 341(6147),732–733.

Kusaba, M., 2004. RNA interference in crop plants. *Curr Opin Biotechnol.* 15(2),139–143. https://doi.org/10.1016/j.copbio.2004.02.004.

Lalani, J., Misra, A., 2011. Gene delivery using chemical methods. In: Ambikanandan Misra (ed) *Challenges in Delivery of Therapeutic Genomics and Proteomics.* Published by: Elsevier, USA. pp. 127–206.

Lelandais-Brière, C., Sorin, C., Declerck, M., Benslimane, A., Crespi, M., Hartmann, C., 2010. Small RNA diversity in plants and its impact in development. *Curr. Genomics.* 11(1),14–23. https://doi.org/10.2174/138920210790217918

Li, M., Li, X., Zhou, Z., Wu, P., Fang, M., Pan, X., Lin, Q., Luo, W., Wu, G., Li, H., 2016. Reassessment of the four yield-related genes Gn1a, DEP1, GS3, and IPA1 in rice using a CRISPR/Cas9 system. *Front Plant Sci.* 7(1),377–390. https://doi.org/10.3389/fpls.2016.00377.

Lim, L.P., Lau, N.C., Garrett-Engele, P., Grimson, A., Schelter, J.M., Castle, J., Bartel, D.P., Linsley, P.S., Johnson, J.M., 2005. Microarray analysis shows that some microRNAs downregulate large numbers of target mRNAs. *Nature*. 433(7027),769–773. https://doi.org/10.1038/nature03315

Losey, J.E., Rayor, L.S., Carter, M.E., 1999. Transgenic pollen harms monarch larvae. *Nature*. 399(6733),214. https://doi.org/10.1038/20338

Low, L.Y., Yang, S.K., Kok, D.X.A., Ong-Abdullah, J., Tan, N.P., Lai, K.S., 2018. Transgenic plants: Gene constructs, vector and transformation method. *New Visions in Plant Science*. 41–61. https://doi.org/10.5772/intechopen.79369

Malik, K.A., Maqbool, A., 2020. Transgenic crops for biofortification. *Front Sustain Food Syst*. 4(571402),1–15. https://doi.org/10.3389/fsufs.2020.571402

Mathur, J., Koncz, C., 1998. PEG-mediated protoplast transformation with naked DNA. *Methods Mol Biol*. 82(1),267–276. https://doi.org/ 10.1385/0-89603-391-0:267

Manshardt, R., 2004. Crop improvement by conventional breeding or genetic engineering: How different are they? *Biotechnology*. 5,1–3.

Martínez-Fortún, J., Phillips, D.W., Jones, H.D., 2017. Potential impact of genome editing in world agriculture. *Emerg Top Life Sci*. 1(2),117–133. https://doi.org/10.1042/ETLS20170010

McBride, K.E., Svab, Z., Schaaf, D.J., Hogan, P.S., Stalker, D.M., Maliga, P., 1995. Amplification of a chimeric Bacillus gene in chloroplasts leads to an extraordinary level of an insecticidal protein in tobacco. *Biotechnology*. 13(4),362–365. https://doi.org/10.1038/nbt0495-362

McDougall, P., 2011. The cost and time involved in the discovery, development and authorisation of a new plant biotechnology derived trait. Consultancy Study for Crop Life International. Midlothian. 1–24.

Mesterházy, Á.,Oláh, J., Popp, J., 2020. Losses in the grain supply chain: Causes and solutions. *Sustainability*. 12(6), 2342–2360. https://doi.org/10.3390/su12062342

Mishra, M., Jalil, S.U., Sharma, N., Hudedamani, U., 2014. An Agrobacterium mediated transformation system of guava (*Psidium guajava* L.) with endochitinase gene. *Crop Breed Appl Biotechnol*. 14(4),232–237. https://doi.org/10.1590/1984-70332014v14n4a36

Morison, J.I., Baker, N.R., Mullineaux, P.M., Davies, W.J., 2008. Improving water use in crop production. *Philos Trans R Soc Lond B Biol Sci*. 363(1491), 639–658. https://doi.org/10.1098/rstb.2007.2175

Nadakuduti, S.S., Buell, C.R., Voytas, D.F., Starker, C.G, Douches, D.S., 2018. Genome editing for crop improvement – Applications in clonally propagated polyploids with a focus on Potato (*Solanum tuberosum* L.). *Front Plant Sci*. 9(1607), 1–11. https://doi.org/10.3389/fpls.2018.01607

Nanjareddy, K., Arthikala, M.K., Blanco, L., Arellano, E.S., Lara, M., 2016. Protoplast isolation, transient transformation of leaf mesophyll protoplasts and improved Agrobacterium-mediated leaf disc infiltration of *Phaseolus vulgaris*: Tools for rapid gene expression analysis. *BMC Biotechnol*. 16(1),1–14. https://doi.org/10.1186/s12896-016-0283-8

Naqvi, S., Maitra, A.N., Abdin, M.Z., Akmal, M., Arora, I., Samim, M., 2012. Calcium phosphate nanoparticle mediated genetic transformation in plants. *J Mater Chem*. 22(8), 3500–3508. https://doi.org/10.1039/c2jm11739h

Nestel, P., Bouis, H.E., Meenakshi, J.V., Pfeiffer, W., 2006. Biofortification of staple food Crops. *J Nutr*. 136(4),1064–1067. https://doi.org/10.1093/jn/136.4.1064

Osakabe, Y., Watanabe, T., Sugano, S.S., Ueta, R., Ishihara, R., Shinozaki, K., Osakabe, K., 2016. Optimization of CRISPR/Cas9 genome editing to modify abiotic stress responses in plants. *Sci Rep*. 6(1),1–10. https://doi.org/10.1038/srep26685.

Passricha, N., Saifi, S.K., Negi, H., Tuteja, R., Tuteja. N., 2020. Transgenic approach in crop improvement. In: Narendra Tuteja, Renu Tuteja, Nishat Passricha , Shabnam Saif (ed) *Advancement in Crop Improvement Techniques*. Published by: Woodhead Publishing. pp. 329–350.

Peer, R., Rivlin, G., Golobovitch, S., Lapidot, M., Gal-On, A., Vainstein, A., Tzfira, T., Flaishman, M.A., 2015. Targeted mutagenesis using zinc-finger nucleases in perennial fruit trees. *Planta* 241(4),941–951. https://doi.org/10.1007/s00425-014-2224-x

Prakash, D., Verma, S., Bhatia, R., Tiwary, B.N., 2011. Risks and precautions of genetically modified organisms. *ISRN Ecol.* 2011(1),1–13. https://doi.org/10.5402/2011/369573

Purchase, I.F., 2005. What determines the acceptability of genetically modified food that can improve human nutrition? *Toxicol Appl Pharmacol.* 207(2 Suppl),19–27. https://doi.org/10.1016/j.taap.2004.12.025

Rakoczy-Trojanowska, M., 2002. Alternative methods of plant transformation--a short review. *Cell Mol Biol Lett.* 7(3),849–858.

Ran, Y., Liang, Z., Gao, C., 2017. Current and future editing reagent delivery systems for plant genome editing. *Sci China Life Sci.* 60(5),490–505. https://doi.org/10.1007/s11427-017-9022-1.

Raney, T., 2006. Economic impact of transgenic crops in developing countries. *Curr Opin Biotechnol.* 17(2),174–178. https://doi.org/10.1016/j.copbio.2006.02.009

Rani, S.J., Usha, R., 2013. Transgenic plants: Types, benefits, public concerns and future. *Journal of Pharmacy Research.* 6(8),879–883.

Rashid, M., Guangyuan, H., Guangxiao, Y., 2014. Agrobacterium-mediated tobacco transformation with wheat dreb2 gene. *Pak J Agri Sci.* 51(4),887–892.

Rasmussen, J.O., Rasmussen, O.S., 1993. PEG mediated DNA uptake and transient GUS expression in carrot, rapeseed and soybean protoplasts. *Plant Sci.* 89(2),199–207. https://doi.org/10.1016/0168-9452(93)90128-m

Risacher, T., Craze, M., Bowden, S., Paul, W., Barsby, T., 2009. Highly efficient Agrobacterium-mediated transformation of wheat via in planta inoculation. In: H. Jones, P. Shewry (ed) *Transgenic Wheat, Barley and Oats. Methods in Molecular Biology™ (Methods and Protocols)*. Published by: Humana Press, USA. pp.115–124. https://doi.org/10.1007/978-1-59745-379-0_7

Redman, M., King, A., Watson, C., King, D., 2016. What is CRISPR/Cas9?. *Arch Dis Child Educ Pract Ed.* 101(4),213–215. https://doi.org/10.1136/archdischild-2016-310459

Ribeiro, T.P., Arraes, F., Lourenço-Tessutti, I.T., Silva, M.S., Lisei-de-Sá, M.E., Lucena, W.A., Macedo, L., Lima, J.N., Santos Amorim, R.M., Artico, S., Alves-Ferreira, M., Mattar Silva, M.C., Grossi-de-Sa, M.F., 2017. Transgenic cotton expressing Cry10Aa toxin confers high resistance to the cotton boll weevil. *Plant Biotechnol J.* 15(8),997–1009. https://doi.org/10.1111/pbi.12694

Ricroch, A., Clairand, P., Harwood, W., 2017. Use of CRISPR systems in plant genome editing: Toward new opportunities in agriculture. *Emerg Top Life Sci.* 1(2),169–182. https://doi.org/10.1042/etls20170085

Roudsari, M.F., Salmanian, A.H., Mousavi, A., Sohi, H.H., Jafari, M., 2009. Regeneration of glyphosate-tolerant *Nicotiana tabacum* after plastid transformation with a mutated variant of bacterial aroA gene. *Iran J Biotechnol.* 7(4),247–253.

Saurabh, S., Vidyarthi, A.S., Prasad, D., 2014. RNA interference: Concept to reality in crop improvement. *Planta.* 239(3),543–564. https://doi.org/10.1007/s00425-013-2019-5

Schaart, J.G., Krens, F.A., Pelgrom, K.T., Mendes, O., Rouwendal, G.J., 2004. Effective production of marker-free transgenic strawberry plants using inducible site-specific recombination and a bifunctional selectable marker gene. *Plant Biotechnol J.* 2(3),233–240.

Schaub, P., Al-Babili, S., Drake, R., Beyer, P., 2005. Why is golden rice golden (yellow) instead of red?. *Plant Physiol.* 138(1),441. https://doi.org/10.1104/pp.104.057927

Schouten, H.J., Krens, F.A., Jacobsen, E., 2006. Do cisgenic plants warrant less stringent oversight? *Nat Biotechnol.* 24(7),753–753. https://doi.org/10.1038/nbt0706-753

Shah, T., Andleeb, T., Lateef, S., Noor, M.A., 2018. Genome editing in plants: Advancing crop transformation and overview of tools. *Plant Physiol Biochem.* 131(1),12–21. https://doi.org/10.1016/j.plaphy.2018.05.009

Shan, Q., Zhang, Y., Chen, K., Zhang, K., Gao, C., 2015. Creation of fragrant rice by targeted knockout of the OsBADH2 gene using TALEN technology. *Plant Biotechnol J.* 13(6),791–800. https://doi.org/10.1111/pbi.12312

Sheldon, I.M., 2002. Regulation of biotechnology: Will we ever 'freely' trade GMOs?. *Eur Rev Agric Econ* 29(1),155–176. https://doi.org/10.1093/erae/29.1.155

Shukla, M., Al-Busaidi, K.T., Trivedi, M., Tiwari, R.K., 2018. Status of research, regulations and challenges for genetically modified crops in India. *GM Crops Food* 9(4),173–188. https://doi.org/10.1080/21645698.2018.1529518

Shukla, V.K., Doyon, Y., Miller, J.C., DeKelver, R.C., Moehle, E.A., Worden, S.E., Mitchell, J.C., Arnold, N.L., Gopalan, S., Meng, X., Choi, V.M., Rock, J.M., Wu, Y.Y., Katibah, G.E., Zhifang, G., McCaskill, D., Simpson, M.A., Blakeslee, B., Greenwalt, S.A., Butler, H.J., Hinkley, S., Zhang, L., Rebar, E., Gregory, P.D., Urnov, F.D., 2009. Precise genome modification in the crop species *Zea mays* using zinc-finger nucleases. Nature. 459(7245),437–441. https://doi.org/10.1038/nature07992

Somssich, M., 2019. A short history of plant transformation. *PeerJ Prepr.* 7(e27556v2), 1–28. https://doi.org/10.7287/peerj.preprints.27556

Sun, L., Liu, Y., Kong, X., Zhang, D., Pan, J., Zhou, Y., Wang, L., Li, D., Yang, X., 2012. ZmHSP16.9, a cytosolic class I small heat shock protein in maize (*Zea mays*), confers heat tolerance in transgenic tobacco. *Plant Cell Rep.* 31(8),1473–1484. https://doi.org/10.1007/s00299-012-1262-8

Sun, Y., Zhang, X., Wu, C., He, Y., Ma, Y., Hou, H., Guo, X., Du, W., Zhao, Y., Xia, L., 2016. Engineering herbicide-resistant rice plants through CRISPR/Cas9-mediated homologous recombination of acetolactate synthase. *Mol Plant.* 9(4),628–631. https://doi.org/10.1016/j.molp.2016.01.001

Tang, G., Qin, J., Dolnikowski, G.G., Russell, R.M., Grusak, M.A., 2009. Golden rice is an effective source of vitamin A. *Am J Clin Nutr.* 89(6),1776–1783. https://doi.org/10.3945/ajcn.2008.27119

Tang, J., Chu, C., 2017. MicroRNAs in crop improvement: Fine-tuners for complex traits. *Nat Plants.* 3(7),1–11. https://doi.org/10.1038/nplants.2017.77

Teotia, S., Singh, D., Tang, X., Tang, G., 2016. Essential RNA-based technologies and their applications in plant functional genomics. *Trends Biotechnol.* 34(2),106–123. https://doi.org/10.1016/j.tibtech.2015.12.001

Tomar, V., Saini, S.S., Juneja, K., Agrawal, P.K., Sircar, D., 2019. Transgenic technologies and their potential applications in horticultural crop improvement. In: Ramalingam Sathishkumar, Sarma Rajeev Kumar, Jagadeesan Hema, Venkidasamy Baskar (ed) *Advances in Plant Transgenics: Methods and Applications*. Published by: Springer, Germany. pp. 189–212.

Tsatsakis, A.M., Nawaz, M.A., Kouretas, D., Balias, G., Savolainen, K., Tutelyan, V.A., Golokhvast, K.S., Lee, J.D., Yang, S.H., Chung, G., 2017. Environmental impacts of genetically modified plants: A review. *Environ Res.* 156(1),818–833. https://doi.org/10.1016/j.envres.2017.03.011

Tseng, M.J., Yang, M.T., Chu, W.R., Liu, C.W., 2014. Plastid transformation in cabbage (Brassica oleracea L. var. capitata L.) by the biolistic process. In: P. Maliga (ed)

Chloroplast Biotechnology. Methods in Molecular Biology (Methods and Protocols). Published by: Humana Press, Totowa. pp.355–366. https://doi.org/10.1007/978-1-62703-995-6_23

Turnbull, C., Lillemo, M., Hvoslef-Eide, T., 2021. Global regulation of genetically modified crops amid the gene edited crop boom – A review. *Front Plant Sci.* 12 (630396), 1–19. https://doi.org/10.3389/fpls.2021.630396

Uzogara, S.G., 2000. The impact of genetic modification of human foods in the 21st century: A review. *Biotechnol Adv.* 18(3),179–206. https://doi.org/10.1016/s0734-9750(00)00033-1

Van Bavel, J., 2013. The world population explosion: Causes, backgrounds and projections for the future. *Facts Views Vis Obgyn.* 5(4),281–291.

Vaucheret, H., 2008. Plant ARGONAUTES. *Trends Plant Sci.* 13(7),350–358. https://doi.org/10.1016/j.tplants.2008.04.007

Wang, H., Wang, H, Shao, H, Tang, X., 2016. Recent advances in utilizing transcription factors to improve plant abiotic stress tolerance by transgenic technology. *Front Plant Sci.* 7(67),1–13. https://doi.org/10.3389/fpls.2016.00067

Weeks, J.T., Ye, J., Rommens, C.M., 2008. Development of an in planta method for transformation of alfalfa (*Medicago sativa*). *Transgenic Res.* 17(4),587–597. https://doi.org/10.1007/s11248-007-9132-9.

Yogindran, S., Rajam, M.V., 2015. RNAi for crop improvement. In: Bir Bahadur, Manchikatla Venkat Rajam, Leela Sahijram, K Krishnamurthy (ed) *Plant Biology and Biotechnology*. Published by: Springer, New Delhi. pp. 623–637. https://doi.org/10.1007/978-81-322-2283-5_31

Zhang, B., 2015. MicroRNA: A new target for improving plant tolerance to abiotic stress. *J Exp Bot.* 66(7),1749–1761. https://doi.org/ 10.1093/jxb/erv013

Zhang, Y., Massel, K., Godwin, I.D., Gao, C., 2018. Applications and potential of genome editing in crop improvement. *Genome Biol.* 19(1),210–221. https://doi.org/10.1186/s13059-018-1586-y

2

Elicitation: A Biotechnological Approach for Enhancement of Secondary Metabolites in In Vitro *Cultures*

Ritu Mahajan
University of Jammu
Jammu (J&K), India

Tania Sagar
University of Jammu
Jammu (J&K), India

Pallavi Billowria
University of Jammu
Jammu (J&K), India

Nisha Kapoor
University of Jammu
Jammu (J&K), India

CONTENTS

DOI: 10.1201/9781003239932-2

2.1 Introduction

Plants are the reservoirs for pharmaceutically important compounds (Aslam et al. 2020). They have been used traditionally against several diseases. At present, due to modernization and rapid development in technologies, pharmaceutical industries are attentively considering such valuable medicinal compounds from plants, including flavonoids, phenolics, terpenoids and saponins present in different parts of the plants (Kaur et al. 2018, 2021). Plants, in response to stress conditions, synthesize an enormous variety of complex chemical compounds called secondary metabolites. These bioactive compounds exhibit specific phytochemical characteristics against several microbial infections (Jawdat, 2016). However, due to the availability of and high demand for synthetic drugs, the use of plants and their extracts was reduced (Karimi et al. 2015). But, at present, the trend is again changing towards the use of these secondary metabolites as medicines, in nutrition, pharmaceuticals, cosmetics, agriculture and numerous other industries of commercial value, as they are cheap, ecofriendly and have no side effects, in comparison with the modern synthetic drugs (Hussain et al. 2011).

Regional and environmental adaptation of plants to their natural milieu has reduced their availability, and hence, there is an urgent need to find better routes for procurement of these crucial organic compounds. Moreover, due to overexploitation, a major concern is depletion of natural resources and the threat of extinction. Production of these highly important compounds from *in vitro* cultures is an economical, efficient, time- and energy-saving method, rather than cumbersome production from whole plants and a tiresome process of exploiting plant material from the natural environment. Using biotechnological approaches, many different *in vitro* culture techniques are being used for the enhanced production of these valuable bioactive metabolites from plants (Mahajan, 2016; Mahajan et al. 2016). As well as the use of callus cell suspension cultures, one of the most efficient tools for increasing the yield of phytochemicals is elicitation, in which diverse metabolic pathways are elicited by the incorporation of agents for optimum production of secondary metabolites (Yue et al. 2016). The agents employed are termed elicitors. Several reports describe the application of elicitors stimulating plant defense mechanisms, which in turn increases the yield of secondary metabolites in cultures grown *in vitro* (Halder et al. 2019; Bajwa et al. 2021). *In vitro* cultures provide an opportunity for increased production of biomass, rapid growth rate, and significant production of desired metabolites (Wawrosch et al. 2021). However, output varies depending upon the nature of the plant, the elicitor, the concentration and stimulation time offered to the plant, and the duration of signal response and production initiation. So, optimization of each factor is needed to accomplish the required target of elicitation (Hashim et al. 2021).

2.2 Elicitation and Secondary Metabolite Production

The use of plant cell cultures for the production of secondary metabolites is important where their synthesis by chemical methods is difficult and there is great demand for important phytochemicals (Wani et al. 2017). Diverse biotechnological approaches, including optimization of medium composition, selection of a high-yielding cell line, elicitation, immobilization, precursor feeding, metabolic engineering, etc., have been evaluated to estimate their efficacy in increasing the production of secondary metabolites by using different *in vitro* plant cell and organ cultures (Gutierrez-Valdes et al. 2020) (Figure 2.1).

The introduction of elicitation into *in vitro* cultures gave the impetus to scale up the process of production in a completely protected and optimized atmosphere. The elicitation technique used today is an evolved practice, based on the natural principle of a plant's response in natural conditions to stress or a defensive signal (Isah, 2019), where plants are artificially induced to activate the defense process (Nabi et al. 2021). The concept of hormesis of a plant defines two types of outcomes from a stress signal. It can be distress, which leads to damage or death of the plant, or it can be eustress, known as good stress, which induces the plant to produce secondary metabolites (Vargas-Hernandez et al. 2017). Stress is the most important factor to accomplish the process of elicitation (Singh and Kumaria, 2021). The stimulation of the plant defense mechanism for biosynthesis of these desired compounds in the laboratory is elicitation (Modarres and Yazdi, 2021). Elicitation has proven itself as an economic strategy for a high yield of these unique low-molecular-weight compounds under optimized conditions of a culture system by inducing various signal molecules (Yazdanian et al. 2021).

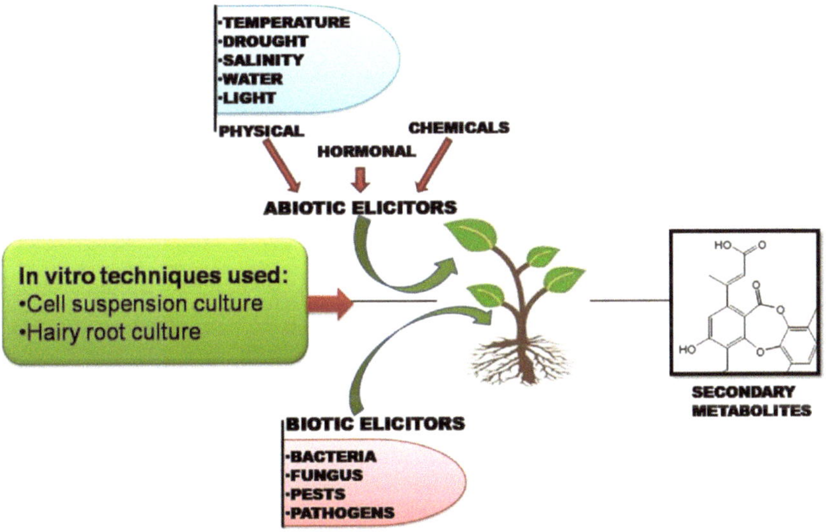

FIGURE 2.1 Effect of elicitors on *in vitro* plant cultures for the enhancement of secondary metabolites.

Several studies indicated that increased target secondary metabolite production in cell suspension cultures as well as in hairy root cultures was possible via the elicitation approach, which will open a new path for the production of various secondary metabolites in the near future (Ramirez-Estrada et al. 2016; Srivastava et al. 2017; Singh et al. 2018). At present, hairy root cultures are mostly preferred over cell cultures or callus culture for elicitation because of their high biosynthetic and genetic stability and their elevated growth rate in the medium without the need for growth regulators (Halder et al. 2018; Srivastava et al. 2018).

2.3 Elicitors and Their Types

Elicitors, as defense-inducing factors, initiate many signal transduction cascades depending upon the plant species (Jan et al. 2021). These secondary metabolism-inducing factors improve secondary metabolite production while switching the plant metabolism towards the defensive mechanism (Gorni et al. 2021). On the basis of their origin, elicitors are either biotic or abiotic (Figure 2.1). Abiotic elicitors include physical factors like exposure to light of different wavelengths, exposure to UV radiation (Ellenberger et al., 2020), temperature stress (Ayoola-Oresanya et al. 2021), nutrient deprivation (Groher et al. 2018), etc. or chemical factors like exposure to nanostructures (Khan et al. 2021), volatile compounds, salts (Hawrylak-Nowak et al. 2021), metal ions or pollutants. Biotic elicitors are of biological origin. They can be exogenous, like bacterial and fungal lysates, chitosan, chitin, yeast extract (Bhaskar et al. 2021), or endogenous elicitors, which include polysaccharides from the plant cell walls (Chandran et al. 2020). Several reports cited in the literature support the use of elicitors for the increased production of required secondary metabolites (Halder et al. 2019) (Table 2.1). However, certain other compounds, such as coronatine and cyclodextrins (Farhadi et al. 2020) or microRNA or abiotic factors of non-biological origin (Sak et al. 2021), are also regarded as potential elicitors.

2.3.1 Abiotic Elicitors

Abiotic elicitors are grouped into three categories: physical, chemical and hormonal factors.

2.3.1.1 Physical Elicitors

The physical elicitors include light, thermal stress, water stress, drought and salt stress.

2.3.1.1.1 Light is an important physical factor that induces secondary metabolite production. In callus culture of *Zingiber officinale*, light enhanced the production of gingerol and zingiberene (Anasori and Asghari 2008). Light irradiation caused anthocyanin production and digitoxin accumulation in cell suspension cultures of *Perilla frutescens* and *Digitalis purpurea*, respectively (Zhong et al. 1993). In *Vitis vinifera*, irradiation of berries with UV-C stimulated stilbene production (Wang et al. 2013). In hairy root cultures of *Fagopyrum tataricum*, the application of UV-B radiation

TABLE 2.1
Types of Elicitors and Secondary Metabolites Produced in Plants

Elicitor (Biotic/Abiotic)	Plant species	Culture nature	Secondary metabolites	Reference
Chitin	*Hypericum perforatum*	Cell suspension	Phenyl propanoid	Gadzovska et al. 2015
Mannan oligosaccharides	*Glycyrrhiza glabra* L.	Hairy root	Glycyrrhizin	Srivastava et al. 2019
Aspergillus niger	*Gymnema sylvestre*	Cell suspension	Gymnemic acid	Chodisetti et al. 2015
AgNPs	*Fagonia indica*	Callus	Phenolics and flavonoids	Begum et al. 2020
CdONPs	*Hordeum vulgare*	Roots and leaves	Saponarin and ferulic acid	Vecerova et al. 2016
AgNPs	*Caralluma tuberculata*	Callus	Phenolics and flavonoids	Ali et al. 2019
Zn and Fe nano-oxides	*Hypericum perforatum*	Cell suspension	Hyperforin and hypericin	Shakya et al. 2019
UV-B irradiation	*Fagopyrum tatarium*	Hairy root	Quercetin and rutin	Huang et al. 2016
Chalcone isomerase and methyl jasmonate	*Plumbago indica*	Hairy root	Plumbagin	Gangopadhyay et al. 2011
Staphylococcus aureus	*Datura metel*	Hairy root	Atropine	Shakeran et al. 2015
Ag, Au and NAA	*Prunella vulgaris*	Cell suspension	Phenolics and flavonoids	Fazal et al. 2019
Light, methyl jasmonate and cyclodextrin	*Scutellaria lateriflora*	Hairy root	Coegonin and baicalein	Marsh et al. 2014
Methyl cyclodextrin and methyl jasmonate	*Arachis hypogaea*	Hairy root	Stilbenoids	Yang et al. 2015
Methyl jasmonate and coronatine	*Salvia sclarea*	Hairy root	Quinone diterpenes	Vaccaro et al. 2017

enhanced the accumulation of flavonoids (Huang et al. 2016), while artemisinin biosynthesis was increased in *Artemisia annua* (Liu et al. 2002).

2.3.1.1.2 Thermal Stress affects plant growth and development. High- or low-temperature stress enhances the production of secondary metabolites. At lower temperature (20 ± 2° C), cell cultures of *Melastoma malabathricum* grew better and with greater production of anthocyanin than those grown at higher temperatures (26 ± 2° C and 29 ± 2° C) (Chan et al. 2010). In *Panax quinquefolius*, higher temperature enhanced senescence in the leaf as well as secondary metabolite content in the roots (Jochum et al. 2007). Chan et al. (2010) observed higher anthocyanin production in *Melastoma malabathricum* cell cultures grown at a lower temperature range (20 ± 2° C) than in those incubated at high temperatures (26 ± 2° C and 29 ± 2° C).

2.3.1.1.3 Water Stress can change the physiological as well as the biochemical properties of plants but also increases the secondary metabolite content in the various plant tissues (Zobayed et al. 2007). Pratibha et al. (2015) observed that in callus and suspension cultures of *Stevia rebaudiana*, osmotic agents such as proline and polyethylene glycol (PEG) induced the production of steviol glycosides. The use of PEG elicited the production of hypericin and pseudohypericin in *Hypericum adenotrichum*, while sucrose enhanced the synthesis of hypericin and hyperforin in *Hypericum perforatum* (Pavlick et al. 2007; Omer and Bengi 2013).

2.3.1.1.4 Drought Stress greatly influences the cellular functions of the plant and enhances the content of secondary metabolites. In *Prunella vulgaris*, moderate drought stress enhanced the production of rosmarinic, oleanolic and ursolic acids (Chen et al. 2011). Furthermore, in roots of *Glycyrrhiza uralensis* and *Salvia miltiorrhiza*, weak water deficit enhanced the production of glycyrrhizic acid and salvianolic acid (Li et al. 2011; Liu et al. 2011).

2.3.1.1.5 Salt Stress also alters physiological as well as metabolic processes. In a study, it has been reported that in the embryogenic tissue of *Catharanthus roseus*, the use of NaCl as an elicitor enhanced the synthesis of vincristine as well as vinblastine (Fatima et al. 2015). Enhanced salt concentration increased the anthocyanin concentration in *Grevillea* (Kennedy and De Filippis 1999), while in *Datura innoxia*, an increase in the total alkaloid content was observed in young leaves (Brachet and Cosson, 1986). Also, enhancement in organic osmolytes such as glycine betaine was recorded under salinity stress in *Triticum aestivum* (Krishnamurthy and Bhagwat 1989).

2.3.1.2 Chemical Elicitors

Heavy metals are environmental pollutants. Their sources are weathering of metal-bearing rocks, volcanic eruptions, and various mining industries and agricultural wastes (Ali et al. 2019). Heavy metals also elicit secondary metabolite production in different plants. Mizukami et al. (1977) observed that Cu^{2+} and Cd^{2+} enhanced shikonin formation in *Lithospermum* callus cultures, while Mg^{2+} increases cardenolide accumulation in *Digitalis lanata* tissue cultures (Ohlsson et al. 1989). In hairy root cultures of *Salvia castanea*, the application of Ag^+ (15 μM) enhanced the content of tanshinone II A 1.8-fold as compared with the untreated control culture (Li et al. 2016). Zaker et al. (2015) observed that $AgNO_3$ enhanced tanshinone production in root cultures of *Perovskia abrotanoides*. Similarly, Cu^{2+} and Co^{2+} enhanced secondary metabolite production in *Beta vulgaris* (Trejo-Tapia et al. 2001), while application of Zn^{2+} (900 μM) enhanced the lepidine content in *Lepidium sativum* cultures (Saba et al. 2000).

2.3.1.3 Hormonal Elicitors

The most studied hormones that affect the plant defense response are salicylic acid and its derivatives as well as jasmonates. For many secondary metabolic pathways in plants, jasmonates make up an important class of elicitors. In *Mentha piperita*, jasmonic acid elicits the production of rosmarinic acid (Krzyzanowska et al. 2012). It has been noted that methyl jasmonate as well as jasmonic acid induced stilbene biosynthesis in hairy root cultures of *Vitis rotundifolia* and in cell cultures of *Vitis vinifera* (Nopo-Olazabal et al. 2014; Taurino et al. 2015). Liang et al. (2013) studied the effect of gibberellin as an elicitor for stimulated production of secondary metabolites, while in cell suspension cultures of *Vitis vinifera*, salicylic acid enhanced the production of stilbene (Xu et al. 2015). Furthermore, in periwinkle, salicylic acid induced the biosynthesis of vinblastine and vincristine (Idrees et al. 2011). In hairy root cultures of *Salvia miltiorrhiza*, methyl jasmonate and salicylic acid along with transgenic technology increased the production of tanshinones (Xiaolong et al. 2015). Brassinosteroids, which are endogenous plant hormones, also play an important role in plant growth and development. These have been used as elicitors to enhance the production of acyl-conjugated metabolites in *Ornithopus sativus* (Kolbe et al. 1999).

2.3.2 Biotic Elicitors

Biotic elicitors are signaling molecules that activate synthesis of many chemicals as defense due to wounding and pathogen attack. Different polysaccharides have been used as biotic elicitors to enhance secondary metabolite production. Taurino et al. (2015) reported that addition of chitosan increased the production of trans-resveratrol (a natural phytoalexin) and viniferins in cell cultures of *Vitis vinifera*. Hu et al. (2003) observed that the application of cell-wall-derived oligogalacturonic acid in cell suspension culture of *Panax ginseng* resulted in enhanced ginseng and saponin content. Biotic elicitors of yeast origin have been utilized to stimulate the production of secondary metabolites (Chandran et al. 2020). Yeast extract induced tanshinone accumulation in *Perovskia abrotanoides* root cultures (Zaker et al. 2015), while it stimulated bacterial resistance in *Phaseolus vulgaris* (Stangarlin et al. 2011).

In bacterial elicitation, maximum increase in glycyrrhizic acid content was observed in cultures challenged with *Bacillus aminovorans*, *Bacillus cereus* and *Agrobacterium rhizogenes*, with no significant enhancement observed in glycyrrhizic acid content in *Agrobacterium tumefaciens*-challenged root cultures (Awad et al. 2014). *Pseudomonas syringae* produced coronatine, a phytotoxin that can significantly enhance the synthesis of taxane, in taxane media cell cultures (Onrubia et al. 2013) and also enhanced the production of viniferins in *Vitis vinifera* cell culture (Taurino et al. 2015).

Fungal cell wall elicitation in cell suspension cultures of *Cathranthus roseus* resulted in a five-fold enhancement of serpentine, catharanthine and ajmalicine production (Zhao et al. 2001). Fungal elicitation in *Taverniera cuneifolia* root cultures resulted in enhanced glycyrrhizic acid yield (Awad et al. 2014). Farhadi et al. (2020) reported that fungal cell wall and chitin, alone or in combination with methyl-β-cyclodextrin, enhanced paclitaxel biosynthesis in culture medium, with 146% more than in control cultures, while Salehi et al. (2020) reported the use of both culture extract and culture filtrate of *in vitro Corylus avellana* culture for paclitaxel biosynthesis. Linh et al. (2021) reported that endophytic fungi isolated from *Catharanthus roseus* improved biosynthesis of vinblastine, vincristine and terpenoid indole alkaloids in callus and cell suspension cultures of *C. roseus*.

2.4 Mechanism of Elicitation

The process of elicitation involves interaction of receptors present on the plasma membrane with the signal molecules. These include receptors from plasma membranes, such as saccharide elicitors, peptide elicitors and glycolipid elicitors (Malik et al. 2020). The interaction among such elicitors is highly specific. The stimulus generated is then transferred to the cells by a signal transduction system, producing changes resulting in the formation of phytoalexins (Ramirez-Estrada et al. 2016).

The interaction of these saccharide, proteinaceous or lipid elicitors with the receptors is reversible and saturable. In spite of this, elicitors have the ability to interact with different species, as plants exhibit common receptors for a number of different signal molecules. Elicitation is a multistep process with several reactions resulting in various responses, depending upon the physiological state and some genetic factors. On receiving the signal from the elicitor molecule, the receptor releases a second messenger, followed by several downstream reactions. The transduction pathways involved show several variations to different signal molecules and also to defensive responses (Liu and Lam 2019). Defensive responses result in reversible phosphorylation and dephosphorylation reactions, change in efflux and influx, and a rise in cytosolic Ca^{2+} levels outside and inside the cell along with the pH changes. Various pathways and reactions result in the production of antimicrobial compounds and some pathogenesis-related proteins to help the plant survive against pathogenic attack. For the synthesis of secondary metabolites, many signaling pathways have to work in collaboration with each other (Ramirez-Estrada et al. 2016). Also, the production of antimicrobial compounds during plant–elicitor interaction may play a potent role in increasing the resistance of host plants against pathogens in the near future (Thakur and Sokal 2013).

2.5 Factors Affecting the Process of Elicitation

Abiotic and biotic elicitors in plants provide an excellent system for the production of secondary metabolites, but the effect of elicitation is also decided by a number of factors like type of elicitor, age of culture used for elicitation, time of exposure to elicitor, and concentration or dose of elicitor.

2.5.1 Time of Elicitor Exposure

Exposure of a plant culture to an elicitor affects the growth of the culture and the yield of secondary metabolites (Chandran et al. 2020). In the cells of *Papaver bracteatum*, the synergistic effects of methyl jasmonate and tyrosine enhanced the content of thebaine yield after six days of treatment. Zaheer et al. (2016) studied the role of jasmonic acid and acetyl salicylic acid in hairy root cultures of *Psoralea corylifolia* as elicitors of daidzin production. Similarly, in *Taxus cuspidata*, production of paclitaxel increased after seven days of culturing suspension cells elicited with phenylalanine, methyl jasmonate, salicylic acid and gibberellins (Wang et al. 2015). Zaheer and Giri (2015) studied the effect of salicylic acid on the production of andrographolides after eight weeks of treatment in *Andrographis paniculata*. Gai et al. (2019), after elicitating the hairy roots of *Isatis tinctoria* by salicylic acid and methyl jasmonate, observed an enhancement in alkaloids and flavonoids after 24 days of culturing. Akhgari et al. (2019) reported a three-fold enhancement in concentrations of eburenine five and seven days after elicitation of hairy roots with methyl jasmonic acid as compared with non-elicited cultures of *Rhazya stricta*.

2.5.2 Concentration of Elicitor

Elicitor concentration is a crucial factor that plays an important role in secondary metabolite production. Different concentrations of biotic and abiotic elicitors like jasmonic acid, chitosan, yeast extract, and bacterial and fungal cultures resulted in variation of secondary metabolite production in various plant cell cultures (Moharrami et al. 2017; Gonçalves et al. 2019; Farjaminezhad and Garoosi, 2021). Chitosan elicitation in basil plant also decreases transpiration under water stress conditions (Malekpoor et al. 2016). Chodisetti et al. (2015) reported that in cell suspension cultures of *Gymnema sylvestre*, different concentrations of salicylic acid and methyl jasmonate were used. However, salicylic acid at 200 μM and methyl jasmonate at 150 μM increased the production of gymnemic acid.

Kang et al. (2006) demonstrated that the methyl jasmonate concentration required for scopolamine and hyoscyamine production varied (1.0 and 0.01 mM, respectively) in adventitious root cultures of *Scopolia parviflora*. However, with one target metabolite, the required optimal concentration of elicitor also varies for enhanced production of the metabolite of interest. Chong et al. (2005) studied the effect of jasmonic acid at different concentrations (10–200 mM) in cell cultures of *Morinda elliptica* on intracellular anthraquinone production. It was found that at a 50 mM concentration of jasmonic acid, anthraquinone content enhanced by 1.6 fold over the control, whereas at 100 or 200 mM concentration, a decrease in anthraquinone content (1.4-fold or 1.2-fold) was observed as compared with the control.

Chu et al. (2017) observed that at the same concentration of elicitor, a two-fold increase in the production of trans-resveratrol was seen in transgenic cell lines of *Vitis vinifera* as compared with the non-transgenic cell lines. Gai et al. (2019) observed that in *Astragalus membranaceus* hairy root cultures, there was a 6.17-fold increase in the yields of formononetin and calycosin as compared with control on treating cultures with 100 mg/L of chitosan for 24 h. Zhao and Tang (2020) observed that methyl jasmonate at a concentration of 100 mg/L increased valtrate production seven days after elicitation by 3.63-fold as compared with that of non-elicited control in rhizomes of *Valeriana jatamansi*.

2.5.3 Age of Culture

Age of culture is also an important parameter in the process of elicitation. Sivanandhan et al. (2013) reported that in a hairy root culture of *Withania somnifera*, the application of salicylic acid and methyl jasmonate enhanced the yield of withanone, withanolide A and withaferin A in a 40-day-old culture. Furthermore, in a six-day-old cell suspension culture of *Catharanthus roseus*, the addition of methyl jasmonate at concentrations of 10 μM and 100 μM enhanced the production of ajmalicine and serpentine. However, a negative effect has been observed on both cell growth as well as alkaloid synthesis during re-elicitation (Lee-Parsons et al. 2004). In another study, on elicitation, higher ajmalicine content was observed in 20-day-old cultures of *Catharanthus roseus* cells (Namdeo et al. 2002). Dowoma et al. (2017) observed increased rosmarinic acid and salvianolic acid A production from 50-day-old seedlings of *Salvia virgata* on elicitation with Ag^+ ions. Huang et al. (2021) reported enhanced production of sanguinarine and chelerythrine from 60-day-old cultures of *Macleaya cordata* using methyl jasmonate and salicylic acid as elicitors.

2.5.4 Elicitor Selection

Selection of the elicitor is imperative because of elicitor specificities in triggering a signaling cascade. However, the elicitor choice also depends on the target secondary metabolites of interest (Wiktorowska et al. 2010). On screening different elicitors such as Ag^+, Co^{2+}, Hg^{2+}, Cu^{2+} and signal molecules like salicylic acid and methyl jasmonate for azadirachtin production in *Azadirachta indica* hairy root cultures, it was observed that salicylic acid enhanced the yield of azadirachtin (Srivastava and Srivastava, 2014). Similarly, in cell suspension culture of *Andrographis paniculata*, among the different metal salts used as elicitors ($CdCl_2$, $AgNO_3$, $HgCl_2$ and $CuCl_2$), $CdCl_2$ was found to be effective, as it maximally enhanced the production of andrographolide (Gandi et al. 2012). These results indicate that judicious elicitor choice is necessary for enhanced production of secondary metabolites. Moreover, elicitor choice also depends on the physiological condition of the plant culture. Moreno-Escamilla et al. (2020) recorded the effect of different concentrations of arachidonic acid, salicylic acid, methyl jasmonate and Harpin protein on red and green butterhead lettuces. They observed that the highest impact of elicitation was with methyl jasmonate, which resulted in enhanced total phenolic and flavonoid content. Jaisi and Panichayupakaranant (2020) studied the dual elicitation effect on production of plumbagin from *Plumbago indica* root cultures and observed that a simultaneous treatment using chitosan and L-alanine increased plumbagin production. Also, sequential additions of methyl-β-cyclodextrin followed by chitosan addition enhanced production of plumbagin to 14.33 mg/g dry weight.

2.6 Elicitors and Enhancement of Valuable Medicinal Compounds

In the 21st century, a novel coronavirus termed the severe acute respiratory syndrome coronavirus 2 (SARS-CoV-2) emerged as the most lethal virus, causing the death of millions of people. Several studies have confirmed the role of phytochemicals against coronaviruses. Since the quality of secondary metabolites isolated from important medicinal plants varies because of seasonal and geographical variations, the enhancement in the production of secondary metabolites is achieved by triggering or eliciting the defense response of *in vitro* plant cultures. An increase in the yield of glycyrrhizin, an important anti-SARS-CoV metabolite from hairy root cultures of *Glycyrrhiza glabra*, is the most significant example of elicitation (Srivastava et al. 2019). It has been reported that in hairy root culture of *Glycyrrhiza glabra*, the application of mannan oligosaccharides obtained from the cell wall of *Saccharomyces cerevisiae* resulted in a 7.8-fold enhancement in the production of glycyrrhizin as compared with the control (De Oliveira et al. 2014). It should be recapitulated here that glycyrrhizin, a saponin, has proven activity against the previously pandemic SARS-CoV. Furthermore, it was proposed that an endogenous biotic elicitor, methyl jasmonate, at a concentration of 100 mM, increased the glycyrrhizin production up to 109 μg/g dry weight. In the same study, the role of other biotic elicitors, including yeast extract and chitosan, in glycyrrhizin production was demonstrated (Wongwicha et al. 2011).

There are several studies in hand on increasing the yields of important phytochemical compounds that possess anti-SARS-CoV-2 activity. Secondary metabolites (replication inhibitors of coronavirus) including reserpine, lycorine, plant lectins, luteolin, apigenin and quercetin (Keyaerts et al. 2007; Ryu et al. 2010) have been triggered via the use of elicitors such as methyl jasmonate and salicylic acid (Ptak et al. 2017; Chandran et al. 2020).

A study on cell suspension cultures of *Hypericumper foratum* reported that the application of chitin resulted in enhancement in the yield of phenylpropanoid and naphthodianthrone (Shakya et al. 2019). Shakeran et al. (2015) proposed enhancement in the production of atropine in hairy root cultures of *Datura metelvia.*

On the other hand, abiotic elicitors increased the yield of flavonoids, including hypericin and hyperforin, in *Hypericum perforatum* (Shakya et al. 2019). It has been reported that in the plantlets of *Hypericum perforatum*, chromium (0.01 mM) enhanced the yield of hypericin by 38% (Tirillini et al. 2006). In *Hypericum perforatum*, the presence of quercetin was found to be very effective as an anti-SARS-CoV-2 agent and elicitation mechanism that could be used as an effective strategy for the enhancement of quercetin (Shakya et al. 2019).

Apart from this, there are also environmental triggers to increase the production of phytochemicals in *in vitro* plant cultures. The use of different spectral lights has increased the production of compounds such as myricetin and apigenin. It has been experimentally shown that myricetin, by affecting ATPase activity, can inhibit the SARS-CoV helicase protein and thus could possess effective potential against SARS-CoV-2 (Yu et al. 2012; White et al., 2020). In another study, it was reported that UV-B irradiation enhanced the biosynthesis of quercetin and rutin in hairy root cultures of *Fagopyrum tatarium* (Huang et al. 2016).

2.7 Nanoparticles and Elicitation

Nanoparticles are materials with size ranging from 1 to 100 nm (Hasan, 2015). Nanomaterials have now become a new addition to the range of abiotic elicitors (Rivero-Montejo et al. 2021; Sagar et al. 2021). Silver nanoparticles enhanced the production of phenolics and flavonoids in callus cultures of *Fagonia indica* (Begum et al. 2020). Similarly, in callus cultures of *Caralluma tuberculata*, silver nanoparticles at a concentration of 90 μg/l increased the production of total phenolics and flavonoid content (Ali et al. 2019). However, in cell suspension cultures of *Hypericum perforatum*, zinc and iron nano-oxides triggered the production of hyperforin and hypericin (Shakya et al. 2019). Karakas (2020) studied the application of silver nanoparticles in *Isatis constricta*, resulting in enhanced production of indigo and tryptanthrin as compared with the control. Furthermore, salinity stress treatment along with nanoparticles (TiO_2 and SiO_2) in *Tanacetum parthenium* enhanced parthenolide production (Khajavi et al. 2019).

Polyphenolic compounds have attracted the attention of scientists as well as consumers due to their medicinal properties in the treatment of various serious diseases, including cancer and cardiovascular disease along with neurodegenerative diseases (Sharifi-Rad et al. 2020). In celery, the foliar application of selenium increased total flavonoids, total phenols, vitamin C and antioxidant capacity (Li et al. 2020). Application of CdO nanoparticles to roots and leaves of barley plants increased the total phenolic content 200 times, particularly saponarin and ferulic acid. The same study resulted in a 183% enhancement of isovitexin content (Vecerova et al. 2016).

2.8 Effect of Combined Elicitors on Production of Secondary Metabolites

In several plant species, it has been observed that the use of combined elicitors can increase the production of secondary metabolites as compared with single-elicitor treatment. In hairy root cultures of *Plumbago indica*, the application of chalcone isomerase and methyl jasmonate more effectively increased the plumbagin (a naphthoquinone) yield than chalcone isomerase alone (Gangopadhyoy et al. 2011). Furthermore, treatment with silver (Ag) and gold (Au) nanoparticles (3:1, 30 μg/l) along with naphthalene acetic acid (NAA) in cell suspension cultures of *Prunella vulgaris L.* enhanced the yield of flavonoids and also increased the DPPH-radical scavenging activity in comparison with the control (Fazal et al. 2019).

Co-treatment with methyl jasmonate or yeast extract and acetylsalicylic acid enhanced the production of glucotropaeolin in hairy root cultures of *Tropaeo lummajus* as compared with use of one of these elicitors alone (Wielanek and Urbanek 2006). The incubation of hairy root cultures of *Scutellaria lateriflora* under continuous light and treatment for a period of 24 hours with 15 mM methyl jasmonate and cyclodextrin significantly increased the yield of flavones, coogonin and baicalein as compared with cultures under darkness (Marsh et al. 2014). In hairy root cultures of *Arachis hypogaea*, the combined application of 9 g/l methyl cyclodextrin and 100 μM methyl jasmonate increased the production of stilbenoids such as resveratrol, arachidin-1, piceatannol and arachidin-3 as compared with treatment using one of these elicitors alone (Yang et al. 2015). Co-treatment with methyl jasmonate and phytotoxin coronatine in hairy roots of *Salvia sclarea* induced the yield of bioactive quinonediterpenes, including abietane and aethiopinone, by transcriptional reprogramming, but long-term exposure to methyl jasmonate impedes the growth of hairy roots (Vaccaro et al. 2017).

2.9 Role of Elicitors in Metabolic Engineering

The biosynthetic pathway is the most challenging issue for efficient yield of secondary metabolites. Some metabolic pathways have been discovered in *in vitro* plant cultures, including the terpenoid pathway, the flavonoid pathway, the iso quinolone alkaloid (berberine, morphine) pathway and the indole alkaloid pathway (Kucht et al. 2004; Shakya et al. 2019). In plant cells, modification of the metabolic pathways is difficult, as these pathways are complicated and at the same time activated by diverse enzymes. As a result, the modification is achieved via metabolic engineering of regulatory genes as well as their transcriptional factors (Fiehn et al. 2000; Oksman-Caldentey and Saito 2005).

In the literature, several examples of metabolic engineering in medicinal plants have been cited. In a recent study, the yield of an isoflavone, genistein, has been studied in three non-leguminous species, viz. *icotiana tabacum, Lactuca sativa* and *Petunia hybrida*, as the compound is not native to these plant species. This was made possible with the introduction of the isoflavone synthase gene from *Glycine max* (Barone et al. 2019). In another study, transformation in *Artemisia annua* was carried out via *Agrobacterium tumefaciens* so as to produce taxane, which is a paclitaxel precursor with anti-cancer and anti-HIV activity (Li et al. 2015; Ryang et al. 2019). Boccalone et al. (2020) reported that glycyrrhizin exhibits antiviral activity against SARS-CoV-2 *in vitro* and biotransformation of this bioactive compound to several other compounds with more emulsifying properties as compared with glycyrrhizin (He et al. 2019).

2.10 Effect of Elicitation Along with Other Strategies

2.10.1 Combined Effect of Precursor Feeding and Elicitation

Simultaneous use of elicitation and precursor feeding enhances the secondary metabolite production in *in vitro* plant cultures and could be used as an effective strategy for economically important metabolites. Qu et al. (2011) reported that the combined application of methyl jasmonate and phenylalanine enhanced the anthocyanin accumulation 4.6-fold as compared with the control in cell suspension cultures of *Vitis vinifera*. Raghavendra et al. (2011) studied the effect of methyl jasmonate, chitin, pectin, yeast extract and precursor L-tyrosine at different concentrations on the production of L-Dopa in suspension cultures of *Mucuna pruriens*. They reported that in comparison to the control suspension cultures, a several-fold enhancement in L-Dopa was observed in elicitor-treated and precursor -fed suspension cultures. Ooi et al. (2016) studied the effect of methyl jasmonate with L-arginine as precursor on hairy root cultures of *Solanum mammosum*, which resulted in a five-fold enhancement in solasodine productivity as compared with the control.

2.10.2 Nutrient Feeding with Elicitation

Zhang et al. (2004) studied the increase in tanshinone yield as compared with the control using a combination of a medium renewal process and a silver elicitation strategy in hairy root cultures of *Salvia miltiorrhiza*. Similarly, Shinde et al. (2009) reported the effect of salicylic acid, phenylalanine, and polyamines such as putrescine and spermidine on accumulation of isoflavones in hairy root cultures of *Psoralea corylifolia*. The addition of phenylalanine enhanced the concentration of daidzein and genistein 1.3-fold. In another study, it has been proposed that the combined effect of a medium renewal process along with a yeast polysaccharides elicitation strategy in hairy root cultures of *Fagopyrum tatarium* enhanced the flavonoid (rutin and quercetin) yield as compared with the control culture (Zhao et al. 2014).

2.11 Conclusions

In *in vitro* plant cultures, several stimulation strategies have been employed to enhance the yield of secondary metabolites. Elicitation has been widely used in plant cell and organ cultures to improve the production of bioactive compounds. The production of such compounds via elicitation varies depending upon several factors, such as nature and concentration of elicitors, physical conditions of growth chamber, types of culture, etc. Hence, more research is needed to optimize the best methods for enhanced secondary metabolite production several-fold. Further studies on various signaling pathways so as to focus on specific signal molecules involved in elicitor enhancement can serve as efficient tools for enhancing the yield of various secondary metabolites in plant cultures.

Acknowledgment

The authors thankfully acknowledge various funding agencies (Govt. of India) such as Department of Science and Technology (DST), the University Grants Commission (UGC) and the National Medicinal Plants Board (NMPB) for research grants in the form of major research projects and School of Biotechnology, University of Jammu, Jammu. In addition, the authors are thankful to Rashtriya Uchchattar Shiksha Abhiyan (RUSA), the University Grants Commission Special Assistance Programme (UGC-SAP), Promotion of University Research and Scientific Excellence (PURSE) grants, Central facility and Department of Biotechnology (DBT) funded Bioinformatics center at School of Biotechnology, University of Jammu for their support.

REFERENCES

Akhgari, A., Laakso, I., Maaheimo, H., Choi, Y.H., Seppanen-Laakso, T., Oksman-Caldentey, K.M., Rischer, H., 2019. Methyl jasmonate elicitation increases terpenoidindole alkaloid accumulation in *Rhazya stricta* hairy root cultures. *Plants*. 8,534.

Ali, A., Mohammad, S., Khan, M.A., Raja, N.I., Arif, M., Kamil, A., Mashwani, Z.U.R., 2019. Silver nanoparticles elicited *in vitro* callus cultures for accumulation of biomass and secondary metabolites in *Caralluma tuberculata* Artif Cells. *Nanomed Biotechnol.* 47, 715–724.

Anasori, P., Asghari, G., 2008. Effects of light and differentiation on gingerol and zingiberene production in callus culture of *Zingiber officinale* Rosc. *Res Pharmaceut Sci.* 3,59–63.

Aslam, L., Kaur, R., Kapoor, N., Mahajan, R., 2020. Phytochemical composition and antioxidant activities of leaf extracts of *Viola odorata* from Kishtwar, Jammu and Kashmir. *J Herbs Spices Med Plants.* 26,77–88.

Awad, V., Kuvalekar, A., Harsulkar, A., 2014. Microbial elicitation in root cultures of *Taverniera cuneifolia* (Roth) Arn. for elevated glycyrrhizic acid production. *Ind Crops Prod.* 54,13–16.

Ayoola-Oresanya, I.O., Sonibare, M.A., et al., 2021. Elicitation of antioxidant metabolites in Musa species in vitro shoot culture using sucrose, temperature and jasmonic acid. *Plant Cell Tissue Organ Cult.* 146,225–236.

Bajwa, M.N., Bibi, A., Idrees, M.Z., Zaman, G., Farooq, U., Bhatti, T.T., 2021. Elicitation, a mechanistic approach to change the metabolic pathway of plants to produce pharmacological important compounds in *in-vitro* cell cultures. *Glob J. Eng.* 8. doi: 10.33552/GJES.2021.08.000678

Barone, R.P., Knittel, D.K., Ooka, J.K., Porter, L.N., Smith, N.T., Owens, D.K., 2019. The production of plant natural products beneficial to humanity by metabolic engineering. *Curr Plant Biol.* 24,100121.

Begum, S., Zahid, A., Khan, T., Khan, N.Z., Ali, W., 2020. Comparative analysis of the effects of chemically and biologically synthesized silver nanoparticles on biomass accumulation and secondary metabolism in callus cultures of *Fagonia indica. Physiol Mol Biol Plants.* 26(8),1739–1750.

Bhasker, R., Xavier, L.S.E., Udayakumaran, G., Kumar, D.S., Venkatesh, R., Nagella, P., 2021. Biotic elicitors: a boon for the in-vitro production of plant secondary metabolites. *Plant Cell Tissue Organ Cult.* 147, 1–18.

Boccalone, E., Lanni, V.M., Magro, V.M., 2020. Role of natural drugs in severe acute respiratory syndrome coronavirus 2: Potential beneficial effects in prevention or prior treatment in elderly and cancer patients. *Geriatr Care.* 6,9016.

Brachet, J., Cosson, L., 1986. Changes in the total alkaloid content of *Datura innoxia* Mill. subjected to salt stress. *J Experiment Bot.* 37(5),650–656.

Chan, L.K., Koay, S.S., Boey, P.L., Bhatt, A., 2010. Effects of abiotic stress on biomass and anthocyanin production in cell cultures of *Melastoma malabathricum. Biol Res.* 43(1),127–135.

Chandran, H., Meena, M., Barupal, T., Sharma, K., 2020. Plant tissue culture as a perpetual source for production of industrially important bioactive compounds. *Biotechnol Rep (Amst).* 26,e00450.

Chen, Y., Guo, Q., Liu, L., Liao, L., Zhu, Z., 2011. Influence of fertilization and drought stress on the growth and production of secondary metabolites in *Prunella vulgaris* L. *J Med Plants Res.* 5(9),1749–1755.

Chodisetti, B., Rao, K., Suryakala, G., Giri, A., 2015. Gymnemic acid enhancement in the suspension cultures of *Gymnema sylvestre* by using the signaling molecules- methyl jasmonate and salicylic acid. *In Vitro Cell Dev Biol Plant.* 51,88–92.

Chong, T.M., Abdullah, M.A., Lai, O.M., Lajis, N.H., 2005. Effective elicitation factors in *Morinda elliptica* cell suspension culture. *Process Biochem.* 40,3397–3405.

Chu, M., Pedreno, M.A., Alburquerque, N., Faize, L., Burgos, L., Almagro, L., 2017. A new strategy to enhance the biosynthesis of trans-resveratrol by overexpressing stilbene synthase gene in elicited *Vitis vinifera* cell cultures. *Plant Physiol Biochem.* 113,141–148.

De Oliveira, C.A.F., Corassin, C.H., Correa, B., Oswald, I.P., 2014. Animal health: Mycotoxins. *Encyclopedia of Agriculture and Food Systems.* pp. 358–377.

Dowoma, S.A., Abrishamchi, P., Radjabian, T., Salami, S.A., 2017. Enhanced phenolic acids production in regenerated shoot cultures of *Salvia virgata* Jacq. after elicitation with Ag+ ions, methyl jasmonate and yeast extract. *Ind Crops Prod.* 103,81–88.

Ellenberger, J., Siefen, N., Krefting, P., Lutum, J.S., Pfarr, D., Remmel, M., Schroder, L., Rohlen-Schmittgen, S., 2020. Effect of UV radiation and salt stress on the accumulation of economically relevant secondary metabolites in bell pepper plants. *Agronomy.* 10,142.

Farhadi, S., Moieni, A., Safaie, N., Sabet, M.S., Salehi, M., 2020. Fungal cell wall and methyl-β–cyclodextrin synergistically enhance paclitaxel biosynthesis and secretion in *Corylus avellana* cell suspension culture. *Sci Rep.* 10,5427.

Farjaminezhad, R., Garoosi, G., 2021. Improvement and prediction of secondary metabolites production under yeast extract elicitation of *Azadirachta indica* cell suspension culture using response surface methodology. *AMB Expr.* 11(1),43.

Fatima, S., Mujib, A., Tonk, D., 2015. NaCl amendment improves vinblastine and vincristine synthesis in *Catharanthus roseus*: A case of stress signalling as evidenced by antioxidant enzymes activities. *Plant Cell Tissue Organ Cult.* 121,445–458.

Fazal, H., Abbasi, B.H., Ahmad, N., Ali, M., Ali, S.S., Khan, A., Wei, D.Q., 2019. Sustainable production of biomass and industrially important secondary metabolites in cell cultures of selfheal (*Prunella vulgaris* L.) elicited by silver and gold nanoparticles. *Artif Cells Nanomed Biotechnol.* 47,2553–2561.

Fiehn, O., Kopka, J., Dörmann, P., Altmann, T., Trethewey, R.N., Willmitzer, L., 2000. Metabolite profiling for plant functional genomics. *Nat Biotech.* 18(11), 1157–1161.

Gadzovska, S., Tusevski, O., Maury, S., Delaunay, A., Laine, E., Joseph, C., Hagege, D., 2015. Polysaccharide elicitors enhance phenylpropanoid and naphtodianthrone production in cell suspension cultures of *Hypericum perforatum. Plant Cell Tissue Organ Cult.* doi: 10.1007/s11240-015-0798-z

Gai, Q.Y., Jiao, J., Wang, X., Liu, J., Wang, Z.Y., Fu, Y.J., 2019. Chitosan promoting formononetin and calycosin accumulation in *Astragalus membranaceus* hairy root cultures via mitogen-activated protein kinase signaling cascades. *Sci Rep.* 9,10367.

Gandi, S., Rao, K., Chodisetti, B., Giri, A., 2012. Elicitation of andrographolide in the suspension cultures of *Andrographis paniculata. Appl Biochem Biotechnol.* 168,1729–1738.

Gangopadhyay, M., Dewanjee, S., Bhattacharya, S., 2011. Enhanced plumbagin production in elicited *Plumbago indica* hairy root cultures. *J Biosci Bioeng.* 111,706–710.

Gonçalves, S., Mansinhos, I., Rodriguez-Solana, R., Perez-Santín, E., Coelho, N., Romano, A., 2019. Elicitation improves rosmarinic acid content and antioxidant activity in *Thymus lotocephalus* shoot cultures. *Ind Crop Prod.* 137,214–220.

Gorni, P.H., Pacheco, A.C., Moro, A.L., Silva, J.F.A., Moreli, R.R., et al. 2021. Elicitation improves the leaf area, enzymatic activities, antioxidant activity and content of secondary metabolites in *Achillea millefolium* L. grown in the field. *Journal of Plant Growth Regulation.* 40,1652–1666.

Groher, T., Schmittgen, S., Noga, G., Hunsche, M., 2018. Limitation of mineral supply as tool for the induction of secondary metabolites accumulation in tomato leaves. *Plant Physiol Biochem.* 130, 105–111.

Gutierrez-Valdes, N., Häkkinen, S.T., Lemasson, C., Guillet, M., Oksman-Caldentey, K.M., Ritala, A., Cardon, F. 2020. Hairy root cultures: A versatile tool with multiple applications. *Front Plant Sci.* 11,33.

Halder, M., Roychowdhury, D., Jha, S., 2018. A critical review on biotechnological interventions for production and yield enhancement of secondary metabolites in hairy root cultures. In: V. Srivastava, S. Mehrotra, S. Mishra (ed), *Hairy Roots*, Springer Nature, Singapore, pp. 21–44.

Halder, M., Sarkar, S., Jha, S., 2019. Elicitation: A biotechnological tool for enhanced production of secondary metabolites in hairy root cultures. *Eng Life Sci.* 19(12), 880–895.

Hasan, S., 2015. A review on nanoparticles: Their synthesis and types. *Res J. Recent Sci.* 2277,2502.

Hashim, M., Ahmad, B., Drouet, S., Hano, C., Abbasi, B.H., Anjum, S., 2021. Comparative effects of different light sources on the production of key secondary metabolites in plants *in vitro* cultures. *Plants* 10,1521.

Hawrylak-Nowak, B., Dresler, S., Jakubas, M.S., Wojciak, M., Sowa I., Matraszek-Gawron, R., 2021. NaCl-induced elicitation alters physiology and increases accumulation of phenolic compounds in *Melissa officinalis* L. *Int J Mol Sci.* 22, 6844.

He, Y., Hu, Z., Li, A., Zhu, Z., Yang, N., Ying, Z., He, J., Wang, C., Yin, S., Cheng, S., 2019. Recent advances in biotransformation of saponins. *Molecules.* 24,2365.

Hu, X., Neill, S., Cai, W., Tang, Z., 2003. Hydrogen peroxide and jasmonic acid mediate oligogalacturonic acid–induced saponin accumulation in suspension–cultured cells of *Panax ginseng. Physiol Plant.* 118,414–421.

Huang, P., Xia, L., Zhou, L., Liu, W., Wang, P., Qing, Z., Zeng, J., 2021. Influence of different elicitors on BIA production in *Macleaya cordata. Sci Rep.* 11,619.

Huang, X., Yao, J., Zhao, Y., Xie, D., Jiang, X., Xu, Z., 2016. Efficient rutin and quercetin biosynthesis through flavonoids related gene expression in *Fagopyrum tataricum* Gaertn. hairy root cultures with UV-B irradiation. *Front Plant Sci.* 7,63.

Hussain, M.I., Gonzalez, L., Souto, C., Reigosa, M.J., 2011. Eco physiological responses of three native herbs to phytotoxic potential of invasive *Acacia melanoxylon* R. *Br Agroforestry Systems* 83(2),149.

Idrees, M., Naeem, M., Aftab, T., Khan, M.M.A., Moinuddin, 2011. Salicylic acid mitigates salinity stress by improving antioxidant defence system and enhances vincristine and vinblastine alkaloids production in periwinkle [*Catharanthus roseus* (L.) G. Don]. *Acta Physiol Plant.* 33,987–999.

Isah, T., 2019. Stress and defense responses in plant secondary metabolites production. *Biol Res.* 52(1),39.

Jaisi, A., Panichayupakaranant, P., 2020. Enhanced plumbagin production in *Plumbago indica* root culture by simultaneous and sequential dual elicitations using chitosan with L-alanine and methyl-β-cyclodextrin. *Bioresour Bioprocess.* 7,10.

Jan, R., Asaf, S., Numan, M., Lubna, Kim, M-K., 2021. Plant secondary metabolite biosynthesis and transcriptional regulation in response to biotic and abiotic stress conditions. *Agronomy.* 11,968.

Jawdat, D., Al-Faoury, H., Odeh, A., Al-Rayan, R., Al-Safadi, B., 2016. Essential oil profiling in callus of some wild and cultivated Daucus genotypes. *Industrial Crops and Products.* 94,848–855.

Jochum, G.M., Mudge, K.W., Thomas, R.B., 2007. Elevated temperatures increase leaf senescence and root secondary metabolite concentration in the understory herb *Panax quinquefolius* (Araliaceae). *Amer J Bot.* 94,819–826.

Kang, S.M., Min, J.Y., Kim, Y.D., Kang, Y.M., Park, D.J., Jung, H.N., Kim, S.W., Choi, M.S. 2006. Effects of methyl jasmonate and salicylic acid on the production of bilobalide and ginkgolides in cell cultures of *Ginkgo biloba. In Vitro Cell Dev Biol Plant.* 42,44–49.

Karakas, O., 2020. Effect of silver nanoparticles on production of indole alkaloids in *Isatis constricta. Iran J Sci Technol Trans A Sci.* 44, 621–627.

Karimi, A., Majlesi, M., Rafieian-Kopaei, M., 2015. Herbal versus synthetic drugs; beliefs and facts. *J Nephropharmacol.* 4(1),27–30.

Kaur, R., Aslam, L., Hussain, S., Kapoor, N., Mahajan, R., 2021. Flavonoid biosynthetic pathway: Genetics and biochemistry. *Biosci Biotechnol Res Asia.* 18(2),271–286.

Kaur, R., Aslam, L., Kapoor, N., Mahajan, R., 2018. Phytochemical analysis and anti-oxidant activity of wild pomegranate collected from Patnitop, Jammu & Kashmir. *Biosci Biotechnol Res Asia.* 15,335–341.

Kennedy, B.F., De Filippis, L.F., 1999. Physiological and oxidative response to NaCl of the salt tolerant *Grevillea ilicifolia* and the salt sensitive *Grevillea arenaria. J Plant Physiol.* 155,746–754.

Keyaerts, E., Vijgen, L., Pannecouque, C., Van Damme, E., Peumans, W., Egberink, H., Balzarini, J., Ranst, M.V., 2007. Plant lectins are potent inhibitors of coronaviruses by interfering with two targets in the viral replication cycle. *Antiviral Res.* 75(3),179–187.

Khajavi, M., Rahaie, M., Ebrahimi, A., 2019. The effect of TiO_2 and SiO_2 nanoparticles and salinity stress on expression of genes involved in parthenolide biosynthesis in Feverfew (*Tanacetum parthenium* L.). *Caryologia Int J Cytol Cytosyst Cytogenet.* 72,3–14.

Khan, A.K., Kousar, S., Tungmunnithum, D., Hano, C., Abbasi, B.H., Anjum, S., 2021. Nano-elicitation as an effective and emerging strategy for *in vitro* production of industrially important flavonoids. *Appl Sci.* 11,1694.

Kolbe, A., Schneider, B., Porze, A., Schmidt, J., Adam, G., 1999. Acyl-conjugated metabolites of brassinosteroids in cell suspension cultures of *Ornitopus sativus. Phytochemistry.* 41,163–167.

Krishnamurthy, R., Bhagwat, K.A., 1989. Polyamines as modulators of salt tolerance in rice cultivars. *Plant Physiol.* 91(2),500–504.

Krzyzanowska, J., Czubacka, A., Pecio, L., Przybys, M., Doroszewska, T., Stochmal, A., Oleszek, W., 2012. The effects of jasmonic acid and methyl jasmonate on rosmarinic acid production in *Mentha piperita* cell suspension cultures. *Plant Cell Tissue Organ Cult.* 108,73–81.

Kucht, S., Groß, J., Hussein, Y., Grothe, T., Keller, U., Basar, S., Konig, WA., Steiner, U., Leistner, E., 2004. Elimination of ergoline alkaloids following treatment of *Ipomoea asarifolia* (Convolvulaceae) with fungicides. *Planta.* 219(4),619–625.

Lee-Parsons, C.W.T., Eeturk, S., Tengtrakool, J., 2004. Enhancement of ajmalicine production in *Catharanthus roseus* cell cultures with methyl jasmonate is dependent on timing and dosage of elicitation. *Biotechnol Lett.* 0,1595–1599.

Li, D., An, Q., Wu, Y., Li, J.Q., Pan, C., 2020. Foliar application of selenium nanoparticles on celery stimulates several nutrient component levels by regulating the α-linolenic acid pathway. *ACS Sustainable Chem Eng.* 8,10502–10510.

Li, B., Wang, B., Li, H., Peng, L., Ru, M., Liang, Z., Yan, X., Zhu, Y., 2016. Establishment of *Salvia castanea* Diels f. tomentosa Stib. hairy root cultures and the promotion of tanshinone accumulation and gene expression with Ag^+, methyl jasmonate, and yeast extract elicitation. *Protoplasma.* 253(1),87–100.

Li, W.D., Hou, J.L., Wang, W.Q., Tang, X.M., Liu, C.L, Xing, D., 2011. Effect of water deficit on biomass production and accumulation of secondary metabolites in roots of *Glycyrrhiza uralensis. Russ J Plant Physiol.* 58,538–542.

Li, Y.J., Guo, Y., Yang, Q., Weng, X.G., Yang, L., Wang, Y.J., Chen, Y., Zhang, D., Li, Q., Liu, X.C., Kan, X.X., Chen, X., Zhu, X.X., Kmoniekova, E., Zidek, Z., 2015. Flavonoids casticin and chrysosplenol D from *Artemisia annua* L. inhibit inflammation *in vitro* and *in vivo. Toxicol Appl Pharmacol.* 286(3),151–158.

Liang, Z., Ma, Y., Xu, T., Cui, B., Liu, Y., Guo, Z., Yang, D., 2013. Effects of abscisic acid, gibberellin, ethylene and their interactions on production of phenolic acids in *Salvia miltiorrhiza* bunge hairy roots. *PLoS ONE.* 8(9),e72806.

Linh, T.M., Mai, N.C., Hoe, P.T., Ngoc, N.T., Thao, P.T.H., Ban, N.K., Van, N.T., 2021. Development of a cell suspension culture system for promoting alkaloid and vinca alkaloid biosynthesis using endophytic fungi isolated from local *Catharanthus roseus. Plants.* 10,672.

Liu, C.Z., Guo, C., Wang, Y., Ouyang, F., 2002. Effect of light irradiation on hairy root growth and artemisinin biosynthesis of *Artemisia annua* L. *Process Biochem.* 38,581–585.

Liu, H., Wang, X., Wang, D., Zou, Z., Liang, Z., 2011. Effect of drought stress on growth and accumulation of active constituents in *Salvia miltiorrhiza. Ind Crops Prod.* 33,84–88.

Liu, J.Z., Lam, H.M., 2019. Signal transduction pathways in plants for resistance against pathogens. *Int J Mo. Sci.* 20(9),2335.

Mahajan, R., 2016. In vitro and cryopreservation techniques for conservation of snow mountain garlic. *Methods Mol Biol.* 1391,335–346.

Mahajan, R., Kapoor, N., Singh, I., 2016. Somatic embryogenesis and callus proliferation in *Picrorhiza kurroa* Royle ex. Benth. *J Experiment Biol Agric Sci.* 4,201–208.

Malekpoor, F., Pirbalouti, A.G., Salimi, A., 2016. Effect of foliar application of chitosan on morphological and physiological characteristics of basil under reduced irrigation. *Res Crops.* 17(2),354–359.

Malik, N.A.A., Kumar, I.S., Nadarajah, K., 2020. Elicitor and receptor molecules: Orchestrators of plant defense and immunity. *Int J Mo. Sci.* 21,963.

Marsh, Z., Yang, T., Nopo-Olazabal, L., Wu, S., et al., 2014. Effect of light, methyl jasmonate and cyclodextrin on production of phenolic compounds in hairy root cultures of *Scutellaria lateriflora. Phytochemistry.* 107,50–60.

Mizukami, H., Konoshima, M., Tabata, M., 1977. Effect of nutritional factors on shikonin derivative formation in *Lithospermum callus* cultures. *Phytochemistry.* 16,1183–1186.

Modarres, M., Yazdi, M.E.T., 2021. Elicitation improves phenolic acid content and antioxidant enzymes activity in *Salvia leriifolia* cell cultures. *Iran J Sci Technol Trans A: Sci.* 45,849–855.

Moharrami, F., Hosseini, B.B., Sharafi, A., Farjaminezhad, M., 2017. Enhanced production of hyoscyamine and scopolamine from genetically transformed root culture of Hyoscyamus reticulatus L. elicited by iron oxide nanoparticles. *In Vitro Cell Dev Biol -Plant.* doi: 10.1007/s11627-017-9802-0

Moreno-Escamilla, J.O., Jimeńez-Hernández, F.E., Alvarez-Parrilla, E., Rosa, L.A., Martínez-Ruiz, N.R., González-Fernández, R., Orozco-Lucero, E., González-Aguilar, G.A., García-Fajardo, J.A., Rodrigo-García, J., 2020. Effect of elicitation on polyphenol and carotenoid metabolism in Butterhead Lettuce (*Lactuca sativa* var. capitata). *ACS Omega.* 5(20),11535–11546.

Nabi, N., Singh, S., Saffeullah, P., 2021. Responses of *in vitro* cell cultures to elicitation: Regulatory role of jasmonic acid and methyl jasmonate: A review. *In Vitro Cell Dev Biol Plant.* 57,341–355.

Namdeo, A., Patil, S., Fulzele, D.P., 2002. Influence of fungal elicitors on production of ajmalicine by cell cultures of *Catharanthus roseus. Biotechnology Progress.* 18,159–162.

Nopo–Olazabal, C., Condori, J., Nopo–Olazabal, L., Medina–Bolivar, F., 2014. Differential induction of antioxidant stilbenoids in hairy roots of *Vitis rotundifolia* treated with methyl jasmonate and hydrogen peroxide. *Plant Physiol Biochem.* 74,50–69.

Ohlsson, A.B., Berglund, T., 1989. Effect of high $MnSO_4$ levels on cardenolide accumulation by *Digitalis lanata* tissue cultures in light and darkness. *J Plant Physiol.* 135,505–507.

Oksman-Caldentey, K.M., Saito, K., 2005. Integrating genomics and metabolomics for engineering plant metabolic pathways. *Curr Opin Biotechnol.* 16(2),174–179.

Omer, Y., Bengi, E., 2013. Effects of sucrose and polyethylene glycol on hypericins content in *Hypericum adenotrichum. Eurasian J Biosci.* 7,101–110.

Onrubia, M., Moyano, E., Bonfill, M., Cusidó, R.M., Goossens, A., Palazón, J., 2013. Coronatine, a more powerful elicitor for inducing taxane biosynthesis in *Taxus* media cell cultures than methyl jasmonate. *J Plant Physiol.* 170,211–219.

Ooi, C.T., Syahida, A., Stanslas, J., Maziah, M., 2016. The influence of methyl jasmonate, cholesterol and L-arginine on solasodine production in hairy root culture of *Solanum mammosum. Eng Life Sci.* 432–442.

Pavlik, M., Vacek, J., Klejdus, B., Kuban, V., 2007. Hypericin and hyperforin production in St. John's wort in vitro culture: Influence of saccharose, polyethylene glycol, methyl jasmonate, and *Agrobacterium tumefaciens. J Agric Food Chem.* 55,6147–6153.

Pratibha, G., Satyawati, S., Sanjay, S., 2015. Biomass yield and steviol glycoside production in callus and suspension culture of *Stevia rebaudiana* treated with proline and polyethylene glycol. *Appl Biochem Biotech.*. doi: 10.1007/s12010-015-1616-0

Ptak, A., Morañska, E., Saliba, S., Zieliñski, A., Simlat, M., Laurain-Mattar, D., 2017. Elicitation of galanthamine and lycorine biosynthesis by *Leucojumae stivum* L. and *L. aestivum* 'Gravety Giant' plants cultured in bioreactor RITA§. *Plant Cell Tissue Organ Cult.* 128,335–345.

Qu, J., Zhang, W., Yu, X., 2011. A combination of elicitation and precursor feeding leads to increased anthocyanin synthesis in cell suspension cultures of *Vitis vinifera. Plant Cell Tissue Organ Cult.* 107(2),261–269.

Raghavendra, S., Ramesh, C.K., Kumar,V., Khan, H.H.M., 2011. Elicitors and precursor induced effect on L-Dopa production in suspension cultures of *Mucuna pruriens* L. *Front Life Sci.* 5,127–133.

Ramirez-Estrada, K., Vidal-Limon, H., Hidalgo, D., Moyano, E., Golenioswki, M., Cusidó, R.M., Palazon, J., 2016. Elicitation, an effective strategy for the biotechnological production of bioactive high-added value compounds in plant cell factories. *Molecules.* 21,182.

Rivero-Montejo, S.J., Vargas-Hernandez, M., Torres-Pacheco, I., 2021. Nanoparticles as novel elicitors to improve bioactive compounds in plants. *Agriculture.* 11,134.

Ryang, J., Yan, Y., Song, Y., Liu, F., Ng, T.B., 2019. Anti-HIV, antitumor and immunomodulatory activities of paclitaxel from fermentation broth using molecular imprinting technique. *AMB Express.* 9,194.

Ryu, Y.B., Jeong, H.J., Kim, J.H., Kim, Y.M., Park, J.Y., Kim, D., et al., 2010. Biflavonoids from *Torreya nucifera* displaying SARS-CoV 3CL(pro) inhibition. *Bioorg Med Chem.* 18,7940–7947.

Saba, P.D., Iqbal, M., Srivastava, P.S., 2000. Effect of $ZnSO_4$ and $CuSO_4$ on regeneration and lepidine content in *Lepidium sativum. Biol Plant.* 43,253–256.

Sagar, T., Kapoor, N., Mahajan, R., 2021. Nano-elicitation: An efficient method for enhancement of secondary metabolite production in plant cultures. *Indian J Applied Pure Bio.* 36(2),561–567.

Sak, M., Dokupilova, I., Ka, S., Markova, M., Mihalik, D., Hauptvogel, P., Kraic, J., 2021. Biotic and abiotic elicitors of stilbenes production in *Vitis vinifera* L. *Cell Culture Plants.* 10,490.

Salehi, M., Moieni, A., Safaie, N., Farhadi, S., 2020. Whole fungal elicitors boost paclitaxel biosynthesis induction in *Corylus avellana* cell culture. *PLoS ONE* 15(7),e0236191.

Shakeran, Z., Keyhanfar, M., Asghari, G., Ghanadian, M., 2015. Improvement of atropine production by different biotic and abiotic elicitors in hairy root cultures of *Datura metel*. *Turk J Biol.* 39,111–118.

Shakya, P., Marslin, G., Siram, K., Beerhues, L., Franklin, G., 2019. Elicitation as a tool to improve the profiles of high-value secondary metabolites and pharmacological properties of *Hypericum perforatum*. *J Pharm Pharmacol.* 71,70–82.

Sharifi-Rad, M., Lankatillake, C., Dias, D.A., Docea, A.O., Mahomoodally, M.F., Lobine, D., Chazot, P.L., Kurt, B., Tumer, T.B., Moreira, A.C., Sharopov, F., Martorell, M., Martins, N., Cho, W.C., Calina, D., Sharifi-Rad, J., 2020. Impact of natural compounds on neurodegenerative disorders: From preclinical to pharmacotherapeutics. *J Clinic Med.* 9(4),1061.

Shinde, A.N., Malpathak, N., Fulzele, D.P., 2009. Enhanced production of phytoestrogenic isoflavones from hairy root cultures of Psoralea corylifolia L. using elicitation and precursor feeding. *Biotechnol Bioproc Eng.* 14,288.

Singh, N.R., Rath, S.K., Behera, S., Naik, S.K., 2018. *In vitro* secondary metabolite production through fungal elicitation: An approach for sustainability. In: R. Prasad, V. Kumar, M. Kumar, S. Wang (ed), *Fungal Nanobionics: Principles and Applications*, Springer, Singapore, pp. 215–242.

Singh, N., Kumaria, S., 2021. Deciphering the role of stress elicitors on the differential modulation of chalcone synthase gene and subsequent production of secondary metabolites in micropropagated Coelogyne ovalis Lindl., a therapeutically important medicinal orchid. *South African J Bot.* 140,336–348.

Sivanandhan, G., Kapil, D.G., Jeyaraj, M., Rajesh, M., Manickavasagam, M., et al., 2013. Increased production of withanolide A, withanone and withaferin A in hairy root cultures of *Withania somnifera* L. Dunal elicited with methyl jasmonate and salicylic acid. *Plant Cell Tissue Organ Cult.* 114,121–129.

Srivastava, M., Singh, G., Sharma, S., Shukla, S., Misra, P., 2019. Elicitation enhanced the yield of glycyrrhizin and antioxidant activities in hairy root cultures of *Glycyrrhiza glabra* L. *J Plant Growth Regul.* 38,373–384.

Srivastava, S., Srivastava, A.K., 2014. Effect of elicitors and precursors on azadirachtin production in hairy root culture of *Azadirachta indica*. *Appl Biochem Biotechnol.* 172,2286–2297.

Srivastava, V., Mehrotra, S., Mishra, S., 2018. *Hairy Roots: An Effective Tool of Plant Biotechnology*, Springer, Singapore.

Srivastava, V., Mehrotra, S., Verma, P.K., 2017. Biotechnological interventions for production of therapeutic secondary metabolites using hairy root cultures of medicinal plants. In: S.K. Dubey, A. Pandey, R.S. Sangwan (ed), *Current Developments in Biotechnology and Bioengineering: Crop Modification, Nutrition, and Food Production*, Elsevier, Netherlands, pp. 259–282.

Stangarlin, J.R., Kuhn, O.J., Assi, L., Schwan–Estrada, K.R.F., 2011. Control of plant diseases using extracts from medicinal plants and fungi. In: A. Mendez–Vilas (ed), Science Against Microbial *Pathogens: Communicating Current Research and Technological Advances*, Formatex Research Center, Badajoz, Spain, pp. 1033–1042.

Taurino, M., Ingrosso, I., D'amico, L., Domenico, S.D., Nicoletti, I., Corradini, D., Santino, A., Giovinazzo, G., 2015. Jasmonates elicit different sets of stilbenes in *Vitis vinifera* cv. Negramaro cell cultures. *SpringerPlus.* 4,49.

Thakur, M., Bhattacharya, S., Khosla, P.K., Puri, S., 2019. Improving production of plant secondary metabolites through biotic and abiotic elicitation. *J Appl Res Med Aromat Plants.* 12,1–1.

Thakur, M., Sohal, B.S., 2013. Role of elicitors in inducing resistance in plants against pathogen infection: A review. *ISRN Biochem.* 2013, 762412.

Tirillini, B., Ricci, A., Pintore, G., Chessa, M., Sighinolfi, S., 2006. Induction of hypericins in *Hypericum perforatum* in response to chromium. *Fitoterapia.* 77,164–170.

Trejo–Tapia, G., Jimenez–Aparicio, A., Rodriguez–Monroy, M., De Jesus–Sanchez, A., Gutierrez–Lopez, G., 2001. Influence of cobalt and other microelements on the production of betalains and the growth of suspension cultures of *Beta vulgaris. Plant Cell Tissue Organ Cult.* 67,19–23.

Vaccaro, M.C., Mariaevelina, A., Malafronte, N., De Tommasi, N., Leone, A., 2017. Increasing the synthesis of bioactive abietanediterpenes in *Salvia sclarea* hairy roots by elicited transcriptional reprogramming. *Plant Cell Rep.* 36,375–386.

Vargas-Hernandez, M., Macias-Bobadilla, I., Guevara-Gonzalez, R.G., Romero-Gomez, S.J., Rico-Garcia, E., Ocampo-Velazquez, R.V., Alvarez-Arquieta, L.L., Torres-Pacheco, I., 2017. Plant hormesis management with biostimulants of biotic origin in agriculture. *Front Plant Sci.* 8,1762.

Vecerova, K., Vecera, Z., Docekal, B., Oravec, M., Pompeiano, A., Triska, J., Urban, O., 2016. Changes of primary and secondary metabolites in barley plants exposed to CdO nanoparticles. *Environ Pollut.* 218,2017–2018.

Wang, L.J., Ma, L., Xi, H.F., Duan, W., Wang, J.F., Li, S.H., 2013.Individual and combined effects of $CaCl_2$ and UV–C on the biosynthesis of resveratrols in grape leaves and berry skins. *J Agric Food Chem.* 61,7135–7141.

Wang, S., Li, C., Wang, H., Zhong, X., Zhao, J., Zhou, Y., 2015. A process optimization study on ultrasonic extraction of paclitaxel from *Taxus cuspidata. Prep Biochem Biotech.* 46,274–280.

Wani, S.H., Kapoor, N., Mahajan, R., 2017. Metabolic responses of medicinal plants to global warming, temperature and heat stress. In: M. Ghorbanpour, A. Varma (ed), *Medicinal Plants and Environmental Challenges*, Springer International Publishing AG. pp. 69–80.

Wawrosch, C., Sergey, B., Zotchev, S.B., 2021. Production of bioactive plant secondary metabolites through *in vitro* technologies—status and outlook. *Applied Microbiol Biotech.* 105,6649–6668.

White, M.A., Lin, W., Cheng, X., 2020. Discovery of COVID-19 inhibitors targeting the SARS-CoV-2 Nsp13 helicase. *J Phys Chem Lett.* 11(21),9144–9151.

Wielanek, M., Urbanek, H., 2006. Enhanced glucotropaeolin production in hairy root cultures of *Tropaeolum majus* L. by combining elicitation and precursor feeding. *Plant Cell Tissue Organ Cult.* 86,177–186.

Wiktorowska, E., Długosz, M., Janiszowska, W., 2010. Significant enhancement of oleanolic acid accumulation by biotic elicitors in cell suspension cultures of *Calendula officinalis* L. *Enzyme Microb Technol.* 46,14–20.

Wongwicha, W., Tanaka, H., Shoyama, Y., Putalun, W., 2011. Methyl jasmonate elicitation enhances glycyrrhizin production in *Glycyrrhiza inflata* hairy roots cultures. *Z Naturforschung C J Biosci.* 66,423–428.

Xiaolong, H., Min, S., Lijie, C., Chao, X., Yanjie, Z., Guoyin, K., 2015. Effects of methyl jasmonate and salicylic acid on tanshinone production and biosynthetic gene expression in transgenic *Salvia miltiorrhiza* hairy roots. *Biotechnol Appl Biochem.* 62(1),24–31.

Xu, A., Zhan, J.C., Huang, W.D., 2015. Effects of ultraviolet C, methyl jasmonate and salicylic acid, alone or in combination, on stilbene biosynthesis in cell suspension cultures of *Vitis vinifera* L. cv. Cabernet Sauvignon. *Plant Cell Tissue Organ Cult.* 122,197–211.

Yang, T., Fang, L., Nopo-Olazabal, C., Condori, J., et al. 2015. Enhanced production of resveratrol, piceatannol, arachidin-1, and arachidin-3 in hairy root cultures of peanut co-treated with methyl jasmonate and cyclodextrin. *J Agric Food Chem.* 63,3942–3950.

Yang, L., Wen, K.S., Ruan, X., Zhao, Y.X., Wei, F., Wang, Q., 2018. Plant-microbial interaction: The mechanism and the application of microbial elicitor induced secondary metabolites biosynthesis in medicinal plants. *Molecules.* 23(4),762.

Yazdanian, E., Golkar, P., Reza, M., Taghizadeh, M., 2021. Elicitation effects on some secondary metabolites and antioxidant activity in callus cultures of *Allium jesdianum* Boiss. & Buhse.: Methyl jasmonate and putrescine. *Appl Biochem Biotech.* doi: 10.1007/s12010-021-03643-4

Yu, M.S., Lee, J., Lee, J.M., Kim, Y., Chin, Y.W., Jee, J.G., et al., 2012. Identification of myricetin and scutellarein as novel chemical inhibitors of the SARS coronavirus helicase, nsP13. *Bioorg Med Chem Letters.* 22,4049–4054.

Yue, W., Ming, Q.L., Lin, B., Rahman, K., Zheng, C.J., et al., 2016. Medicinal plant cell suspension cultures: Pharmaceutical applications and high-yielding strategies for the desired secondary metabolites. *Crit Rev Biotechnol.* 36(2),215–232.

Zaheer, M., Giri, C.C., 2015. Multiple shoot induction and jasmonic versus salicylic acid driven elicitation for enhanced andrographolide production in Andrographis paniculata. *Plant Cell Tiss. Organ Cult.* 122, 553–563.

Zaheer, M., Reddy, V.D., Giri, C.C., 2016. Enhanced daidzin production from jasmonic and acetyl salicylic acid elicited hairy root cultures of Psoralea corylifolia L. (Fabaceae). *Nat. Prod. Res.* 30, 1542–1547.

Zaker, A., Sykora, C., Gössnitzer, F., Abrishamchi, P., Asili, J., Mousavi, S.M., Wawrosch, C., 2015. Effects of some elicitors on tanshinone production in adventitious root cultures of *Perovskia abrotanoides* Karel. *Ind Crops Prod.* 67,97–102.

Zhang, C.H., Yan, Q., Cheuk, W., Wu, J., 2004. Enhancement of tanshinone production in *Salvia miltiorrhiza* hairy root culture by Ag^+ and nutrient feeding. *Planta Medica.* 700(2),147–151.

Zhao, J.L., Zou, L., Zhang, C.Q., Li, Y.Y., et al., 2014. Efficient production of flavonoids in *Fagopyrum tataricum* hairy root cultures with yeast polysaccharide elicitation and medium renewal process. *Pharmacogn Mag.* 10,234–240.

Zhao, J., Zhu, W., Hu, Q., 2001. Selection of fungal elicitors to increase indole alkaloid accumulation in *Catharanthus roseus* suspension cell culture. *Enzyme Microb Technol.* 28,666–672.

Zhao, S., Tang, H., 2020. Enhanced production of valtrate in hairy root cultures of *Valeriana jatamansi* Jones by methyl jasmonate, jasmonic acid and salicylic acid elicitors. *Notulae Botanicae Horti Agrobotanici Cluj-Napoca.* 48(2),839–848.

Zhong, J.J.T., Seki, S.I., Kinoshita Yoshida, T., 1993. Effect of light irradiation on anthocyanin production by suspended culture of *Perilla frutescens. Biotechnol Bioeng.* 38,653–658.

Zobayed, S.M.A., Afreen, F., Kozai, T., 2007. Phytochemical and physiological changes in the leaves of St. John's wort plants under a water stress condition. *Environ Experiment Bot.* 59,109–116.

3

Tissue Culture of Rare and Endangered Forest Plant Species of India

Radheshyam Sharma
Jawaharlal Nehru Krishi Vishwa Vidhalya
Jabalpur
Madhya Pradesh, India

Vikram Singh Gaur
College of Agriculture
Waraseoni
Jawaharlal Nehru Krishi Vishwa Vidhalya
Jabalpur
Madhya Pradesh, India

Varsha Kumari
Sri Karan Narendra Agriculture University
Jobner-Jaipur
Rajasthan, India

S.R. Maloo
Pacific University
Udaipur
Rajasthan, India

CONTENTS

DOI: 10.1201/9781003239932-3

3.1 Introduction

Nowadays, one of the most urgent needs of human beings is the sustainable utilization and conservation of the existing biodiversity of all living forms. Continuous loss of biodiversity is a big threat to humans and will weaken our capacity for poverty reduction, food and nutritional security, and human health. A sustainable agriculture system looks to utilize natural resources in such a way that they can regenerate and also to minimize harmful impacts on ecosystems beyond the field's edge. In the recent past, much attention has been paid to the conservation of the genetic stocks of many plant species. It is important to preserve the gene pools not only of crops, but also of economically important rare and endangered forest tree species, which often have several medicinal, culinary, decorative, forage, and other useful properties. Tremendous success has been achieved in conserving and introducing farmer- and consumer-preferred traits into many cultivated crops and plant species (Corlett, 2017; Miolkanova et al. 2015).

Many forest plant species and their products are the main sources for livelihood of many tribals. In a study of a household survey of tribals of Central India, it was found that ~30% of their livelihood income was obtained from forest produce. Apart from this, forest plants are an important component of green vegetation, essential to maintain ecological diversity and to act as a major carbon sink of an ecosystem. Loss of forest vegetation leads to the annihilation of the essence of the biological flora and fauna, which is of global concern, irrespective of regional and local importance (Kumari et al. 2019). Destruction of plant species from their natural habitat eliminates the source of economic gain and increases the risk of flood and other natural calamities. A study indicates that there is a close association between disease outbreaks and the degradation of natural vegetation. Various forest tree species in Asia, Europe, Australia, Africa, South America, North America, and Antarctica; particularly in India, are currently under threat of extinction. Tropical forests of India are the most productive as well as the most threatened, with an extreme rate of deforestation. Over-exploitation, anthropogenic load in the form of grazing, plowing land, shifting cultivation, rapid industrialization, urbanization, logging, and geo-mining are the main reasons for the extinction of important forest tree species from their natural habitat. Other factors influencing the reduced reproduction of these tree species are low seed germination capacity, poor seed viability, lack of alternative methods of propagation, relict species, harsh climatic conditions, being eaten by animals and birds, and lack of government support for conservation (Marchese, 2015; Miolkanova et al. 2015).

Several strategies have been utilized for the preservation of the gene pool, such as (a) establishment of botanical gardens and nurseries, (b) creation of reservoirs and other specific protected forests, and (c) recent biological tools to create seed banks, gene banks, genetic stock, cells, tissues, pollen storage, and national and international plant repositories. In India, the Indian Council of Agricultural Research (ICAR)–National Bureau of Plant Genetic Resources (NBPGR) is the key institute for the collection and conservation of germplasm of agricultural and horticultural crops. Apart from NBPGR, the Botanical Survey of India-Kolkata, Forest Research Institute-Dehradun, and National Botanical Research Institute-Locknow are the main organizations involved in the exploration, collection, identification, and documentation of plant resources in the country. Modern biotechnological tools play an important role in the multiplication and conservation of the endangered elite genotypes of many

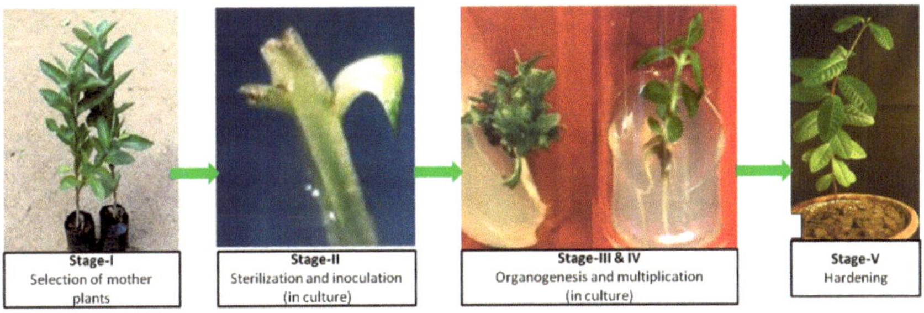

FIGURE 3.1 Stages of micro-propagation for effective *in vitro* regeneration of plantlets.

forest plants. Therefore, these tools can be successfully applied in the conservation of the gene pool of rare and endangered forest plants (Coelho et al. 2020). Plant tissue culture is a very effective approach for rapid multiplication and genetic improvement of plants. It is widely used for *ex situ* germplasm conservation and restoration of the gene pool of many critical forest tree species. This approach consists of five major steps: selection of elite mother plants, sterilization of explants, inoculation into suitable medium, organogenesis, and acclimatization (Figure 3.1). Thus, the development of plants through plant tissue culture is an effective method for *ex situ* biodiversity conservation of plant species (Grigoriadou et al. 2019).

3.2 Endangered Plants in India

Forest is a conditional renewable resource which can be reproduced but requires a specific time and climate to maintain its sustainable functioning. In India, many forest plant species are being continuously depleted, which is an alarming situation for the nation. The number of Indian plants in the International Union for Conservation of Nature (IUCN) Red List is steadily increasing, much to the dismay of conservationists. In 2018, the Red List featured 4,537 endangered species globally, while in 2019–20, this number went up to 4,993. In 2019, around 176 endangered forest plant species were recorded from India only. The major threats noted were habitat destruction, climate change, urbanization, global warming, and other anthropogenic activities. According to the IUCN Red List, recent biotechnological tools are essential to conserve rare and endangered plant species that do not reproduce easily. It is a fact that many useful forest plant species are critical, extinct, or near to extinction, and may vanish in the near future. A list of major endangered forest plants of India is shown in Table 3.1.

3.3 Tissue Culture Techniques for Some Endangered Plants

Biodiversity is an essential component of the earth's climate and offers innumerable services to human beings. Biodiversity boosts ecosystem productivity; therefore, the study of biodiversity and the identification and assessment of new resources are essential. An ecosystem continuously destabilized by the hammering of biodiversity is less likely to deliver those services, especially the daily needs of an ever-growing

TABLE 3.1

List of Some Major Endangered Forest Plants Species in India

Name of the plants	Family	Distribution	Status
Vateria macrocarpa	Dipterocarpaceae	Endemic to southern Western Ghat	CR
Hopea erosa (Bedd.)	Dipterocarpaceae	Endemic to Western Ghats, in Tamil Nadu and Kerala.	CR
Aglaia malabarica	Meliaceae	Endemic to Western Ghats, in Tamil Nadu and Kerala	CR
Cynometra beddomei	Leguminosae	Endemic to southern Kerala and Wayanad	CR
Dipterocarpus indicus	Dipterocarpaceae	Endemic to Western Ghats, throughout	E
Hopea ponga	Dipterocarpaceae	Endemic, throughout Western Ghats, semi-evergreen forests	E
Kingiodendron pinnatum	Leguminosae	Endemic to southern Western Ghats, throughout	E
Cynometra travancorica	Leguminosae	Endemic to Western Ghats in Karnataka, Tamil Nadu, Kerala	E
Atuna travancorica (Bedd.)	Chrysobalanceae	Endemic to Western Ghats, Tamil Nadu, Kerala	E
Madhuca bourdillonii	Sapotaceae	Endemic to southern Western Ghats	E
Vateria indica L.	Dipterocarpaceae	Endemic to Western Ghats	CR
Arenga wightii	Arecaceae	Endemic to southern Western Ghats, throughout, in evergreen forests along stream sides	V
Bentinckia condapanna	Arecaceae	Endemic to southern Western Ghats, south of Palakkad Gap, in Tamil Nadu and Kerala, always on steep slopes of high altitudes	V
Garcinia wightii	Clusiaceae	Endemic to southern Western Ghats	V
Syzygium occidentale	Myrtaceae	Endemic to Southern Western Ghats, in riverine vegetation	V
Myristica malabarica Lam	Myristicaceae	Endemic to Western Ghats in riverine evergreen forests and Myristica swamps	V
Cinnamomum riparium Gamble	Lauraceae	Endemic to southern Western Ghats, along stream sides	V
Ochreinauclea missionis	Rubiaceae	Endemic to southern Western Ghats, in riverine vegetation	V
Buchanania lanceolata Wt.	Anacardiaceae	India, Myanmar	V
Gymnacranthera canarica	Myristicaceae	Endemic to southern Western Ghats near streams and Myristica swamps	V
Hydnocarpus macrocarpa	Flacourtiaceae	Endemic to Western Ghats	V
Garcinia travancorica Bedd	Clusiaceae	Restricted endemic to Agasthyamala Biosphere Reserve	V

TABLE 3.1 (Continued)

List of Some Major Endangered Forest Plants Species in India

Name of the plants	Family	Distribution	Status
Garcinia imberti Bourd	Clusiaceae	Restricted endemic to Agasthyamal Biosphere Reserve	E
Garcinia indica	Clusiaceae	Endemic to northern Western Ghats, introduced to many parts of Asia, Europe	V
Myristica beddomei subsp. *sphaerocarpa*	Myristicaceae	Endemic to southern Western Ghats	E
Psydrax dicoccos Gaertn.	Rubiaceae	Indo-Malaysia, China	V
Calophyllum apetalum Willd.	Clusiaceae	Endemic to Western Ghats on river banks	V
Santalum album L.	Santalaceae	India, Malaysia	V
Saraca asoca (Roxb.)	Caesalpinioideae	Indo-Malaya	V
Polygala irregularis	Polygalaceae	Endemic to Rann of Kutch Gujarat (rare)	E
Lotus corniculatus	Fabaceae	Gujarat	E
Amentotaxus assamica	Taxaceae	Arunachal Pradesh	CR
Psilotum nudum	Psilotaceae	Karnataka	E
Diospyros celibica	Ebenaceae	Karnataka	CR
Actinodaphne lawsonii	Lauraceae	Kerala	CR
Acacia planifrons	Fabaceae	Endemic to Western Ghat of Tamil Nadu	E
Abutilon indicum	Malvaceae	Endemic to Western Ghat of Tamil Nadu	E
Chlorophytum tuberosum	Asparagaceae	Endemic to Western Ghat of Tamil Nadu	E
Nymphaea tetragona	Nymphaea	Jammu	E
Belosynapsis vivipara	Commelinaceae	Madhya Pradesh	E
Colchicum luteum	Colchicaceae	Himachal Pradesh	CR
Pterospermum reticulatum	Malvaceae	Kerala	CR
Ceropegia odorata	Asclepiadaceae	Gujarat	E
Buchanania lanzan	Anacardiaceae	Endemic to Central MP, CG and Gujarat	E

CR: Critical; E: Endangered; V: Vulnerable

human population. Different types of pollution, climate change, anthropogenic load, and industrialization are all threats to biodiversity. These threats are responsible for an unprecedented rise in the rate of species extinction. Scientists have predicted that approximately one-third of all species on earth will be wiped out within the next century. Therefore, biodiversity conservation efforts are essential to preserving existing biodiversity as well as protecting endangered species in their natural habitats. Many species of forest trees are being depleted from their natural habitat, and researchers are making efforts to counter this. The main approaches for the regeneration and conservation of rare and endangered plants are *in situ*, *ex situ*, and modern *in vitro* approaches. The *in situ* approach will take place in natural existing ecosystems through the construction of uniquely protected natural territories: nature reserves, national parks, natural monuments, etc. This approach allows continuous evolution in the area of their natural habitat, while *ex situ* conservation involves the off-site conservation of the wild genetic resources/genetic diversity. *Ex situ* conservation takes place outside the natural surroundings and includes collection, preservation, and maintenance of selected genetic resources from the wild; development of botanical gardens, gene banks, and DNA banks; techniques involved in tissue culture and cryopreservation; integration of biotic and abiotic stress tolerance traits through genetic transformation; and ecological restoration of rare and endangered species of plants and their populations. These two groups of techniques have basic differences: in the *ex situ* conservation process, the taxon of interest is taken out from its natural domain and grown under artificially developed conditions that provide a better degree of protection to the germplasm, whereas *in situ* conservation involves determining the natural habitat and observing plant growth and development (Maxted et al. 1997). The major problems in developing protected areas and setting up living collections of rare and endangered species are the creation, regular tracking, and protection of natural habitats, which require large areas, as well as injury to the plants by wild animals and insect pests (Laslo et al. 2011). It is noteworthy that *in situ* conservation of biodiversity is recommended in most cases but is not always appropriate for the protection of individual plant species. Hence, *ex situ* gene pool conservation approaches are becoming more attractive and popular. They originally depended on creating large-scale collections of rare and endangered plants with the establishment of field gene banks and seed banks (Pence et al. 2017). Therefore, *ex situ* approaches gained international recognition with their inclusion in Article 9 of the Convention on Biological Diversity (Kapai et al. 2010).

Currently, deposition under slow-growth conditions at low temperatures (+2–15 °C), storage in conditions of active growth, and cryopreservation in liquid nitrogen (−196 °C) are the three preferable approaches and have proved successful in many vegetatively propagated crops. Rhizomes, tubers, corns, roots, and cuttings of many perennial and woody plants are shipped and stored under such conditions. Seeds that have very low germination rates and require specific conditions for growth and development are generally preserved, reproduce, and are reintroduced in active growth conditions. This approach is broadly used for *in vitro* conservation of both monocotyledonous and dicotyledonous plants. Additionally, *in vitro* culture of apical shoots and axillary buds is preferred to obtain virus- and disease-free planting materials (Reed et al. 2011; Matsumoto, 2017).

3.4 *In vitro* Propagation Techniques for Endangered Forest Plant Species

Rare and endangered plant species are highly specialized for survival in a particular environment. Such species require either natural habitat or artificially developed controlled conditions to regenerate them. *In vitro* propagation of such plant species is usually undertaken to augment the biomass and conserve the germplasm, particularly when the population is very low in the wild. This approach has been successfully utilized in many plants where conventional ways are not effectively working and the population has decreased due to over-exploitation by destructive harvesting. Additionally, the approach can effectively be used to meet the growing demand for clonally uniform elite plants. It is also providing an alternative to produced by-products of a plant species through suspension culture and alleviates pressures on wild populations. *In vitro* micropropagation with advanced biotechnological tools offers avenues for conservation, genetic improvement, and efficient use of endangered plant resources and products (Bapat et al. 2008).

New and updated *in vitro* micropropagation of endangered plant species requires an effective combination of various macro and micro elements, vitamins, phytohormones, and amino acids, and sometimes specific antibiotics, activated charcoal, antioxidants, and woody plant medium, to regenerate them. However, microspore and anther culture, protoplast culture, embryo culture, bud culture, callus culture, and meristem culture are commonly used today for regeneration, not only species-specific but also for conservation of many endangered plant species.

Many rare and endangered plant species are grown in plant cell bioreactors. The design of these bioreactors is based on plant cell and tissue culture characteristics. These systems provide quick and well-organized propagation of many forestry and agricultural plant species through utilizing liquid and semi-liquid mediums. This approach is more efficient and faster than using a solid medium due to maximum supply and absorbance of nutrients and hormones to explants, better contact of plant tissue with the culture medium, and provision of aeration and circulation for maximum growth in scaling-up processes. But the whole process of optimal plant regeneration will be based on a proper understanding of plant responses to signals from the microenvironment and on manipulation in culture to control the morphogenesis of plants in liquid cultures. Automation in a bioreactor has been advanced as a possible way of minimizing costs and has given good results. Several plant species, such as *Anoectochilus*, guava, apple, garlic, *Chrysanthemum*, pomegranate, ginseng, grape, *Phalaenopsis*, *Lilium*, and potato, have been propagated using bioreactor systems. Due to automation and advances in technological properties, bioreactors not only promote a quick and top-quality micropropagation process but also have the facility to rapidly accumulate substances valuable for medicine from the roots of rare and endangered plant species (Yoon et al. 2007).

However, the cells, tissues, and organs of some medicinal and forest tree species do not respond to tissue culture manipulations. Such types of species are very difficult at one or more stages of micropropagation. These species are considered as "recalcitrant" plants. For "recalcitrant" plant species, it is essential to search for suitable explants, nutrient medium compositions, and environmental adaptation methods, which are very difficult with many tree species and aquatic plants. The physiology of the donor plant, *in vitro* manipulations of media, and the cultural environment are the main factors that influence recalcitrance in plants. Therefore, integrated approaches to whole-plant physiology with an understanding of tissue culture responses are essential for overcoming recalcitrance (Benson, 2000).

In the case of several aquatic plant species, *in vitro* culture techniques face a high degree of contamination of plants as well as poor seed germination (Nguyen, 2016). Therefore, to overcome these problems, efficient surface sterilization and different types of explants must be sought. In the case of orchids, difficulties occur when using seed material devoid of stored food material or endosperm as well as when adapting them *ex situ* (Harrap and Harrap 2005). In nature, orchids cannot utilize their own scanty lipid reserves, break down starch, or photosynthesize. After the uptake of water, seed swelling and turning green may occur but fail to develop further in the absence of mycorrhizal fungi infection (symbiotic germination). Lacking nutrients in the orchid seed and without the ability to utilize reserves, these plants are forced to enter into a symbiotic relationship with mycorrhizal fungi. Therefore, without the symbiotic association of mycorrhizal fungi, even with micropropagation, the propagation of orchids is almost impossible. Thus, in the case of orchids, the introduction of fungal culture into the *in vitro* culture of explants is employed to obtain successful regeneration.

Protoplast culture is another approach to *in vitro* micropropagation of many endangered plant species. Plant protoplasts are totipotent and can regenerate into various organs. In addition, protoplasts easily take up foreign genetic material and have become important in diverse fields of plant biotechnology such as genetic manipulation, gene expression, functional characterization, genome editing, and transcriptome studies (Mitrofanova and Moroz 2018).

A new approach to the micropropagation of both rare and many endangered plant species is the biotization of endophytic microorganisms. This is a bio-hardening technique whereby endophytes are used under both *in vitro* and *ex vitro* conditions to stimulate growth, reduce stress, and increase plant immunity (Kanani et al. 2020). Generally, spores of arbuscular mycorrhiza, Trichoderma, and plant growth promoting rhizobacteria (PGPR) are inoculated in either liquid or powder formulations. Recent research has shown several beneficial effects of many microorganisms on the growth of the vegetative part of plants, callus growth, seed maturation, resistance to pathogens, and increased tolerance to low temperature. Microplant biotization is an emerging field of science aiming to reduce chemical input and increase plant fitness and productivity for sustainable agriculture.

Micrografting is another technique that allows the massive propagation of several plant taxa of wild and endangered species. It involves the *in vitro* grafting of small shoot apices or lateral buds onto decapitated rootstock seedlings. However, *in vitro* grafting is influenced by scion size and rootstock age and requires great skill (Vidoy-Mercado et al. 2021).

3.5 Examples of Micropropagation Protocol Development of Some Rare and Endangered Plant Species in India

Raju and Divya (2020) carried out a study at the Department of Biology, The Gandhigram Rural Institute, Gandhigram, India, and developed an efficient micropropagation protocol of an endangered tree species, *Syzygium densiflorum*. The species belongs to the Myrtaceae family and has many medicinal values. *S. densiflorum* species has been overexploited due to the presence of several medicinal compounds in every part, such as bark, leaves, root, and seeds, and therefore categorized under severe threat of extinction. The over-exploitation, anthropogenic load, habitat degradation, shifting cultivation, irregular phenological events, geo-mining, low viability of the seed, and reduced adaptability of the seedling into the natural habitat are the key factors for the disappearing population of the species.

3.5.1 Explant Selection and Sterilization

Fresh and healthy nodal, inter-nodal, leaf, and seed explants were collected from wild trees for sterilization and further inoculation into the medium. The sterilization of explants included washing of explants with tap water for 5–10 min, treatment with 0.1% mercuric chloride for 5 min and 2–4% sodium hypochlorite for 5 min, followed by treatment with 70% ethanol for 5 min to remove both bacterial and fungal contaminants. Additionally, explants were washed twice with double sterile distilled water and finally, washed four times with autoclaved distilled water.

3.5.2 Shoot Formation from Nodal Explants

Woody Plant Medium (WPM) with pH 5.8 gave the best response for organogenesis when compared with B_5 and Murashige and Skoog (MS) medium. For shoot induction, explants were inoculated on WPM supplemented with different concentrations (0.5–2.5 mg/l) of phyto-hormones such as 6-benzylaminopurine (BAP), Kn alone, and combinations of BAP + indole-3-acetic acid (IAA), BAP + indole-3-butyric acid (IBA), and BAP + 2,4-dichlorophenoxyacetic acid (2,4-D). In all combinations, bud break was observed after 28–35 days, and the number and length of shoots were observed, with considerable differences at various concentrations in the induction media. Multiple shoot induction was observed in BAP alone as well as in the combination of BAP with IAA, IBA, or 2,4-D. For kinetin, the maximum number of shoots (2.2 ± 0.23) was observed in WPM containing 1.0 mg/l kinetin. Similarly, for BAP, the maximum number of shoots (3.5 ± 0.08) and frequency (45.43%) of shoot regeneration were observed after the eighth week in the nodal explants inoculated on WP medium containing BAP (0.5 mg/l). In combination with different plant growth regulators, BAP + IAA (1.0 g/l) showed the maximum number of shoots (6.3 ± 0.17) and frequency (81.77%). Likewise, in combinations of BAP + IBA, 1.5 mg/l showed the highest number of shoots (7.7 ± 0.08) with 100% frequency, while in the combinations of BAP + 2,4-D, 2.0 mg/l showed the maximum number of shoots (2.8 ± 0.03), with 36.34% frequency after the eighth week of inoculation (Table 3.2).

3.5.3 Root Formation from Shootlet Development from Nodal Explants

Auxin is essential for artificial root induction in *in vitro* culture of explants. It causes rapid cell division and is involved in the organization of defined organs. Healthy developed 4–5-cm long shoots with six to eight nodes from nodal explants were

TABLE 3.2

Shoot Proliferation from Nodal Explants of *S. densiflorum* on Woody Plant Media with Different Plant Growth Regulators

WP ;+ 1 mg/ l Kinetin	WP + 0.5 mg/ l BAP	WP + 1 mg/l each BAP and IAA	WP + 1.5 mg/l each BAP and IBA	WP + 2 mg/l each BAP and 2,4-D
2.2 ± 0.23	3.5 ± 0.08	6.3 ± 0.17	7.7 ± 0.08	2.8 ± 0.03

TABLE 3.3

Root Induction of *S. densiflorum* on Woody Plant Media with IAA and IBA

	WP + 0.5 mg/l IBA	WP + 105 mg/l IAA
Frequency of roots (%)	100	62.64
Mean number of roots	3.83 ± 0.53	2.4 ± 0.23
Mean length of roots (cm)	3.4 ± 0.05	1.9 ± 0.17

inoculated individually on rooting media containing half-strength WPM with different concentrations of auxins. The maximum response for root induction was observed in IBA when compared with the concentrations of IAA. The highest frequency (100%) of root formation in nodal explants, with 3.83 ± 0.53 roots and 3.4 ± 0.05 cm root length, was seen on WPM fortified with 1.0 mg/l IBA. Similarly, with IAA, the maximum frequency (62.64%) was noticed with 1.5 mg/l on WPM (Table 3.3).

Sanjay et al (2006), working at Tree Improvement and Propagation Division, Institute of Wood Science and Technology, Malleshwaram, Bangalore, India, developed a micropropagation protocol for endangered Indian sandalwood (*Santalum album* L). Sandalwood is an economically important plant harvested for heartwood oil. India is the main exporter of sandalwood oil and its products to various countries.

Fresh and healthy nodal segments were collected from 50–60-year-old trees and inoculated on MS medium supplemented with 0.53 μM α-naphthalene acetic acid (NAA) and 11.09 μM 6-benzyl amino purine (BA). Later, *in vitro* differentiated shoots were transferred into multiplication medium supplemented with 4.44 μM BA, 0.53 μM NAA, and the following additives: 283.93 μM ascorbic acid, 118.10 μM citric acid, 104.04 μM cystine, 342.24 μM glutamine, and 10% (v/v) coconut milk. Repeated subculture was done on fresh medium at four-week intervals. After 30–40 days of inoculation, micro shoots had developed and were transferred into the root induction medium. For root induction, micro shoots were inoculated on MS supplemented with 98.4 μM IBA for 48 h and produced roots on growth-regulator-free, quarter-strength MS basal salts medium with vitamin B_5 and 2% sucrose. *In vitro* root induction was obtained from micro shoots pulsed with 1230 μM IBA for 30 min in soilrite rooting medium. The percentage of rooting in soilrite was higher than in agar medium, and *in vitro* raised plants were established in the field and showed normal growth.

Singh and Sharma (2020) developed an efficient *in vitro* micropropagation protocol of an economically important endangered plant, chironji (*Buchanania lanzan* Spreng). The species is a medium-sized deciduous forest tree, native to the Indian subcontinent, and belongs to the family Anacardiaceae. It is found naturally in the dry deciduous forests of Madhya Pradesh, Chhattisgarh, Jharkhand, South East Uttar Pradesh, Gujarat, and Rajasthan. Seven species of *Buchanania* have been reported in India; only *Buchanania lanzan* and *Buchanania axillaries* (Syn. *Angustifolia*) produce edible fruits, while other species are not edible. The establishment of an efficient *in vitro* mass multiplication protocol is an essential prerequisite to conserve the depleted population of chironji.

3.6 Explant Selection and Sterilization

Young leaves and nodal segments of chironji were chosen. Initially, the explants were thoroughly washed with tap water followed by 0.15 (v/v) Tween-20 solution for 15 minutes and rinsed with double distilled water three times. Explants were transferred to the laminar airflow and treated with 0.5% bavistin for 30 minutes, followed by 0.1% mercuric chloride (for different time durations, ranging from 3 to 6 minutes depending on the type of explants taken), and washed with sterile double distilled water. MS and WPM medium (pH 5.8) were used for callus, shoot, and root induction. Explants were incubated at 24–28 °C with a relative humidity of 55–65%, light intensity of 40 mmol/m^2/s, and a 16-h light/8-h dark period. Cultures were transferred to a fresh medium every three weeks.

3.6.1 Shoot Formation from Nodal Explants

For direct shoot induction from nodal segments, WPM medium was used with different concentrations of TDZ, BAP, and kinetin as well as WPM medium without growth regulator as a control. Sterilized explants were placed vertically on the medium and incubated at 25 ± 2 °C with a photosynthetic photon intensity of 50 μmol/m^2/s under 16/18h photoperiod and 60–70% humidity for 21–24 days. After 21–24 days of incubation, cultures were transferred to the same medium for further growth. The number of shoots per explant was counted after 45 days of incubation. Nodal segments showed an initial response of swelling after 15–20 days of incubation. Maximum shoot initiation response (34%) was observed in ½WPM fortified with 2.5 mg/l TDZ, while minimum shoot initiation response (3%) was observed in ½WPM fortified with 0.5 mg/l TDZ. Activated charcoal (0.1%) was also added to the medium to prevent phenolic secretions from the explants. Further, maximum shoot multiplication (78%) was observed in ½WPM supplemented with 2.5 mg/l TDZ and 0.5 mg/l GA_3, with 5.7 shoots per explant.

3.6.2 Root Induction

For root formation, WPM with various concentrations of IBA (0–3 mg/l) enriched with activated charcoal was used. A maximum of 8 roots was observed in 13 inoculated *in vitro* regenerated shoots with ½WPM supplemented with 2.0 mg/l IBA enriched with 0.2% activated charcoal. Further, plants with roots 3–4 cm in length were acclimatized and transferred to pots containing an autoclaved mixture of soil, sand, and manure (2:1:1). A survival rate of approximately 70% was recorded from *in vitro* grown plantlets on transfer to pots.

Acknowledgment

The authors gratefully acknowledge the Director, Biotechnology Centre and staff members for their support and suggestions for the improvement of the chapter.

REFERENCES

Bapat, V.A., Yadav, S.R., Dixit, G.B., 2008. Rescue of endangered plants through biotechnological applications. *Natl Acad Sci Lett.* 31,201–210.

Benson, E.E., 2000. *In-vitro* plant recalcitrance: An introduction. In Vitro *Cell Dev Biol Plant.* 36,141–148.

Coelho, N., Gonçalves, S., Romano, A., 2020. Endemic plant species conservation: Biotechnological approaches. *Plants.* 9,345.

Corlett, R.T., 2017. A bigger toolbox: Biotechnology in biodiversity conservation. *Trends Biotechnol.* 35,55–65.

Grigoriadou, K., Krigas, N., Sarropoulou, V., Papanastasi, K., Tsoktouridis, G., Maloupa, E., 2019. *In-vitro* propagation of medicinal and aromatic plants: The case of selected Greek species with conservation priority. In *Vitro Cell Dev Biol Plant.* 12.

Harrap, A., Harrap, S., 2005. *Orchids of Britain and Ireland: A Field and Site Guide.* A&C Black Publishers Ltd., London.

Kanani, P., Modi, A., Kumar, A., 2020. Biotization of endophytes in micropropagation: A helpful enemy. In: Ajay Kumar, Vipin Kumar Singh (ed), Woodhead Publishing Series in Food Science, *Technology and Nutrition, Microbial Endophytes*, Woodhead Publishing, pp. 357–379, ISBN 9780128187340.

Kapai, V.Y., Kapoor, P., Rao, I.U., 2010. *In-vitro* propagation for conservation of rare and threatened plants of India: A review. *International Journal of Biological Technology.* 1(2),1–14.

Kumari, R., Banerjee, A., Kumar, R., Kumar, A., Saikia, P., Khan, M.L., 2019. Deforestation in India: Consequences and sustainable solutions. In: Mohd Nazip Suratman, Zulkiflee Abd Latif, Gabriel De Oliveira, Nathaniel Brunsell, Yosio Shimabukuro and Carlos Antonio Costa Dos Santos (eds), *Forest Degradation around the World*, Intech Open, doi:10.5772/intechopen.85804

Laslo, V., Zăpârţan, M., Agud, E., 2011. *In-vitro* conservation of certain endangered and rare species of Romanian spontaneous flora. *An Univ Oradea Fasc Protectia Mediu.* XVI,252–261.

Marchese, C., 2015. Biodiversity hotspots: A shortcut for a more complicated concept. *Glob Ecol Conserv.* 3,297–309.

Matsumoto, T., 2017. Cryopreservation of plant genetic resources: Conventional and new methods. *Rev Agric Sci.* 5,13–20.

Maxted, N., Ford-Lloyd, B.V., Hawkes, J.G., 1997. Complementary conservation strategies. In: N. Maxted, B.V. Ford-Lloyd, J.G. Hawkes (ed), *Plant Genetic Conservation. The In Situ Approach*, Chapman, London, pp. 15–41.

Mitrofanova, I., Moroz, L., 2018. Development of the protocol for protoplast isolation from lavender and lavandin plants cultured *in vitro. J Biotechnol.* 280,83.

Miolkanova, O.I., Vasilyeva, O.G., Konovalova, L.N., 2015. The scientific basis for conservation and sustainable reproduction of plant gene fond in culture *in vitro. Bull Udsu Biol Earth Sci.* 25,95–100.

Nguyen, H., 2016. In vitro physiology of recalcitrant tissue cultured plants in the Nymphaeaceae, Alismataceae, and Orchidaceae. A Dissertation Presented to the Graduate School of The University of Florida in Partial fulfillment of the Requirements for the Degree of Doctor of Philosophy, University of Florida, Gainesville, FL.

Pence, V.C., Finke, L.R., Chaiken, M.F., 2017. Tools for the *ex situ* conservation of the threatened species, *Cycladenia humilis* Var. Jonesii. *Conserv Physiol.* 5,cox053.

Raju, R., Divya, C., 2020. Micropropagation of Syzygium densiflorum wall. Ex Wight & arn: An endemic and endangered semi-evergreen tree species of western Ghats, India. *Trees, Forests and People*. 2,100037.

Reed, B.M., Sarasan, V., Kane, M., Bunn, E., Pence, V.C., 2011. Biodiversity conservation and conservation biotechnology tools. *In Vitro Cell Dev Biol Plant.* 47,1–4.

Sanjay, M.B., Rathore, T.S., Ravishankar, R.V., 2006. Micropropagation of endangered Indian sandalwood (*Santalum album* L). *Journal of Forest Research.* 11,203–209.

Singh, S.V., Sharma, R.S. 2020. Standardization and development of in-vitro mass multiplication protocol of Chironji (*Buchanania lanzan* Spreng). Thesis submitted at JNKVV, Jabalpur 2020.

Vidoy-Mercado, I., Narvaez, I., Palomo-Rios, E., Litz, R.E., Barcelo-Munoz, A., Pliego-Alfaro, F., 2021. Reinvigoration/rejuvenation induced through micrografting of tree species: Signaling through graft union. *Plants.* 10,1197.

Yoon, Y.J., Murthy, H.N., Hahn, E.J., Paek, K.Y., 2007. Biomass production of *Anoectochilus formosanus hayata* in a bioreactor system. *J. Plant Biol* 50,573–576.

4

Enhancement of Nutritional, Pharmaceutical and Industrial Value of Crops through Genetic Modification with Carotenoid Pathway Genes

Amar A. Sakure
Anand Agricultural University
Anand
Gujarat, India

CONTENTS

4.1 Introduction

Value addition is the key aspect for improving the quality of agriculturally important crops, and metabolic engineering has become a very important approach for such modification. Cultivation of plants with improved quality traits may bring more economic return to the pharmaceutical industries in the preparation of various drugs and related products, perfumery industries and ultimately, farmers. Carotenoids are the most important isoprenoids found in photosynthetic bacteria, algae and plants. These pigments produce a variety of red, bright yellow and orange colours in different parts of plants, vegetables and fruits. More than 600 different types of carotenoids have been identified. Among these, alpha-carotene, beta-carotene, beta-cryptoxanthin, lutein, zeaxanthin and lycopene are the most common. In addition to chlorophyll pigments present in plants, carotenoids, as secondary light-absorbing pigments, are present in the thylakoid membranes of chloroplasts. Basically, carotenoids are hydrocarbons categorized according to their structural forms: cyclic include α- and β-carotene, while acyclic include lycopene and phytoene (Faure et al. 1999) and xanthophylls such as zeaxanthin, lutein, β-cryptoxanthin, bixin and capsanthin (Khachik et al. 1997, Meena et al. 2019). These carotenoids are used as food, as feed, and in cosmetic industries

DOI: 10.1201/9781003239932-4

(Jaswir et al., 2011). Astaxanthin has been identified as having potent antioxidant activity and helping to promote immune response, reduce eye fatigue and elevate muscle performance (Kidd, 2011). Its effect on the production of nitric acid (NO) in macrophages has also been examined *in vitro* and *in vivo*. It was identified that astaxanthin strongly suppresses the level of proinflammatory mediators such as NO, prostaglandin (PGE2), tumour necrosis factor (TNF-alpha) and interleukin-1beta (IL-1beta) in lipopolysaccharide-administered mice (Lee et al. 2003). Due to their potential antioxidant activity and as precursors for vitamin A, carotenoids are always in high demand. Therefore, enhancement of carotenoid content in crop plants and modification of metabolite pathways are always desirable, helping to alleviate vitamin A deficiencies and health-related ailments in malnourished populations around the world. Carotenoids and their derivatives, such as apocarotenoids, are protective against various cancers and age-related macular degeneration. Carotenoids are reported to play an important role as quenchers for light to protect cells from superoxide radicals and UV light (Ong and Tee, 1992). Also, oxidative cleavage of carotenoids produces apocarotenoids, which serve crucial functions in signalling, as ROS scavengers, giving aroma to flowers and fruits, and as antifungal and antibacterial agents.

Some carotenoids are precursors of vitamins, and they also present anti-inflammatory, antioxidant, immunomodulatory and anticancer activities, for cardiovascular therapy and neurodegenerative diseases (Shahidi and Ambigaipalan 2015). Carotenoids, acting as antioxidants eliminating free radicals, can modulate the risk of developing chronic diseases by inhibiting reactions mediated by ROS. Reactive species are produced during cellular metabolism as a defense against infectious and chemical agents; they may cause damage to DNA, proteins and tissues, contributing to the development of chronic diseases such as diabetes, Parkinson's, Alzheimer's, cardiovascular diseases and cancer (Bakan et al. 2014).

In addition to their antioxidant properties, carotenoids exhibit anti-inflammatory activities due to the protective effects of lutein and astaxanthin. Astaxanthin has been shown to inhibit the production of pro-inflammatory mediators such as nitric oxide (NO) in macrophages, to increase the level of inflammatory cytokines and to reduce oxidative stress. Neuroprotective effects, reduced neuroinflammation, improvement of insulin signals and reduction of lipid levels were also verified (Lu and Yen 2015). In addition to β-carotene, other highly valued carotenoids, including astaxanthin, ketocarotenoids, adonirubin, canthaxanthin, echinenone, adonixanthin and β-cryptoxanthin, are utilised in the food industry as feed supplements and colorants.

4.2 Biosynthesis of Carotenoids

In nature, the biosynthesis of carotenoids occurs by two independent pathways (Figure 4.1). One pathway starts in the plastid via methylerythritol 4-phosphate (MEP) (Eisenreich et al. 2004), and the other is the mevalonate pathway in cytoplasm (Miziorko, 2011). The MEP pathway starts with a condensation reaction between glyceraldehyde-3-phosphate and pyruvate and ends up with dimethylallyl pyrophosphate (DMAPP). In the MVA pathway, a condensation reaction occurs with two molecules of acetyl-CoA to form acetoacetyl-CoA by the action of *acetyl-CoA acetyltransferase* (atoB). Acetoacetyl-CoA is then converted into mevalonate via various intermediates into the final product, *isopentenyl pyrophosphate* (IPP). IPP is then transported into the plastid, where three molecules of IPP and one molecule of DMAPP undergo a condensation reaction via *geranyl-geranyl diphosphate* (GGPP) *synthase* to produce C_{20} GGPP (Wise, 2007). Further condensation of two molecules of GPPP via *phytoene synthase* (*PSY*) enzyme forms a C_{40} colourless compound with three conjugated bonds, named phytoene. Two to three *PSY* genes have been identified: *PSY1* showed expression in fruits, *PSY2* in leaves and *PSY3* in the roots of tomato and citrus (Peng et al. 2013 and Fantini et al. 2013). All C_{40} carotenoids are synthesized from the colorless precursor phytoene, accounting for 90% of total carotenoids (Yabuzaki 2017). The

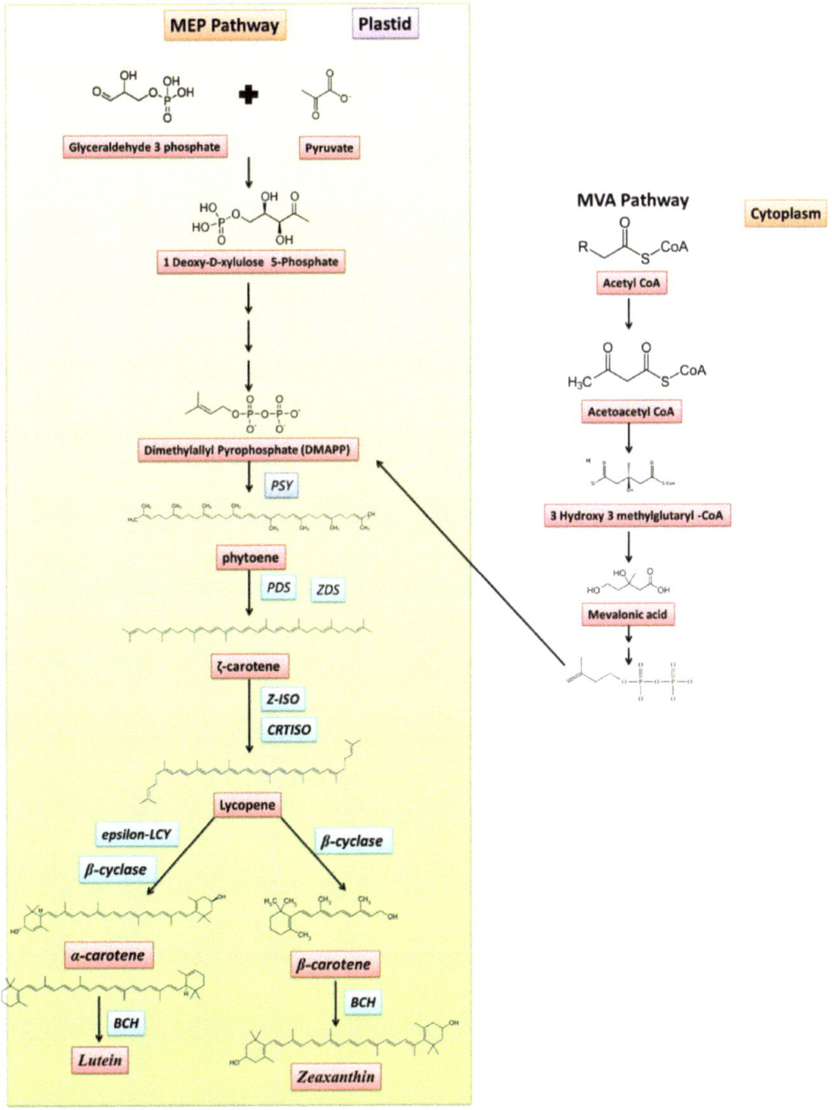

FIGURE 4.1 Schematic representation of biosynthetic pathway of carotenoids in plants.

colourless phytoene undergoes four sequential reactions to produce the red-coloured pigment lycopene. Initially, phytoene is converted via a series of desaturations by *phytoene desaturase* (*PDS*) and *ζ-carotene desaturase* (*ZDS*) into ζ-carotene. Then, the combined activity of *ζ-carotene isomerase* (*Z-ISO*) and *carotenoid isomerase* (*CRTISO*) results in the red-coloured pigment named trans-lycopene. Trans-lycopene further undergoes cyclisation to give rise to carotenoid diversity. Carotenoids can be differentiated based on the presence of either a beta or an epsilon ring. The reaction catalysed by *epsilon-lycopene cyclase* (*ε-LCY*) and *beta-lycopene cyclase* (*β-LCY*) leads to the formation of α-carotene and β-carotene, respectively. β-carotene is then further converted into zeaxanthin, referred to as an isoform of lutein, via *β-carotene hydroxylase* (*BCH*) enzyme (Figure 4.1).

4.3 Approaches for Enhancing Beta-carotene in Crops

Metabolic engineering for enhancing plant carotenoid biosynthesis has placed more emphasis on β-carotene, a precursor of nutritional vitamin A, for alleviating vitamin A deficiency in the diet. To enhance the carotenoids in crop plants, different approaches can be used, such as breeding approaches, gene transfer or overexpression of genes, gene silencing and genome editing.

4.3.1 Breeding Approaches

Biofortification through plant breeding is possible when there is availability of genetic diversity in the primary, secondary or tertiary gene pool of the targeted crop in utilizable form (Garg et al, 2018). Quantitative trait loci (QTLs) responsible for accumulation of α- and β-carotene have been identified by bi-parental mating of two inbred lines of carrot, P50006 and HCM A.C. (Ou et al, 2010). Similarly, amplified fragment length polymorphic (AFLP) loci, AACCAT178-Q and AAGCAG233-Q, have been identified, which are associated with the carotenoid pathway on linkage group 5 and explain 17.8%, 22.8% and 23.5% of total phenotypic variation for zeta-carotene, phytoene and beta-carotene, respectively.

A different approach, target induced local lesions in genomes (TILLING), was applied to enhance the production of β-carotene in the grains of wheat. A key enzyme involved in the carotenoid pathway generating β-carotene, namely *ε-LCY*, was targeted. The null mutant line showed a robust reduction in the expression of the *ε-LCY* gene and also showed pleiotropic effects. Biochemical profiling of total carotenoids mutant lines showed an upsurge of 75% β-carotene as compared with the control (Sestili et al. 2019).

A single nucleotide change in different genes may change carotenoid accumulation in plant tissues. A number of single nucleotide polymorphisms (SNPs) in carotenoid-biosynthesizing genes have been identified. In maize, an SNP was identified for increasing carotenoid content in *ε-LCY* (Harjes et al. 2008) and in *BCH* genes of maize (Yan et al. 2010). Similarly, in red and yellow watermelon, SNPs were identified in the *β-LCY* (Bang et al. 2007). A mutation in *PSY* gene product has been studied in maize and rice, which showed significant changes in carotenoid accumulation (Shumskaya et al. 2012); in tomato, mutations in *PSY* gene showed changes in carotenoid accumulation during fruit development (Gady et al. 2012); and likewise in carrot and canola (Arango et al. 2014 and López-Emparán et al. 2014). Cassava presents in its genome three psy genes, known as *psy1*, *psy2* and *psy3* (Arango et al. 2014).

Some breeding research has been focused on developing high-carotenoid breeding lines. Several dominant alleles of HIGH-BETA were isolated following the introgression of the *CycB* (B) gene from wild tomato species. This allele was characterized in the heirloom tomato line 'Jaune Flamme' of an unknown genetic background with an indeterminate growth habit and orange-coloured fruit (Karniel et al. 2020). A cross was made between the mutant B^{Sh} line and the 'wild-type' tomato variety M82. The high β-carotene phenotype was found co-segregated with the B^{Sh} allele in F_2 offspring at a ratio of 1:3. A non-transgenic tomato line, named 'Xantomato', was generated. Tomatoes of this line were shown to accumulate zeaxanthin at a concentration of 39 μg/g fresh weight (FW) or 577 μg/g dry weight (DW), which was around 50% of total fruit carotenoids compared with zero in the wild type. This is the highest concentration of zeaxanthin reached in a primary crop. The developed line named Xantomato can serve as the richest source of zeaxanthin in the human diet and may serve as a raw material for industrial applications. In tomato, many different breeding lines have been developed for high lycopene content. Advance breeding (HLT-F51 and HLT-F52) lines were developed, which exhibited 2.65-, 2.62- and 3.57-fold higher total carotenoids, lycopene and flavonoids, respectively, than control (Ilahy et al. 2009).

4.3.2 Gene Transfer or Overexpression of Genes

Research on biofortification of crop plants with carotenoid pathway genes is currently needed in agriculture. Much research has been done into value addition to crop plants. Multiplex transgenics were developed by the combinatorial nuclear approach for the enhancement of β-carotene as well as other nutrients in white maize (Naqvi et al. 2009). This was achieved by transforming white maize with five carotenoid genes expressed under different endosperm-specific promoters: *phytonene synthase* (*Zmpsy1*), *phytonene desaturase* (*Pacrt1*), *lycopene β-cyclase* (*Gllycb*), *β-carotene hydroxylase* (*Glbch*) and *β-carotene ketolase* (*ParacrtW*) from maize, *Pantoea ananatis*, *Gentiana lutea, G. lute* and *Paracoccus*, respectively. The resultant multi-transgenic maize showed a 169-fold increased level of β-carotene, a six-fold increase in ascorbate, and a two-fold increase in folate.

Besides β-carotene, other carotenoids, such as lutein and zeaxanthin, have also been identified as playing an important role in human health. Lutein has been reported to have extraordinary anti-inflammatory properties for eye health. It helps to improve or prevent age-related macular disease, which causes blindness and vision impairment (Buscemi et al. 2018). Among the carotenoids, only zeaxanthin and its isomer lutein can cross the blood–retina barrier and accumulate macular pigment in the retina of the eye (Stringham et al. 2019).

For the enhancement of this highly valued carotenoid, overexpression of *PSY* gene in Arabidopsis under a seed-specific promoter showed a 43-fold (260 μg/g FW) increase in β- carotene and a substantial increase in lutein (Lindgren et al. 2003). It was also reported that the overexpression of *ε-LCY* of celery (*Apium graveolens* L.) in *Arabidopsis* resulted in a substantial increase in β-carotene and lutein with enhanced salt tolerance (Yin et al. 2010). Similarly, overexpression of *ZDS* enhancing the β-carotene and lutein content has been reported in sweet potato (Li et al. 2017). In another report in sweet potato, expression of *Orange* (*Or*) gene encoding a 313-amino-acid protein having a cysteine-rich zinc finger domain showed strong expression in orange-fleshed storage roots of sweet potato (Kim et al. 2013). The role of *neoxanthin synthase* gene (*BoaNXS*) in the accumulation of carotenoids in Chinese kale (*Brassica oleracea*) was also studied. The average carotenoid content in three transgenic lines was recorded as 4.99 mg/g DW. Among these carotenoids, especially neoxanthin and lutein showed significantly higher levels in all three overexpressed plants (Jian et al. 2021).

Enhancement of carotenoids has also been exploited in eggplant (*Solanum melongena* L.), which is globally cultivated, especially in Asian countries, as a regular diet item of developing countries. This crop has important nutrients but a low provitamin A content. An attempt has been made to improve the carotenoid content by expressing carotenoid genes in eggplants (Mishiba et al. 2020). An experiment was conducted by introducing *phytoene synthase* (crtB) gene of bacterial origin under the control of the promoter region of eukaryotic elongation factor 2 (*EEF2*). Among different transgenic lines produced, one line showed 1.50 μg/g fresh weight (FW) of β-carotene, which was 30-fold higher than the level in the untransformed fruits (0.05 μg/g FW).

With the creation of 'golden rice', biofortification is a promising strategy followed in different crops for the mitigation of nutritional deficiencies. Similarly to golden rice, quite a successful creation has also been done using potato. Potato is extremely poor in provitamin A carotenoids. To achieve the objective of biofortification of potato with β-carotene content, a group of scientists worked on transformation of potato with genes of bacterial origin. Three genes, encoding *phytoene synthase* (*CrtB*), *phytoene desaturase* (*CrtI*) and *lycopene β-cyclase* (*CrtY*) from *Erwinia*, were transformed in potato under tuber-specific promoter (Diretto et al. 2007). The expression of all these genes under tuber-specific promoter showed a deep yellow ('golden') phenotype with no abnormalities in leaf morphology. Tubers showed accumulation of approximately 20-fold (114 μg/g DW) total carotenoids and 3,600–fold (47 μg/g DW) of β-carotene. This transformed potato version showed higher carotenoid and β-carotene as compared with 'golden rice 2', with 31 μg/g DW beta-carotene. This is sufficient to supplement 50% of the Recommended Daily Allowance of Vitamin A with 250 g (FW) of 'golden' potatoes.

A novel transformation method termed combinatorial nuclear transformation has been devised to produce multiplex transgenic plants (Zhu et al. 2008). To demonstrate this principle, they established a carotenoid pathway by transferring five genes into white maize. All these genes were driven by different endosperm-specific promoters. A number of diverse populations of transgenic plants with different levels of enzyme expression were identified. Three different phenotypes were identified. High-performance liquid chromatography (HPLC) analysis of these phenotypes showed varied accumulation of metabolites, confirming a direct relation between genotype and carotenoid accumulation. Independent gene events such as Phenotype 1 (*Zmpsy1* alone) showed a 53-fold increase in total carotenoids, predominantly zeaxanthin (18.25 μg/g DW), lutein (14.95 μg/g DW) and beta-carotene (7.10 μg/g DW), whereas Phenotype 3 showed high beta-carotene (57.35 μg/g DW) and lycopene (26.69 μg/g DW). Phenotype 2 (*crtI* alone) and the combination of both 1 and 2 (*Zmpsy1* and *PacrtI*) showed 2.5- and 142-fold increases, respectively.

Soybean oil is very sensitive to oxidation when stored at room temperature and also undergoes polymerization at high temperatures during frying because of the presence of high levels of polyunsaturated fatty acids. This problem can be solved by accumulating antioxidant in soybean seeds. A transgenic soybean was developed to accumulate β-carotene, which exhibits enhanced protein and oleate content traits (Schmidt et al. 2015). Overexpression of carotenoid pathway genes such as seed-specific bacterial phytoene synthase gene from *Pantoea ananatis* was modified and targeted to plastids, where it accumulated about 845 μg/g DW of β-carotene in dry seed weight with a desirable 12:1 ratio of β to α. The β-carotene-accumulating seeds showed changes in oil composition, with an increase in oleic acid and a decrease in linoleic acid. The seed protein content was also shown to increase by 4% (w/w). The effects of antioxidants such as β-carotene were tested in soybean salad oil to study the effect of light exposure on flavour deterioration. When oil was treated with 20 ppm β-carotene, it was found to be more stable to light exposure (Warner and Frankel 1987). Therefore, an increase in carotenoids in the oil of biofortified soybeans helps to improve oil quality, reducing oxidation and thereby, rancidity.

The carotenoid zeaxanthin provides numerous health benefits to humans due to its antioxidant properties. It protects the retina in the human eye by filtering harmful blue light and thus delaying the progression of age-related macular degeneration (Roberts and Dennison, 2015). Despite its high nutritional value, zeaxanthin is not available in large amounts as compared with other carotenoids in the diet. To solve this issue, a transgenic tomato from a mutant breeding line (*Solanum lycopersicum* L.) was developed (Karniel et al. 2020). A gene named *BCH2* isolated from *Citrus clementine* was expressed under the constitutive promoter CaMV35S in the double-mutant hp3/BSh tomato line. The resulting transgenic T_1 plants showed accumulation of various xanthophylls in the ripe fruits of the tomato, including mainly zeaxanthin and a small amount of β-cryptoxanthin.

Carrot (*Daucus carota* L.) roots are an extraordinary source of dietary α-carotene, β-carotene (provitamin A), zeaxanthin and lutein. The carrot is an excellent feeder for the nutraceutical industries to produce molecular farming products. As well as routine carotenoids, ketocarotenoids such as canthaxanthin and astaxanthin are reported to have strong antioxidant effects; these have been chemically synthesized and used as dietary supplements in aquaculture and industry. Ketocarotenoids have been successfully synthesized in carrot by introducing β-carotene ketolase gene from the alga *Haematococcus pluvialis* under a constitutive promoter. The gene product was fused with the small subunit of pea Rubisco peptide, which helps to target the enzyme to the plastids of leaf and root. Expression of this gene leads to 70% conversion of total carotenoids into novel ketocarotenoids (2,400 μg/g root DW). In addition, production of astaxanthin, adonirubin, canthaxanthin, echinenone, adonixanthin and β-cryptoxanthin was also detected in transgenic carrot (Jayaraman et al. 2008). This makes carrot a perfect biopharmaceutical source for production of carotenoids. Wheat is one of the most important extensively grown staple food crops across the world. Being a staple food, but lacking nutrients like vitamin A, iron and quality proteins, wheat is always considered for biofortification. Efforts have been made to enhance the provitamin A content by expressing maize *PSY1* driven under the control of endosperm-specific 1Dx5 promoter and bacterial-origin *phytoene desaturase* (*CrtI*) from *Erwinia uredovora* under the constitutive promoter CaMV 35S in wheat (Cong et al. 2009). The developed transgenic wheat line showed a 10.8-fold increase in total carotenoids as compared with an elite non-transgenic line.

4.3.3 Gene Silencing and Genome Editing

Carotenoid content in *Brassica napus* has been significantly increased by down-regulating the gene *lycopene epsilon-cyclase*. The RNA interference (RNAi) technique was employed for down-regulating the expression of *lycopene epsilon-cyclase.* Down-regulation of *lycopene epsilon-cyclase* gene leads to accumulation of beta-carotene, zeaxanthin, violaxanthin and lutein in seeds of *B. napus* (Yu et al. 2008). But an increased level of these carotenoids was accompanied by reduction in various fatty acids. The presence of carotenoid content in oil provides an antioxidant advantage that aids in quality improvement and reducing the oxidation of the oil (Taghvaei and Jafari, 2015). In the case of potato, tubers contain mainly xanthophylls, namely lutein, violaxanthin, and antheraxanthin, of which xanthophyll esters are present at a very low level. An attempt has been made to improve thc beta-carotenoid levels by silencing the *ε-LCY* gene via Agrobacterium-mediated transformation (Diretto et al. 2006). An antisense construct expressed under the control of patatin promoter revealed the tuber-specific silencing of *ε-LCY* gene. Silencing of this gene resulted in a 14-fold increase in beta-carotenoid content and a 2.5-fold increase in other carotenoids.

With recent new advances in genetic engineering, a very popular and valuable tool comes into the picture, which has revolutionised research in biotechnology: genome editing. Genome editing serves not only to mutate genes of interest but also to improve the existing genes. With this technique, a number of crops have been modified to alter the carotenoid pathway. The CRISPR-Cas9 system was used in rice to increase beta-carotene content with the advantage of being marker free (Dong et al. 2020). The authors optimised the CRISPR construct to integrate a carotenoid biosynthesis cassette of around 5.2 kb into the rice genome. Dehusked seeds derived from genome-edited rice 48A-7 showed golden colour, indicating the accumulation of 7.90 μg/g DW of carotenoids in rice endosperm. As an alternative to the golden rice approach, genome editing has been successfully used to produce beta-carotene in scutellum-derived rice calli using directed gene modification of the Osor (rice ortholog) gene via CRISPR-Cas9 (Endo et al. 2019). CRISPR was employed to disrupt the junction between the third exon and intron in the Osor.

Lycopene is a red-coloured pigment reported to reduce the risk of a variety of cancers and cardiovascular disease (Li and Xu, 2014; Tang et al. 2014). Lycopene is among the top six commercial carotenoid pigments found in many ripening fruits. It is a good source of nutritional and pharmaceutical products, where its activity has been studied for reducing the risk of prostate cancer (Ilic et al. 2011 and Barber and Barber, 2002). In tomato, lycopene has been successfully increased to its highest level using CRISPR-Cas9 technology. CRISPR-based mutation was carried out using different single guide RNAs (sgRNA) involving *chloroplast stay-green protein 1* (*SGR1*) for enhancing lycopene synthesis and others (*lycopene ε-cyclase* (*LCY-E*), *beta-lycopene cyclase* (*Blc*), *lycopene β-cyclase 1* and *2* (*LCY-B1, B2*)) to catalyse the cyclisation of lycopene. Knocking out *LCY-E* activity prevents the cyclisation from lycopene to α-carotene, and *LCY-B1* and *LCY-B2* were used to prevent the cyclisation from lycopene to β-carotene. HPLC analysis of mutant edited lines showed that the contents of lycopene and β-carotene were significantly higher as compared with wild-type plants. Mutation in *SGR1* gene showed the highest (5.1-fold) increase in lycopene content when compared with a previous study of RNAi-based gene silencing of *SGR1* gene in tomato (Luo et al. 2013). On the other hand, mutation of the *Blc* did not cause a noticeable improvement in lycopene accumulation.

Interestingly, knocking out of *ε-LCY* gene by CRISPR was done to enrich beta-carotene in banana. For inhibition of *ε-LCY*, sgRNA was selected from the fifth exon and was tested for its ability to create indels in the genome of cv. Grand Naine. Metabolic profiling of edited plants' fruit pulp showed a six-fold (~24 μg/g) increase in beta-carotene content when compared with the unedited plants without disturbing any agro-morphological parameters.

4.4 Cleaved Product of Carotenoids: Apocarotenoids

Besides carotenoids, their derivatives, apocarotenoids, have immense pharmaceutical properties, which have been reported by many workers to fight cancer and other ailments. To date, more than 1,117 natural carotenoids and apocarotenoids have been reported, which range from C_{30} to C_{40}, C_{45} and C_{50} carotenoids (Yabuzaki, 2017). Of these, C_{40} are the most common, with 1,093 different structures.

Different oxygenase enzymes have been reported for conversion of carotenoids to apocarotenoids. These enzymes are named *carotenoid cleavage dioxygenases* (*CCD*). The genes that encode *CCD* are divided mainly into two types: the *nine-cis-epoxycarotenoid dioxygenases* (NCEDs), which catalyse the synthesis of xanthoxin, the precursor of ABA, from neoxanthin and violaxanthin (Seo and Koshiba, 2002; Walter et al. 2010), and *CCDs*, which catalyse an array of different cleavages, leading to the formation of various apocarotenoids. Carotenoid cleavage byproducts such as abscisic acid regulate many biological processes in plants. In vascular plants, two types of carotenoids are observed: unoxygenated carotenoids, known as carotenes (e.g., β-carotene and lycopene), and oxygenated carotenoids, named xanthophylls (such as lutein and zeaxanthin). All these carotenoids play vital roles in the protection of the plant photosystem. Next to *NCEDs*, *CCD1* is the best-studied *CCD*. Its involvement has been confirmed in the conversion of C_{13} apocarotenoid in fruit aroma biosynthesis. The role of *Petunia hybrid CCD1* (Ph*CCD1*) in the formation of an aromatic compound named beta-ionone was confirmed by down-regulating its activity in transgenic petunia (Simkin et al. 2004). These apocarotenoids derived from the action of *CCDs* are now the target for production of many aromatic scents, which are in high demand in perfumery industries. The violet-coloured compound β-ionone is in high demand in the foodstuff and beverage industries due to its flavour and fragrance characteristics. There are two well-established methods for the production of this compound; one is based on nonspecific cleavage of carotenoids by lipoxygenase enzyme (Wu and Robinson 1999) and the other on direct cleavage by fungal peroxidase (Zorn et al. 2003).

Important unique red-coloured glycosylated apocarotenoids and monoterpene glycosides, such as crocins and picrocrocin, respectively, have been reported in saffron (*Crocus sativus*), making this spice highly in demand and with a high price. In saffron, around 150 volatile and non-volatile compounds have been identified by various workers (Pfander and Schurtenberger 1982; Zargani and Heinz, 1971; Winterhalter and Straubinger, 2000). All of them have concluded that saffron contains three major carotenoid derivatives: crocin, which is composed of unusual water-soluble carotenoids (mono- and diglycosyl esters of a polyene dicarboxylic acid named crocetin), picrocrocin and safranal, which are responsible for its intense colour, bitter taste and aroma, respectively.

Crocin has a vast number of pharmaceutical properties and has been tested in animal models for various ailments. The effect of crocin on anxiety was tested in rats, and it was found that crocins at a 50 mg/kg dose did not influence the animals' motor activity but significantly increased latency to enter the dark compartment and prolonged the time spent by the rats in the lit chamber. This result indicated that treatment with these active constituents of *Crocus sativus* L. induced anxiolytic-like effects in the rats (Pitsikas et al. 2008). The effect of crocin on learning and memory was also tested (Manuchair, 2006), and it showed a preventive effect on ethanol-induced impairment of learning and memory. It has also been reported to have strong antioxidant activity in scavenging free radicals, especially superoxide anions, thereby providing protection to cells against oxidative stress (Shinji et al. 2007). It is also reported that crocetin, an intermediate product of crocin biosynthesis, may exert a beneficial effect in preventing diabetes-associated vascular complications (Xiang et al. 2006). Another byproduct of carotenoid present in saffron is a flavouring volatile essential oil named safranal. Its bio-pharmacological activities have been studied in the last decades. An increasing number of papers have been published on the neuropsychological effects of safranal on the central nervous system. Research has been carried out to study its pharmaceutically important effects as an antioxidant (Assimopoulou et al. 2005), a protective agent against indomethacin-induced gastric ulcers (Kianbakht and Mozaffari, 2009) and as protection against PTZ-induced status epilepticus (Pathan et al. 2009).

The availability and cultivation of crops that contain these precious apocarotenoids are very limited. In the case of saffron, the novel apocarotenoid crocin is present, but due to its restricted location in hilly areas of Kashmir in India and limited production, it has a high price. Due to its high price, it is always subject to adulteration. Therefore, a system of production of these metabolites needs to be established by genetic transformation in various host systems so that it can be extracted at the highest possible level. A similar attempt has been made to produce crocins and picrocrocin in *Nicotiana benthamiana* using a virus-driven system. A novel carotenoid cleavage dioxygenase has been identified, named *CCD2L*, which catalyses zeaxanthin into crocetin dialdehyde via *CCD2L* enzyme action and subsequently into crocetin by endogenous aldehyde dehydrogenase enzyme. This crocetin is then catalysed by glycosyl transferase enzyme into crocin (Frusciante et al. 2014). The recombinant virus expressing *CCD2L* showed accumulation of 0.2% crocin and 0.8% picrocrocin DW (Marti et al. 2020). This system opens the door for future production of these compounds in industry on a large scale.

The *CCD4* family has so far only been identified in flowering plants, located in the plastids. Similarly to *CCD1*, *CCD4* is involved in the formation of volatile compounds at the cleavage of 9,10 (9′,10′) or the 7,8 (7′,8′) positions (Huang et al. 2009; Rodrigo et al. 2013). *In vitro* analysis of *CCD4s* showed cleavage of β-carotene, β-apo-8′-carotenal, β-cryptoxanthin or zeaxanthin at the 9,10 and/or 9′,10′ double bond to produce β-ionone (Rubio et al. 2008; Huang et al. 2009; Bruno et al. 2015).

Generally, *CCD7* and *CCD8* contribute to the synthesis of strigolactone in the plastids. Strigolactone has been reported as a novel plant hormone involved in the inhibition of shoot branching and acts as a signal molecule for *arbuscular mycorrhizal* (AM) symbiosis (Al-Babili and Bouwmeester, 2015).

REFERENCES

Al-Babili, S., Bouwmeester, H.J. (2015). Strigolactones, a novel carotenoid-derived plant hormone. *Annu Rev Plant Biol.* 66,161–186.

Arango, J., Jourdan, M., Geoffriau, E., Beyer, P., Welsch, R. (2014) Carotene hydroxylase activity determines the levels of α-carotene and total carotenoids in orange carrots. *Plant Cell* 26, 2223–2233.

Assimopoulou, A.N., Sinakos, Z., Papageorgiou, V.P. (2005). Radical scavenging activity of *Crocus sativus* L. extract & its bioactive constituents. *Phytother Res.* 19(11),997–1000.

Bakan, E., Akbulut, Z.T., Inanç, A.L. (2014). Carotenoids in foods and their effects on human health. *Akademik Gıda.* 12,61–68.

Bang, H., Kim, S., Leskovar, D., King, S. (2007). Development of a codominant CAPS marker for allelic selection between canary yellow and red watermelon based on SNP in lycopene β-cyclase (LCYB) gene. *Mol Breeding* 20,63–72.

Barber, N.J., Barber, J. (2002). Lycopene and prostate cancer. *Prostate Cancer and Prostatic Diseases.* 2002(5),6.

Bruno, M., Beyer, P., Al-Babili, S. (2015). The potato carotenoid cleavage dioxygenase 4 catalyzes a single cleavage of β-ionone ring-containing carotenes and non-epoxidated xanthophylls. *Archives of Biochemistry and Biophysics.* 572,126–133.

Buscemi, S., Corleo, D., Di Pace, F., Petroni, M.L., Satriano, A., Marchesini, G. (2018). The effect of lutein on eye and extra-eye health. *Nutrients.* 10(9),1321.

Cong, L., Wang, C., Chen, L., Liu, H., Yang, G., He, G. (2009). Expression of *phytoene synthase1* and *carotene desaturase crtI* genes results in an increase in the total carotenoids content in transgenic elite wheat (*Triticum aestivum* L.). *J Agric Food Chem.* 57(18),8652–8660.

Diretto, G., Tavazza, R., Welsch, R., Pizzichini, D., Mourgues, F., Papacchioli, V., Beyer, P., Giuliano, G. (2006) Metabolic engineering of potato tuber carotenoids through tuber-specific silencing of lycopene epsilon cyclase. *BMC Plant Biol.* 26(6),13.

Diretto, G., Al-Babili, S., Tavazza, R., Papacchioli, V., Beyer, P., Giuliano, G. (2007). Metabolic engineering of potato carotenoid content through tuber-specific overexpression of a bacterial mini-pathway. *PLoS ONE.* 2(4),e350.

Dong, O.X., Yu, S., Jain, R., et al. (2020). Marker-free carotenoid-enriched rice generated through targeted gene insertion using CRISPR-Cas9. *Nat Commun.*11,1178.

Eisenreich, W., Bacher, A., Arigoni, D., Rohdich, F. (2004). Biosynthesis of isoprenoids via the non-mevalonate pathway. *Cellular and Molecular Life Sciences.* 61, 1401–1426.

Endo, A., Saika, H., Takemura, M., et al. (2019). A novel approach to carotenoid accumulation in rice callus by mimicking the cauliflower Orange mutation via genome editing. *Rice.* 12,81.

Fantini, E., Falcone, G., Frusciante, S., Giliberto, L., Giuliano, G. (2013). Dissection of tomato lycopene biosynthesis through virus-induced gene silencing. *Plant Physiol.* 163,986–998.

Faure, H., Galabert, G., Le Moel, G., Nabet, F. (1999). Carotenoids: Metabolism and physiology. *Ann de Biologie Clinique.* 57,169–183.

Frusciante, S.G., Diretto, M., Bruno, P., Ferrante, M., Pietrella, A., Prado-Cabrero, A., Rubio-Moraga, P., Beyer, L., Gomez-Gomez, S., Al-Babili, et al. (2014). Novel carotenoid cleavage dioxygenase catalyzes the first dedicated step in saffron crocin biosynthesis. *Proc Natl Acad Sci.* 111,12246–12251.

Gady, A.L.F., Vriezen, W.H., Van de Wal, M.H.B.J., Huang, P., Bovy, A.G., Visser, R.G.F., Bachem, C.W.B. (2012). Induced point mutations in the phytoene synthase 1 gene cause differences in carotenoid content during tomato fruit ripening. *Mol Breeding.* 29,801–812.

Garg, M., Sharma, N., Sharma, S., Kapoor, P., Kumar, A., Chunduri, V., Arora, P. (2018). Biofortified crops generated by breeding, agronomy, and transgenic approaches are improving lives of millions. *Frontiers in Nutrition.* 5,12.

Harjes, C.E., Rocheford, T.R., Bai, L., Brutnell, T.P., Kandianis, C.B., Sowinski, S.G., Stapleton, A.E., Vallabhaneni, R., Williams, M., Wurtzel, E.T., Yan, J., Buckler, E.S. (2008). Natural genetic variation in lycopene epsilon cyclase tapped for maize biofortification. *Science.* 319,330–333.

Huang, F.C., Molnár, P., Schwab, W. (2009). Cloning and functional characterization of carotenoid cleavage dioxygenase 4 genes. *J of Experimental Botany.* 60(11),3011–3022.

Ilahy, R., Hdider, C., Tlili, I. (2009). Bioactive compounds and antioxidant activity of tomato high lycopene content advanced breeding lines. *The African Journal of Plant Science and Biotechnology.* (Special Issue 1), 1–6. Global Science Books.

Ilic, D., Forbes, K.M., Hassed, C. (2011). Lycopene for the prevention of prostate cancer. *Cochrane Database of Systematic Reviews.* 11,CD008007.

Jaswir, I., Noviendri, D., Hasrini, R.F., Octavianti, F. (2011). Carotenoids: Sources, medicinal properties and their application in food and nutraceutical industry. *Journal of Medicinal Plants Research.* 5,7119–7131.

Jayaraman, J., Devlin, R., Punja, Z. (2008). Metabolic engineering of novel ketocarotenoid production in carrot plants. *Transgenic Research.* 17,489–501.

Jian, Y., Zhang, C., Wang, Y., Li, Z., Chen, J., Zhou, W., Huang, W., Jiang, M., Zheng, H., Li, M., Miao, H., Zhang, F., Li, H., Wang, Q., Sun, B. (2021). Characterization of the role of the neoxanthin synthase gene *BoaNXS* in carotenoid biosynthesis in Chinese kale. *Genes.* 12(8),1122.

Karniel, U., Koch, A., Zamir, D., Hirschberg, J. (2020). Development of zeaxanthin-rich tomato fruit through genetic manipulations of carotenoid biosynthesis. *Plant Biotechnol J.* 18(11),2292–2303.

Kianbakht, S., Mozaffari, K. (2009). Effects of saffron and its active constituents, crocin and safranal, on prevention of indomethacin induced gastric ulcers in diabetic and nondiabetic rats. *J Med Plants.* 5,30–38.

Khachik, F., Beecher, G.R., Smith, J.C. (1997). Lutein, lycopene, and their oxidative metabolites in chemoprevention of cancer. *J Cell. Biochem.* 22,236–246.

Kidd, P. (2011). Astaxanthin, cell membrane nutrient with diverse clinical benefits and anti-aging potential. *Alternative Medicine Review.* 16,355–364.

Kim, S.H., Ahn, Y.O., Ahn, M.J., Jeong, J.C., Lee, H.S., Kwak, S.S. (2013). Cloning and characterization of an Orange gene that increases carotenoid accumulation and salt stress tolerance in transgenic sweetpotato cultures. *Plant Physiol Biochem.* 70,445–454.

Lee, S.J., Bai, S.K., Lee, K.S., Namkoong, S., Na, H.J., Ha, K.S., Han, J.A., Yim, S.V., Chang, K., Kwon, Y.G., Lee, S.K., Kim, Y.M. (2003). Astaxanthin inhibits nitric oxide production and inflammatory gene expression by suppressing I(kappa)B kinase-dependent NF-kappaB activation. *Mol Cells.* 16(1),97–105.

Li, R., Kang, C., Song, X., Yu, L., Liu, D., He, S., Zhai, H., Liu, Q. (2017). A ζ-carotene desaturase gene, IbZDS, increases β-carotene and lutein contents and enhances salt tolerance in transgenic sweetpotato. *Plant Sci.* 262,39–51.

Li, X., Xu, J. (2014). Meta-analysis of the association between dietary lycopene intake and ovarian cancer risk in postmenopausal women. *Scientific Reports.* 4,4885.

Lindgren, L.O., Stålberg, K.G., Höglund, A.S. (2003). Seed-specific overexpression of an endogenous Arabidopsis phytoene synthase gene results in delayed germination and increased levels of carotenoids, chlorophyll, and abscisic acid. *Plant Physiology.* 132(2),779–785.

López-Emparán, A., Quezada-Martinez, D., Zúñiga-Bustos, M., Cifuentes, V., Iñiguez-Luy, F., Federico, M.L. (2014). Functional analysis of the *Brassica napus* L. phytoene synthase (PSY) gene family. *PLoS ONE.* 9,1–19.

Lu, C.C., Yen, G.C. (2015). Antioxidative and anti-inflammatory activity of functional foods. *Current Opinion in Food Science.* 2,1–8.

Luo, Z., Zhang, J., Li, J., Yang, C., Wang, T., Ouyang, B., et al. (2013). A STAY-GREEN protein SlSGR1 regulates lycopene and beta-carotene accumulation by interacting directly with SlPSY1 during ripening processes in tomato. *New Phytol.* 198,442–452.

Manuchair, E. (2006). *Pharmacodynamic basis of herbal medicine.* Taylor & Francis, 2nd edition, 525–527.

Marti, M., Diretto, G., Aragones, V., Frusciante, S., Ahrazem, O., Gomez, G.L., Daros, J-A. (2020). Efficient production of saffron crocins and picrocrocin in *Nicotiana benthamiana* using a virus-driven system, *Metabolic Engineering.* 61,238–250.

Meena, M., Swapnil, P., Zehra, A., Dubey, M.K., Aamir, M., Patel, C.B., Upadhyay, R.S. (2019). Virulence factors and their associated genes in microbes, in: H.B. Singh, V.K. Gupta, S. Jogaiah (ed), *New and Future Developments in Microbial Biotechnology and Bioengineering*, Elsevier, pp. 181–208.

Mishiba, K.I., Nishida, K., Inoue, N., et al. (2020). Genetic engineering of eggplant accumulating β-carotene in fruit. *Plant Cell Rep.* 39, 1029–1039.

Miziorko, H.M. (2011). Enzymes of the mevalonate pathway of isoprenoid biosynthesis. *Archives of Biochemistry and Biophysics.* 505(2),131–143.

Naqvi, S., Zhu, C., Farre, G., Ramessar, K., Bassie, L., Breitenbach, J., et al. (2009). Transgenic multivitamin corn through biofortification of endosperm with three vitamins representing three distinct metabolic pathways. *Proc Natl Acad Sci USA.* 106(19),7762–7767.

Ong, A.S.H., Tee, E.S. (1992). Natural sources of carotenoids from plants and oils. *Meth Enzymol.* 213,142–167.

Ou, C.G., Deng, B.T., Bao, S.Y., Zhao, Z.W., Hu, H., Zhuang, F.Y., Mao, S.M. (2010). QTL mapping for contents of main carotenes and lycopene in carrot (*Daucus carota* L.). *Chuan.* 32(12),1290–1295.

Pathan, S.A., Zaidi, S.M.A., Jain, G.K., Vohora, D., Ahmad, F.J., Khar, R.K. (2009). Anticonvulsant evaluation of safranal in pentylenetetrazole-induced status epilepticus in mice. *Int J Essential Oil Therapeutics.* 3,106–108.

Peng, G., Wang, C., Song, S., Fu, X., Azam, M., et al. (2013). The role of 1-deoxy-d-xylulose-5-phosphate synthase and phytoene synthase gene family in citrus carotenoid accumulation. *Plant Physiol Biochem.* 71,67–76.

Pfander, H., Schurtenberger, H. (1982). Biosynthesis of C20-carotenoids in *Crocus sativus. Phytochemistry.* 21(5),1039–1042.

Pitsikas, N.A., Boultadakis, G., Georgiadou, P.A., Tarantilis, N.S. (2008). Effects of the active constituents of *Crocus sativus* L., crocins, in an animal model of anxiety. *Phytomedicine.* 15,1135–1139.

Roberts, J.E., Dennison, J. (2015). The photobiology of lutein and zeaxanthin in the eye. *Journal of Ophthalmology.* 2015,687173.

Rodrigo, M.J., Alquézar, B., Alós, E., Medina, V., Carmona, L., Bruno, M., Al-Babili, S., Zacarías, L. (2013). A novel carotenoid cleavage activity involved in the biosynthesis of citrus fruit-specific apocarotenoid pigments. *J of Experimental Botany.* 64(14),4461–4478.

Rubio, A., Rambla, J.L., Santaella, M., Gomez, M.D., Orzaez, D., Granell, A., Gomez-Gomez, L. (2008). Cytosolic and plastoglobule-targeted carotenoid dioxygenases from *Crocus sativus* are both involved in beta-ionone release. *J Biol Chem.* 283(36),24816–24825.

Schmidt, M.A., Parrott, W.A., Hildebrand, D.F., Berg, R.H., Cooksey, A., Pendarvis, K., He, Y., McCarthy, F., Herman, E.M. (2015). Transgenic soya bean seeds accumulating β-carotene exhibit the collateral enhancements of oleate and protein content traits. *Plant Biotechnol J.* 13(4),590–600.

Seo, M., Koshiba, T. (2002). Complex regulation of ABA biosynthesis in plants. *Trends Plant Sci.* 7,41–48.

Sestili, F., Garcia-Molina, M.D., Gambacorta, G., Beleggia, R., Botticella, E., De Vita, P., Savatin, D.V., Masci, S., Lafiandra, D. (2019). Provitamin A biofortification of durum wheat through a TILLING approach. *Int J Mol Sci.* 20(22),5703.

Shahidi, F., Ambigaipalan, P. (2015). Novel functional food ingredients from marine sources. *Current Opinion in Food Science.* 2,123–129.

Shinji, S., Ochiai, T., Shimeno, H., Saito, H., Abe, K., Tanaka, H., Shoyama, Y. (2007). Pharmacological activities of crocin in saffron. *Journal of Natural Medicine.* 61,102–111.

Shumskaya, M., Bradbury, L.M.T., Monaco, R.R., Wurtzel, E.T. (2012). Plastid localization of the key carotenoid enzyme phytoene synthase is altered by isozyme, allelic variation, and activity. *Plant Cell.* 24,3725–3741.

Simkin, A.J., Underwood, B.A., Auldridge, M., Loucas, H.M., Shibuya, K., Schmelz, E., Clark, D.G., Klee, H.J. (2004). Circadian regulation of the PhCCD1 carotenoid cleavage dioxygenase controls emission of beta-ionone, a fragrance volatile of petunia flowers. *Plant Physiol.* 136(3),3504–3514.

Stringham, J.M., Johnson, E.J., Hammond, B.R. (2019). Lutein across the lifespan: From childhood cognitive performance to the aging eye and brain. *Current Developments in Nutrition.* 3(7),2475–2991.

Taghvaei, M., Jafari, S.M. (2015). Application and stability of natural antioxidants in edible oils in order to substitute synthetic additives. *Journal of Food Science and Technology.* 52(3),1272–1282.

Tang, L., Lee, A.H., Su, D., et al. (2014). Fruit and vegetable consumption associated with reduced risk of epithelial ovarian cancer in southern Chinese women. *Gynecologic Oncology.* 132(1),241–247.

Walter, M.H., Floss, D.S., Strack, D. (2010). Apocarotenoids: Hormones, mycorrhizal metabolites and aroma volatiles. *Planta.* 232,1–17.

Warner, K., Frankel, E.N. (1987). Effects of β-carotene on light stability of soybean oil. *J Am Oil Chem Soc.* 64,213–218.

Winterhalter, P, Straubinger, M. (2000). Saffron: Renewed interest in an ancient spice. *Food Rev Int.* 16(1),39–59.

Wise, R.R. (2006). *The Diversity of Plastid Form and Function.* Advances in Photosynthesis and Respiration Book 23. Springer, Berlin, 3–26.

Wu, Z., Robinson, D.S. (1999). Co-oxidation of beta-carotene catalyzed by soybean and recombinant pea lipoxygenase. *J Agric Food Chem.* 47,4899–4906.

Xiang, M., Yang, M., Chenghua, Z., Juan, L., Wenna, L., Zhiyu, Q. (2006) Crocetin prevents AGEs-induced vascular endothelial cell apoptosis. *Pharmacological Research.* 54, 268–274.

Yabuzaki, J. (2017). Carotenoids database: Structures, chemical fingerprints and distribution among organisms. *Database: The Journal of Biological Databases and Curation.* 2017,bax004.

Yan, J., Kandianis, C.B., Harjes, C.E., Bai, L., Kim, E-H., Yang, X., et al. (2010). Rare genetic variation at *Zea mays* crtRB1 increases β-carotene in maize grain. *Nat Genet.* 42,322–327.

Yin, L., Liu, J.X., Tao, J.P., Xing, G.M., Tan, G.F., Li, S., Duan, A.Q., Ding, X., Xu, Z.S., Xiong, A.S. (2010). The gene encoding lycopene epsilon cyclase of celery enhanced lutein and β-carotene contents and confers increased salt tolerance in Arabidopsis. *Plant Physiol Biochem.* 157,339–347.

Yu, B., Lydiate, D.J., Young, L.W., Schäfer, U.A., Hannoufa, A. (2008) Enhancing the carotenoid content of *Brassica napus* seeds by downregulating lycopene epsilon cyclase. *Transgenic Res.* 17(4),573–585. doi: 10.1007/s11248-007-9131-x. Epub 2007 Sep 13. PMID: 17851775.

Zargani, N.S., Heinz, D.E. (1971). The volatile constituents of saffron. *Lebensm Wiss Technol.* 4(2),43–45.

Zhu, C., Naqvi, S., Breitenbach, J., Sandmann, G., Christou, P., Capell, T. (2008). Combinatorial genetic transformation generates a library of metabolic phenotypes for the carotenoid pathway in maize. *Proc Natl Acad Sci U S A.* 105(47),18232–18237.

Zorn, H., Langhoff, S., Schreibner, M., Nimtz, M., Berger, R.G. (2003). A peroxidase from *Lepista irina* cleaves beta,beta-carotene to flavor compounds. *Biol Chem.* 384,1049–1056.

5

Factors Influencing Somatic Embryogenesis and Regeneration with Particular Reference to Carica papaya *L.*

Manish Shukla
Amity University Uttar Pradesh
Lucknow Campus
Lucknow, India

Mala Trivedi
Amity University Uttar Pradesh
Lucknow Campus
Lucknow, India

Rajesh K. Tiwari
Amity University Uttar Pradesh
Lucknow Campus
Lucknow, India

CONTENTS

DOI: 10.1201/9781003239932-5

5.1 Introduction

Somatic embryogenesis is a process to produce an embryo or plant from a single somatic cell of the plant. Somatic embryos (SEs) thus produced are devoid of any seed coat or endosperm. Totipotency in cells of higher plants is the foremost responsible factor for somatic embryogenesis. Through somatic embryogenesis, plants can regenerate bipolar structures from a somatic cell, which is converted into a complete plant (Mendez-Hernandez et al. 2019). The transition of a somatic cell into an embryo cell is the critical step in the somatic embryogenesis process (Guan et al. 2016). Clonal propagation of plants, production of synthetic seeds, cryopreservation, germplasm conservation, and regeneration of genetically transformed plants are the most important applications of somatic embryogenesis (Guan et al. 2016). Conventional plant breeding for the improvement of crops has benefitted significantly from somatic embryogenesis and gene transfer methods (Litz and Grey 1995). The plant kingdom has the unique property of production of developmentally and morphologically normal SEs and complete plants through somatic embryogenesis (Zimmerman, 1993). Steward (1958) reported the first description of SE production from carrot root cells, and since then, this process has become an important pathway for the production of SEs and regeneration of plants using different types of plant cells. SEs are induced from cultured callus cells by a simple modification of culture conditions, and this development process closely resembles that of natural zygotic embryos (Zimmerman, 1993). Somatic embryogenesis has already been widely used in many woody and non-woody plants species for crop improvement programs and germplasm conservation purposes (Guan et al. 2016). It is imperative to develop a robust somatic embryogenesis and *in vitro* regeneration system in order to attempt clonal propagation of elite cultivars, to obtain virus-free plants, and for genetic modification of crops for virus resistance, herbicide tolerance and insect resistance traits. *In vitro* regeneration systems using somatic embryogenesis pathways in several woody and non-woody plants have been developed by many researchers worldwide (Fitch, 1991; Montalbán et al. 2015; Guan et al. 2016); however, somatic embryogenesis protocols for several other plant species are still under development. Detailed studies on the factors influencing somatic embryogenesis and *in vitro* regeneration are highly desirable to establish improved and efficient regeneration systems through somatic embryogenesis.

5.2 Types of Somatic Embryogenesis

There are two different routes for inducing somatic embryogenesis in plants (Sharp 1980): direct somatic embryogenesis and indirect somatic embryogenesis (Yang and Zhang, 2010). It will be difficult to distinguish between direct and indirect somatic embryogenesis, as both processes have been observed to occur simultaneously in the same tissue culture conditions (Gaj, 2004).

5.2.1 Direct Somatic Embryogenesis

Inducing SEs directly from the explant under certain culture conditions without any intermediate callus stage is called direct somatic embryogenesis (Guan et al. 2016). Direct somatic embryogenesis is rarer than indirect somatic embryogenesis (Gholami et al. 2013).

5.2.2 Indirect Somatic Embryogenesis

Indirect somatic embryogenesis occurs through an intermediate callus stage and has been observed in most plant species (Montalbán et al. 2015; Guan et al. 2016).

5.3 Characteristics and Stages of Somatic Embryogenesis

Somatic embryogenesis is a process of embryo development from a somatic cell in which a bipolar structure has been formed without establishing any vascular connection with the original tissue, and resembling most of the steps of zygotic embryogenesis (Von Arnold et al. 2002). Formation of proembryogenic masses from the explant is the first step of the multi-step process of somatic embryogenesis. SE formation, regeneration of secondary SEs, maturation, desiccation and plant regeneration are the subsequent steps of the somatic embryogenesis process (Thomas, 1993). SEs can develop directly from isolated protoplasts or microspores as explants from an organized tissue or indirectly from suspension cells or callus aggregates (Williams and Maheswaran, 1986). These embryos may develop as single structures in suspension culture, or attached to each other with callus tissue in solidified agar medium or liquid culture (Emons, 1994). The somatic cells that produce embryos are called embryogenic cells, and only a few cells of explants, microspores, callus aggregates and protoplasts are able to produce embryogenic structures that exhibit embryo development (Emons, 1994). Globular stage SEs appear as the first recognizable stage after placing the callus cluster in culture medium for somatic embryogenesis (Zimmerman, 1993). Thereafter, the heart or torpedo stage, the cotyledonary stage and regeneration of whole plants occur subsequently (Rao et al. 2006). All these stages of somatic embryogenesis process show parallels with zygotic embryogenesis (Zimmerman, 1993). The somatic embryogenesis process in plants represents the basic concept of totipotency and involves various proteins, a modified gene expression system and a complex signaling network regulated for a special purpose (Mendez-Hernandez et al. 2019). Exogenous stimuli in the form of plant growth regulators and stress conditions such as high or low temperature, drought and osmotic shock were used to regulate the gene expression patterns in artificial conditions (Nic-Can and Loyola-Vargas 2016).

5.4 Factors Affecting Somatic Embryogenesis with Special Reference to Papaya

Somatic embryogenesis and plant regeneration in papaya has been developed by several researchers (Fitch and Manshardt, 1990; Chen et al. 1987; Fitch, 1995; Heringer et al. 2013) using many explant types, genotypes and variable combinations of plant growth regulators, and continuous improvement and modification has been attempted for different genotypes by assessing the effects of various factors actively involved in somatic embryogenesis in plants. Somatic embryogenesis in plants is known to be significantly influenced by several factors: genotype of explant, explant type, explant wounding, plant growth regulators, light, temperature and several other media components (Karami, 2008). Figure 5.1 represents a schematic representation of somatic embryogenesis and in vitro regeneration in papaya.

FIGURE 5.1 Somatic embryogenesis and *in vitro* regeneration in papaya: (a) papaya fruits of different cultivars, (b) papaya seed, (c) immature zygotic embryo, (d) callus, (e) developing somatic embryos, (f) germination of embryos, (g) regenerated papaya plantlet, (h) rooting of regenerated shoots, (i) acclimatization of plantlets in cocopeat, (j) establishment of plants in pots.

5.4.1 Explant Type

Explants have a vital role in the induction of robust and efficient somatic embryogenesis in any plant. Among four explants, i.e. leaf, shoot tip, immature zygotic embryo and root tip, evaluated from various papaya cultivars, immature zygotic embryos showed the highest somatic embryogenesis and regenerated the maximum number of papaya plants (Shukla et al. 2019). Along with this, several other researchers also observed that immature zygotic embryos were the best explants to induce somatic embryogenesis in papaya (Tsay and Su, 1985; Li et al. 2017). The axillary buds of a three-year-old field-grown papaya plant of a dioecious variety were used as explants for tissue culture propagation of papaya (Shlesinger et al. 1987), while apical and lateral buds from seedling-stage and mature plants of the Pusa Nanha variety were also used as explants (Podikunju, 2017) for *in vitro* propagation studies in papaya. Apical shoot tips and young leaves as explants delivered a low frequency of somatic embryogenesis (Shukla, 2020), while higher somatic embryogenesis was reported in Eksotika cultivar using immature zygotic embryos as explants (Al-Shara et al. 2020). Shoot buds were induced from epicotyle segments in CO7 variety (Anandan et al. 2011), while young leaf segments from *in vitro* grown plants of Shahi *cv.* were utilized to induce embryogenic callus (Roy et al. 2016). A high frequency of embryogenesis was achieved from cotyledonary leaves of THB papaya cultivar as explants (Cipriano et al. 2018). Immature zygotic embryos from unripe papaya fruits from Washington and Honey dew cultivars were also used as explants to produce SEs and plant regeneration (Bhattacharya et al. 2002). Induction of shoot bud differentiation was successfully achieved while culturing young inflorescence tips of male and female papaya plants in the presence of different plant growth regulators (Agnihotri et al. 2004). A high frequency of friable embryogenic calli (FEC) was produced from the leaf explants of hermaphrodite papaya plants in a culture medium supplemented with 2,4 dichlorophenoxy acetic acid (2,4-D) (Koehler et al. 2013). The SEs produced will be beneficial and used for regenerating a large number of disease-free high-quality papaya plantlets as well as target tissue for genetic transformation experiments.

5.4.2 Genotype of Explant

The genetic background of plant species and explant development stages during somatic embryogenesis also affects somatic embryogenesis in several plants, including papaya (Krishnan, 2009). The effect of different papaya genotypes on frequency of somatic embryogenesis has been evaluated by many researchers. Zygotic embryos, hypocotyl and leaf explants of Rathna cultivars have been utilized for induction of embryogenic callus by applying various combinations of plant growth regulators and basal media (Farzana et al. 2008). A mixed response by different genotypes to induction of embryogenic tissues from immature zygotic embryos as explants from different papaya cultivars was reported by Malabadi et al. (2011). Furthermore, the variable embryogenic response from different genotypes of papaya was studied with hypocotyl sections as explants from four Hawaiian cultivars (Fitch, 1993). The frequency of initiation of somatic embryogenesis is directly dependent on the genotype of the mother plant, and the culture medium used also affects the genotype's competence for somatic embryogenesis in plants (Guerra et al. 2000). The difference in efficiency of embryogenic response in various genotypes was well established by Fitch (1993) while using many varieties of papaya for the induction of somatic embryogenesis. Somatic embryogenesis and regeneration systems for some Philippine papaya genotypes were developed by Magdalita and San Pascual (2019). Commercial propagation is also hampered severely by difficulty in rooting in plantlets raised in tissue culture and low survival rate after transplantation (Agnihotri et al. 2004). Thus, wide variation in the *in vitro* responses of different cultivars of same species has been reported. This underlines that the development of high-frequency genotype-independent somatic embryogenesis and *in vitro* regeneration of plantlets is a pre-requisite for a clonal propagation and plant breeding program.

5.4.3 Role of Plant Growth Regulators (PGRs)

The optimum concentration of the PGR 2,4-D has a significant effect on the response of somatic embryogenesis in papaya (Bhattacharya et al. 2002; Bukhori, 2013). Maximum induction of callusing and somatic embryogenesis from immature zygotic embryos was achieved when explants were cultured in induction medium containing 10 mg/l 2,4-D (Shukla et al., 2019). Similarly, a high frequency of embryogenic callus (77.5%) was produced using immature zygotic embryo explants of Eksotika variety cultured in Murashige and Skoog (MS) medium supplemented with 10 mg/l 2,4-D and 260 mg/l carbenicillin (Bukhori, 2013). However, in another study, well-developed shoots were regenerated from Pusa Nanha cultivar in MS medium under light conditions without using any PGRs (Podikunju, 2017). The auxins dicamba and picloram, chitosan, and coconut water were also tested for callus induction in papaya tissue culture (Da Silva, 2016), while Shukla (2020) was able to produce embryogenic calli from immature zygotic embryos of papaya in MS medium containing 10 mg/l 2,4-D, 60 g/l sucrose and 400 mg/l glutamine. An efficient protocol for micropropagation of Indian papaya cv CO7 was developed using epicotyl segment as explant on MS medium (Anandan et al. 2011), and high regeneration frequency occurred on MS medium supplemented with 2.5 μM thidiazuron (TDZ) (Al-Shara et al. 2020). Ramesh et al. (2018) were able to establish an efficient micropropagation system for Surya cultivar of papaya and reported callus induction with TDZ at 0.1 mg/l and shoot regeneration in 1.0 mg/l gibberellic acid (GA3). Fitch and Manshardt (1990) reported efficient somatic embryogenesis and plant regeneration from immature zygotic embryos of papaya and were able induce embryogenic culture in 0.1–25 mg/l 2,4-D, 400 mg/l glutamine and 6% sucrose. The highest percentage of SEs was produced in MS medium augmented with 2,4-D 5.0 mg/l, glutamine 400 mg/l and sucrose 60 g/l by Kabir et al. (2016) from immature zygotic embryos of papaya cv Red lady. Germination of developing SEs was also obtained on MS medium devoid of any PGRs (Kabir et al. 2016).

5.4.4 Polyamines and Amino Acids

The polyamines spermidine, spermine and putrescine are nitrogenous bases found in all living organisms and considered as hormonal messengers due to their role in the modulation of signal response (Nagar and Sharma, 2008). It has been reported that polyamine metabolism has an immense role in growth, development and stress responses in plants and acts as a crucial factor in somatic embryogenesis in plants

(Minocha and Minocha, 1995). Addition of the amino acids glycine, arginine, glutamine and asparagine to tissue culture media is known to modulate the somatic embryogenesis process positively and accelerate the cell division process during embryogenesis (Kamada and Harada, 1979). Amino acid glutamine has been shown to increase plant biomass in several *in vitro* regeneration experiments, including somatic embryogenesis (Carlsson et al. 2017). A somatic embryogenesis protocol for papaya cv Rathna from hypocotyl and zygotic embryos was developed by Farzana et al. (2008); it was found that casein hydrolysate is most suitable for maturation of calli, and 0.02 mg/l naphthalene acetic acid (NAA) and 0.5 mg/l benzylaminopurine (BAP) produced higher germination rates (Farzana et al. 2008). The addition of arginine, spermidine, or a combination of the polyamines spermidine, spermine and putrescine increased the embryogenic potential of callus in the rubber tree (El Hadrami and D'Auzac, 1992). Davis and Ying (2004) induced SEs from immature seeds placed on Fitch's liquid medium with half MS and vitamins, 50 mg/l myo-inositol, 6% sucrose, 10 mg/l 2,4-D and 400 mg/l glutamine for genetic transformation in papaya.

5.4.5 Carbon Source

The choice and optimum concentration of the carbon source play a critical role in somatic embryogenesis in papaya. Sucrose and maltose have been utilized by most researchers as a carbon source, while some have also tried fructose as a carbon source in induction of somatic embryogenesis in plants. Sucrose at 6% concentration was proved to be a better carbon source than maltose when induction of somatic embryogenesis and plant regeneration was attempted from immature embryos of Eksotika papaya (Vilasini et al. 2000). Fitch (1993) obtained high-frequency somatic embryogenesis in papaya using explant hypocotyl sections with 6% sucrose and 400 mg/l glutamine along with other essential media components. Significantly higher SE production was reported with the carbon source maltose during the induction of somatic embryogenesis in *Hevea brasiliensis* (Blanc et al. 1999). An increased number of SEs were obtained when 6% sucrose was added to the culture medium while studying somatic embryogenesis and plant regeneration of papaya cv Shahi (Roy et al, 2016).

5.4.6 Somatic Embryogenesis as a Result of Stress

The osmotically active compounds polyethylene glycol (PEG), mannitol and sorbitol have been established to have significant effects on the conversion and maturation of developing SEs in different plants. The use of PEG or ABA (abscisic acid) or activated charcoal has been recommended for maturation of SEs to provide a stress condition for synchronization in development and rapid conversion of globular SEs (Shukla et al. 2016). Significantly higher percentages of SEs (68.86%) matured and plantlets (35.25%) were regenerated in 45 mg/l PEG in MS medium (Shukla et al. 2019). However, this finding is not consistent with the results of Heringer et al. (2013), who recorded the highest numbers of matured SEs in 60 g/l PEG (Heringer et al. 2013), while germinated normal papaya seedlings have been recovered in a medium containing PEG, ABA and activated charcoal for maturation (Schmidt et al. 2005). Similarly, matured soybean somatic embyos were obtained with high conversion ability in a maturation medium containing 3% sorbitol (Li et al. 1998). Also, high embryo maturation efficiency was found in a maturation medium supplemented with 6% PEG and 4% maltose (Vilasini et al. 2000). Globular SEs were formed from the FEC continuously, and cotyledonary SEs were matured in a medium containing PEG, activated charcoal and abscisic acid (Koehler et al. 2013). An improved and modified protocol for *in vitro* propagation of papaya was developed by Magdalita and San Pascual (2019), and successful rooting in germinating plantlets was achieved by soaking the bases of the germinating plantlets in a 5 mg/l indole-3-butyric acid (IBA) solution for one hour and transferring them to a sucrose-free vermiculite medium (McCubbin et al. 2003).

5.4.7 Importance of Signaling for Plant Somatic Embryogenesis

Signaling plays a pivotal role in somatic embryogenesis in plants from the induction stage to the regeneration stage. PGRs, mainly auxins, cytokinins, ethylene and abscisic acid, interact during the induction of somatic embryogenesis (Mendez-Hernandez et al. 2019). The role of signaling during somatic embryogenesis was studied extensively by many researchers, who examined its interaction with various regulators from initial cell differentiation during induction to the regeneration stage of somatic embryogenesis. Since clonal propagation via somatic embryogenesis is widely exploited for many plants of agronomic interest, understanding the mechanisms controlling somatic embryogenesis, signaling in the embryogenic pathway, and its regulation is crucial in plant science (Salaün et al. 2021). Hormonal control and plant tolerance to stress conditions are determining factors for the onset of somatic embryogenesis (Mendez-Hernandez et al. 2019). The LAFL gene regulatory pathway, which consists of LEC1, LEC2, ABI3 and FUS3 genes, mainly shows specificity with somatic embryogenesis and is regulated by transcriptional factors specific to somatic embryogenesis (Salaün et al. 2021). Although zygotic embryogenesis and somatic embryogenesis have similar hormonal and transcriptional control systems, they exhibit differences in some specificities (Leljak-Levanić et al. 2015). It is well understood that various genetic and physiological factors are involved in activation of *in vitro* embryogenesis for different types of somatic cells. The initial stage of somatic embryogenesis is characterized by the induction of several genes and activation of the signal transduction pathway to determine the fate of embryogenesis. The developmental mechanism of embryogenesis in plants requires precise gene expression regulation during the early stages of embryogenesis (Karami et al. 2009). Several genes that are involved in somatic embryogenesis in plants have been identified and characterized by proteome and transcriptome analysis (Karami et al. 2009). SERK genes, LEC genes, BBM genes, AGL15 gene, MtSERF1 gene, MtSK1 gene, GST gene, WUS gene and PKL gene are some of the genes that have been identified as being involved in the signal transduction pathway of somatic embryogenesis in plants (Karami et al. 2009).

5.5 Regeneration of Plants from Somatic Embryos

A high-frequency *in vitro* regeneration system for papaya is essential for mass multiplication and genetic manipulation experiments. Regeneration of complete rooted plantlets from matured SEs is a critical step. The PGRs kinetin and BAP will be required for the conversion of matured SEs into complete plantlets. Thereafter, treatment with the optimum concentration of IBA will induce rooting in regenerated plantlets. Root regeneration was also achieved in papaya cultivar Eksotika by Al-Shara et al (2020) on MS medium containing 7.9 mg/l phloroglucinol and vermiculite. Regeneration of matured SEs of papaya occurred when transferred to full-strength MS medium, 3% sucrose and a combination of the PGRs NAA (0.1 mg/l) and 6-benzylaminopurine (6-BAP) (0.1 mg/l) (Vilasini et al. 2000). The transfer of the developing SEs to hormone-free medium before regeneration was beneficial in decreasing the incidence of abnormal plants. Plants were successfully rooted and transferred to a vermiculite:sand:soil mixture (1:1:1) for further growth and development (Vilasini et al. 2000).

5.6 Factors Influencing *in vitro* Regeneration of Plantlets from Somatic Embryos

5.6.1 Plant Growth Regulators

Papaya plantlets were regenerated from matured SEs in 4–8 weeks of incubation in regeneration media. The germinating SEs of papaya were regenerated and elongated in MS medium with 1 mg/l gibberellic acid, 0.5 mg/l IBA, 100 mg/l myo-inositol and 3.76 mg/l riboflavin (Al-Shara et al. 2020). A higher germination rate from matured SEs was obtained by Farzana et al (2008) in a medium supplemented with 0.02 mg/l NAA and 0.5 mg/l BAP (Farzana et al. 2008). Rooting medium containing 2 mg/l IBA and 8 g/l powdered cocopeat along with basal MS medium showed significantly best rooting in 21 days in regenerated papaya plantlets, as established by Shukla et al. (2019). Maximum numbers of primary and secondary roots emerged in plantlets cultured in this rooting medium. In another study, 3 mg/l IBA, 30 g/l sucrose and 0.05% activated charcoal were used along with full-strength MS medium for *in vitro* rooting in papaya (Podikunju, 2017). Pulse treatment with 10 mg/l IBA for 24 hours was used to initiate rooting in regenerated papaya plantlets and achieved 80% transplant success when rooted and hardened plants were transferred to a mixture of garden soil and soilrite (1:1) in earthen pots (Agnihotri et al. 2004). Half-strength MS medium augmented with 1.0 mg/l IBA in culture medium was used to produce rooting (Rohman et al. 2007), while MS medium supplemented with 0.5 mg/l IBA also produced well-developed shoots and roots (Bukhori, 2013). A fast and robust rooting system was developed in 30 days in a medium containing 2 mg/l IBA and 8 g/l powdered cocopeat along with basal MS medium (Shukla et al. 2019).

5.6.2 Media Components

In addition to PGRs, many other culture media components, such as carbon source, amino acids and polyamines, and stress conditions play significant roles in regeneration of plantlets from germinating matured SEs. Sucrose is the primary carbon source even in the regeneration stage of embryogenesis. A few researchers have reported efficient regeneration in plant hormone-free MS medium (Farzana et al. 2008; Vilasini et al. 2000). Casein hydrolysate along with BAP in MS medium was used for obtaining highly efficient regeneration in germinating cotyledonary-stage SEs of papaya (Shukla et al. 2019). Silver thiosulphate was also tested for somatic embryo proliferation, maturation and germination of plantlets in papaya (Adkins et al, 1997).

5.6.3 Culture Growth Conditions (Light, Temperature, Humidity)

Optimum temperature range, light quality and duration, and relative humidity also affect *in vitro* regeneration of plantlets after somatic embryogenesis. Relative humidity of 55–60% along with a 12-h light and dark cycle for one month was found to be effective for regeneration of papaya plantlets from SEs (Shukla, 2020). A photoperiod of 12 h with 48% relative humidity was recommended for the multiplication and germination of papaya SEs (Solórzano-Cascante et al. 2018). Papaya plantlets were regenerated from matured SEs under a 16-h photoperiod (50 $\mu mol/m^2/s$) and 25 ± 3 °C temperature range with a relative humidity of 55–60% in a growth room (Malabadi et al. 2011). Farzana et al. (2008) regenerated papaya plantlets cv Rathna under 16 h light/8 h dark conditions, light intensity of 55 $\mu mol/m^2/s$ at 25 °C. Temperature, humidity, and light quality and duration are important factors for the successful regeneration of plantlets from papaya SEs.

5.7 Applications of Somatic Embryogenesis

SEs produced through the process of somatic embryogenesis by using different explant types are used widely for the clonal propagation of elite cultivars, for regeneration of different crops and as a target tissue for transformation experiments (Salaün et al. 2021). Globular SEs of papaya were utilized as target tissue for *Agrobacterium*-mediated genetic transformation in papaya to obtain virus-resistant plants (Shukla, 2020). SEs may also be used as artificial seeds and are very suitable for long-term storage such as cryopreservation (Farzana et al, 2008). Cryopreservation of SEs is an effective strategy for germplasm storage and germplasm conservation (Tessereau et al. 1994). Somatic embryogenesis is also used widely in the propagation of adult woody plants, which is difficult otherwise (Guan et al. 2016). Somatic embryogenesis is an important tool in plant biotechnology, which has wide applications in many ways for crop improvement and cryopreservation of germplasm. It is used for the commercial production of several high-value plants and as a tool for the advancement of embryology and plant biology.

REFERENCES

Adkins, S.W., Samosir, Y.M., Ernawati, A., Godwin, I.D., & Drew, R.A. (1997). Control of ethylene and use of polyamines can optimise the conditions for somatic embryogenesis in coconut (*Cocos nucifera* L.) and papaya (*Carica papaya* L.). In *International Symposium on Biotechnology of Tropical and Subtropical Species Part 2 461*, September (pp. 459–466).

Agnihotri, S., Singh, S. K., Jain, M., Sharma, M., Sharma, A. K., & Chaturvedi, H. C. (2004). In vitro cloning of female and male *Carica papaya* through tips of shoots and inflorescences. Indian Journal of Biotechnology, 3, 235–240.

Al-Shara, B., Taha, R. M., Mohamad, J., Elias, H., & Khan, A. (2020). Somatic embryogenesis and plantlet regeneration in the *Carica papaya* L. cv. Eksotika. *Plants*, *9*(3), 360.

Anandan, R., Thirugnanakumar, S., Sudhakar, D., & Balasubramanian, P. (2011). In vitro organogenesis and plantlet regeneration of (*Carica papaya* L.). *Journal of Agricultural Technology*, *7*(5), 1339–1348.

Bhattacharya, J., Khuspe, S. S., Renukdas, N. N., & Rawal, S. K. (2002). Somatic embryogenesis and plant regeneration from immature embryo explant of papaya (*Carica papaya* L. cv. washington and honey dew).

Blanc, G., Michaux-Ferrière, N., Teisson, C., Lardet, L., & Carron, M. P. (1999). Effects of carbohydrate addition on the induction of somatic embryogenesis in *Hevea brasiliensis*. *Plant Cell, Tissue and Organ Culture*, *59*(2), 103–112.

Bukhori, M. F. (2013). Improved protocol for high frequency plant regeneration through somatic embryogenesis in Carica papaya. *Research in Biotechnology*, 4(5).

Carlsson, J., Svennerstam, H., Moritz, T., Egertsdotter, U., & Ganeteg, U. (2017). Nitrogen uptake and assimilation in proliferating embryogenic cultures of Norway spruce—Investigating the specific role of glutamine. *PLoS ONE*, *12*(8), e0181785.

Chen, M. H., Wang, P. J., & Maeda, E. (1987). Somatic embryogenesis and plant regeneration in *Carica papaya* L. tissue culture derived from root explants. *Plant Cell Reports*, *6*(5), 348–351.

Cipriano, J. L., Cruz, A. C. F., Mancini, K. C., Schmildt, E. R., Lopes, J. C., Otoni, W. C., & Alexandre, R. S. (2018). Somatic embryogenesis in *Carica papaya* as affected by auxins and explants, and morphoanatomical-related aspects. *Anais da Academia Brasileira de Ciências*, *90*, 385–400.

Da Silva, J. A. T. (2016). In vitro response of papaya (*Carica papaya*) to plant growth regulators. *Nusantara Bioscience*, *8*(1).

Davis, M. J., & Ying, Z. (2004). Development of papaya breeding lines with transgenic resistance to Papaya ringspot virus. *Plant Disease*, *88*(4), 352–358.

El Hadrami, I., & D'Auzac, J. (1992). Effects of polyamine biosynthetic inhibitors on somatic embryogenesis and cellular polyamines in *Hevea brasiliensis*. *Journal of Plant Physiology*, *140*(1), 33–36.

Emons, M. C. (1994). Somatic embryogenesis: Cell biological aspects. *Acta Botanica Neerlandica*, *43*(1), 1–14.

Farzana, A. R. F., Palkadapala, P. G. V. N., Meddegoda, K. M. M. N., Samarajeewa, P. K., & Eeswara, J. P. (2008). Somatic embryogenesis in papaya (*Carica papaya* L.) cv. Rathna. *Journal of the National Science Foundation of Sri Lanka*, *36*(1).

Fitch, M. M. (1993). High frequency somatic embryogenesis and plant regeneration from papaya hypocotyl callus. *Plant Cell, Tissue and Organ Culture*, *32*(2), 205–212.

Fitch, M. M. M. (1991). *Development of genetic transformation systems for papaya*. University of Hawai'i at Manoa.

Fitch, M. M. M. (1995). Somatic embryogenesis in papaya (*Carica papaya* L.). In *Somatic Embryogenesis and Synthetic Seed I* (pp. 260–279). Springer, Berlin, Heidelberg.

Fitch, M. M., & Manshardt, R. M. (1990). Somatic embryogenesis and plant regeneration from immature zygotic embryos of papaya (*Carica papaya* L.). *Plant Cell Reports*, *9*(6), 320–324.

Gaj, M. D. (2004). Factors influencing somatic embryogenesis induction and plant regeneration with particular reference to *Arabidopsis thaliana* (L.) Heynh. *Plant Growth Regulation*, *43*(1), 27–47.

Gholami, A. A., Alavi, S. V., Majd, A., & Fallahian, F. (2013). Plant regeneration through direct and indirect somatic embryogenesis from immature seeds of citrus. *European Journal of Experimental Biology*, *3*, 307–310.

Guan, Y., Li, S. G., Fan, X. F., & Su, Z. H. (2016). Application of somatic embryogenesis in woody plants. *Frontiers in Plant Science*, *7*, 938.

Guerra, M. P., Silveira, V., dos Santos, A. L., Astarita, L. V., & Nodari, R. O. (2000). Somatic embryogenesis in *Araucaria angustifolia* (Bert) O. Ktze. In *Somatic embryogenesis in woody plants* (pp. 457–478). Springer, Dordrecht.

Heringer, A. S., Vale, E. M., Barroso, T., Santa-Catarina, C., & Silveira, V. (2013). Polyethylene glycol effects on somatic embryogenesis of papaya hybrid UENF/CALIMAN 01 seeds. *Theoretical and Experimental Plant Physiology*, *25*(2), 116–124.

Kabir, M. H., Rahman, M. Z., & Mamun, A. N. K. (2016). Somatic embryogenesis and plant regeneration from zygotic embryo in *Carica papaya* L., cv. Red-Lady. *J. Plant*, *4*(6), 45.

Kamada, H., & Harada, H. (1979). Studies on the organogenesis in carrot tissue cultures II. Effects of amino acids and inorganic nitrogenous compounds on somatic embryogenesis. *Zeitschrift für Pflanzenphysiologie*, *91*(5), 453–463.

Karami, O. (2008). Factors affecting Agrobacterium-mediated transformation of plants. *Transgenic Plant J.*, *2*(2), 127–137.

Karami, O., Aghavaisi, B., & Pour, A. M. (2009). Molecular aspects of somatic-to-embryogenic transition in plants. *Journal of Chemical Biology*, *2*(4), 177–190.

Koehler, A. D., Carvalho, C. R., Abreu, I. S., & Clarindo, W. R. (2013). Somatic embryogenesis from leaf explants of hermaphrodite *Carica papaya*: A new approach for clonal propagation. *African Journal of Biotechnology*, *12*(18).

Krishnan, S. (2009). Genetic transformation of *Carica papaya* L.- Indian cultivar CO7. Doctoral Research, TERI School of Advanced Studies, New Delhi.

Leljak-Levanić, D., Mihaljević, S., & Bauer, N. (2015). Somatic and zygotic embryos share common developmental features at the onset of plant embryogenesis. *Acta Physiologiae Plantarum*, *37*(7), 1–14.

Li, S., Cong, Y., Liu, Y., Wang, T., Shuai, Q., Chen, N., ... & Li, Y. (2017). Optimization of Agrobacterium-mediated transformation in soybean. *Frontiers in Plant Science*, *8*, 246.

Li, X. Y., Huang, F. H., Murphy, J. B., & Gbur, E. E. (1998). Polyethylene glycol and maltose enhance somatic embryo maturation in loblolly pine (*Pinus taeda* L.). *In Vitro Cellular & Developmental Biology-Plant*, *34*(1), 22–26.

Litz, R. E., & Gray, D. J. (1995). Somatic embryogenesis for agricultural improvement. *World Journal of Microbiology and Biotechnology*, *11*(4), 416–425.

Magdalita, P. M., & San Pascual, A. O. (2019). Somatic embryogenesis, regeneration, phenotypic and cytological evaluation of selected Philippine papaya (*Carica papaya* L.) genotypes. *Philippine Journal of Crop Science (PJCS)*, *44*(3), 20–30.

Malabadi, R. B., Kumar, S. V., Mulgund, G. S., & Nataraja, K. (2011). Induction of somatic embryogenesis in papaya (*Carica papaya*). *Research in Biotechnology*, 2(5), 40–55.

McCubbin, M. J., Van Staden, J., Debergh, P. (2003). A modified technique for in vitro propagation of papaya (*Carica papaya* L.). *South African Journal of Botany*, *69*(3), 287–291.

Mendez-Hernandez, H. A., Ledezma-Rodriguez, M., Avilez-Montalvo, R. N., Juarez-Gomez, Y. L., Skeete, A., Avilez-Montalvo, J., De-la-Pena, C., & Loyola-Vargas, V. M. (2019). Signaling overview of plant somatic embryogenesis. *Frontiers in Plant Science*, *10*, 77.

Minocha, S. C., & Minocha, R. (1995). Role of polyamines in somatic embryogenesis. In *Somatic Embryogenesis and Synthetic Seed I* (pp. 53–70). Springer, Berlin, Heidelberg.

Montalbán, I. A., García-Mendiguren, O., Goicoa, T., Ugarte, M. D., & Moncaleán, P. (2015). Cold storage of initial plant material affects positively somatic embryogenesis in *Pinus radiata*. *New Forests*, *46*(2), 309–317.

Nagar, P. K., & Sharma, M. (2008). Role of polyamines during somatic embryogenesis. *Advances in Plant Physiology-A Treatise*, 275–294.

Nic-Can, G. I., & Loyola-Vargas, V. M. (2016). The role of the auxins during somatic embryogenesis. In *Somatic embryogenesis: fundamental aspects and applications* (pp. 171–182). Springer, Cham.

Podikunju, B. (2017). Tissue culture protocol for in-vitro propagation of papaya (*Carica papaya* L.). *Journal of Krishi Vigyan*, *6*(1), 205–212.

Ramesh, A. N., Santhosh, D. B., Suresh, D., Radha, K., Rashmi, K., & Hajira, K. (2018). In vitro regeneration of papaya (*Carica papaya* L.) Variety Surya. *International Journal of Pure & Applied Bioscience*, *6*(4), 456–461.

Rao, A. Q., Hussain, S. S., Shahzad, M. S., Bokhari, S. Y. A., Raza, M. H., Rakha, A., ... & Riazuddin, S. (2006). Somatic embryogenesis in wild relatives of cotton (*Gossypium* Spp.). *Journal of Zhejiang University Science B*, *7*(4), 291–298.

Rohman, M. M., Islam, M. N., Alam, M. S., Ahmad, M. R., & Paul, T. K. (2007). Lateral bud culture of papaya (*Carica papaya*) for clonal propagation. *Biotechnology*, *6*(3), 339–343.

Roy, P. K., Roy, S. K., Hakim, L., & Mamun, A. N. K. (2016). Somatic embryogenesis and plant regeneration of papaya (*Carica papaya* L. Cv. Shahi). *Nucl. Sci. App*, *25*.

Salaün, C., Lepiniec, L., & Dubreucq, B. (2021). Genetic and molecular control of somatic embryogenesis. *Plants*, *10*(7), 1467.

Schmidt, M. A., Tucker, D. M., Cahoon, E. B., & Parrott, W. A. (2005). Towards normalization of soybean somatic embryo maturation. *Plant Cell Reports*, *24*(7), 383–391.

Sharp, W. R. (1980). The physiology of in vitro asexual embryogenesis. *Horticultural Reviews*, *2*, 268–310.

Shlesinger, D., Reuveni, O., & Lavi, U. (1987). Tissue culture propagation of papaya. In *Symposium on In Vitro Problems Related to Mass Propagation of Horticultural Plants*, *212*, 570–570.

Shukla, M. (2020). Development of an efficient Agrobacterium mediated genetic transformation system in papaya (Carica papaya L.) (Unpublished doctoral dissertation). Amity Institute of Biotechnology, Lucknow Campus, Amity University Uttar Pradesh, India.

Shukla, M., Al-Busaidi, K. T., Al-Blushi, G. S., Al-Burashdi, A. H., Al-Hasani, H. D., Al-Jabri, M. H., & AL-Kalbani, B. S. (2016). Nucellar embryogenesis and plantlet regeneration in monoembryonic and polyembryonic mango (*Mangifera indica* L.) cultivars. *African Journal of Biotechnology*, *15*(52), 2814–2823.

Shukla, M., Tiwari, R. K., & Al-Busaidi, K. T. (2019). High-frequency somatic embryogenesis and plantlet regeneration in papaya. *Research Journal of Biotechnology*, 14(3), 100–108.

Solórzano-Cascante, P., Sánchez-Chiang, N., & Jiménez, V. M. (2018). Explant type, culture system, 6-benzyladenine, meta-topolin and encapsulation affect indirect somatic embryogenesis and regeneration in *Carica papaya* L. *Frontiers in Plant Science*, *9*, 1769.

Steward, F. C. (1958). Growth and organized development of cultured cells. III. Interpretations of the growth from free cell to carrot plant. *American Journal of Botany*, *45*(10), 709–713.

Tessereau, H., Florin, B., Meschine, M. C., Thierry, C., & Pétiard, V. (1994). Cryopreservation of somatic embryos: A tool for germplasm storage and commercial delivery of selected plants. *Annals of Botany*, *74*(5), 547–555.

Thomas, T. L. (1993). Gene expression during plant embryogenesis and germination: An overview. *The Plant Cell*, *5*(10), 1401.

Tsay, H. S., & Su, C. Y. (1985). Anther culture of papaya (*Carica papaya* L.). *Plant Cell Reports*, *4*(1), 28–30.

Vilasini, P., Latipah, Z., & Salasiah, A. (2000). Induction of somatic embryogenesis and plant regeneration from immature embryos of Eksotika papaya (*Carica papaya*). *Journal of Tropical Agriculture and Food Science, 28*, 121–126.

Von Arnold, S., Sabala, I., Bozhkov, P., Dyachok, J., & Filonova, L. (2002). Developmental pathways of somatic embryogenesis. *Plant Cell, Tissue and Organ culture*, *69*(3), 233–249.

Williams, E. G., & Maheswaran, G. (1986). Somatic embryogenesis: Factors influencing coordinated behaviour of cells as an embryogenic group. *Annals of Botany*, *57*(4), 443–462.

Yang, X., & Zhang, X. (2010). Regulation of somatic embryogenesis in higher plants. *Critical Reviews in Plant Science*, *29*(1), 36–57.

Zimmerman, J. L. (1993). Somatic embryogenesis: A model for early development in higher plants. *The plant Cell*, *5*(10), 1411.

6

Application of Plant Tissue Culture for Improvement of Centella asiatica

Shweta Kumari
Patna University
Patna
Bihar, India

Nitish Kumar
Central University of South Bihar
Gaya
Bihar, India

Maheshwar Prasad Trivedi
Patna University
Patna
Bihar, India

CONTENTS

6.1 Introduction

The use of medicinal plants for curing various diseases either originates from ancestors or has been developed by human intervention. The application of medicinal plants for treating diseases has not changed in spite of the Cultural Revolution. Ayurvedic

DOI: 10.1201/9781003239932-6

medicine, also called Indian medicine, is known as the mother of all remedies and is the oldest therapeutic system on the earth. The exact translation of Ayurveda is "knowledge of life" (Khan, 2014). Some literature states that the medicinal use of plants dates from approximately 4000–5000 BC. However, the Indian Rig-veda, which was written between 1600 and 3500 BC, refers to the use of plants as medicine. Ancient literature such as the Rig-veda and Atharva-veda mentions the Ayurvedic system of therapies.

The Charak Samhita was the earliest sacred book to be completely focused on the application and notion of Ayurveda as well as its therapeutic prospects for the wellness of human beings. The first application of natural herbs as medicine was by the Chinese. In India, the basic foundation for the medical sciences was laid by ancient physicians, who also studied in detail the use of medicinal plants for therapeutic purposes. Medicinal plants have played an important part in indigenous medical systems worldwide. Ethnobotany facilitates the traditional use of plants by human beings for natural drug development and research (Schippman et al. 2002; Hosseinzadeh et al. 2015). There are various well-established herbal and medicinal plants centers in India, which is also recognized as an eminent Ayurvedic medicine center in different parts of the world. The World Health Organization (WHO) reported that 80% of citizens of developing countries use medicinal herbs for their primary treatment (Sofowora et al. 2013; Roy and Bharadvaja 2017; Aziz et al. 2018). In developed countries such as the UK, 25% of citizens use medicinal plants for their primary health treatment (Lemma et al. 2020). Some compounds cannot be synthesized or are not economically feasible to synthesize in the pharmaceutical industry; 40% of these are derived from medicinal plants to use in the pharmaceutical industry (Bajaj et al. 1988; Yoshimatsu 2008). Various parts of herbal plants— stem, root, leaf, flowers, seeds or sometimes the complete plant—are used for treatment. These parts of medicinal plants possess certain molecules called bioactive molecules, which influence the physiological system of an organism (Kia et al. 2018). Because of overexploitation, habitat destruction, industrialization and urbanization lead to the loss of valuable medicinal plants, and they are also listed under threatened categories (Chokheli et al. 2020).

Centella asiatica, also called *Hydrocotyle asiatica* or Indian pennywort, is an important valuable medicinal plant with many medicinal properties. It is called "Brain Food of India" because it re-energizes as well as rebuilds age-related injury, and is notably used to repair nerve and brain cells. It belongs to the family Apiaceae, earlier called Umbelliferae. *Centella asiatica* is a perennial, prostrate, stoloniferous plant with height up to 6 inches. The plant possesses orbicular–reniform-shaped leaves attached with nodes, one to three in number. *Centella* is mostly found in subtropical and tropical region of India. This plant is native to South Africa, south east Asia, eastern South America, south east USA, Venezuela, Columbia, Mexico, Madagascar, some parts of China, Sri Lanka and India (Jamil et al. 2007).

It contains a broad variety of bioactive molecules, which are also called secondary metabolites. These chemical constituents are categorized into two groups: triterpenes and saponins. Both these bioactive compounds take part in therapeutic activities along with nutraceutical applications. *Centella* triterpenes include asiaticoside, madecassoside, brahmoside, brahminoside, thankiniside, isothankunisode, asiatic acid, madecassic acid, centic acid, cenellic acid, centelloside and madasiatic acid. Out of these triterpenes, the four most important bioactive molecules are asiaticoside, madecassic acid, asiatic acid and madecassoside. *Centella asiatica* also possesses high amounts of phenolics and flavonoids. The leaves have a higher quantity of phytochemicals as compared with stem, root and petiole (Aziz et al. 2007; Zainol et al. 2008; Bhavna and Jyoti 2011; Chong and Aziz 2011). This valuable medicinal plant shows many pharmacological properties: neuroprotective (Gray et al. 2018), antibacterial (Zaidan et al. 2005), antifungal (Jagtap et al. 2009), antioxidant (Seevaratnam et al. 2012), cardiac (Gnanapragasam et al. 2004), antiaging (Lee et al. 2004), wound healing (Rao et al. 2005; Ruksiriwanich et al. 2020), anti-inflammatory (Abdulla et al. 2010) and anti-tumor activity (Bunpo et al. 1997).

Due to the broad spectrum of uses of this valuable medicinal plant, it has been listed as a principal herbal plant all over the world by the export and import bank of India. Unimpeded exploitation of *Centella asiatica* has led to reduction of its wild stock, and coupled with limited cultivation and incompetent approach towards replacement, this has caused it to be registered as a threatened and endangered species by the IUCN. The preservation, restoration and regeneration of this costly medicinal plant are a serious challenge. A constant supply of *Centella* species is required, on the one hand, and pressure on natural sources needs to be relieved, on the other hand. The plant tissue culture technique appears to be an important tool to conserve this costly medicinal plant (Naidu et al. 2010). Over time, advances in plant tissue culture techniques have led to protoplast, cell and embryo culture along with regeneration of whole plant, proving plant tissue culture as a technique.

Plant tissue culture is also applied by plant breeders, pathologists, biochemists and geneticist researchers. For large-scale production of desired molecules, tissue culture may provide an alternative way to ensure a continuous supply of material (Ghareeb and Taha 2018). Plant tissue culture is a systematic tool for the production of secondary metabolites such as naphthoquinones and shikonin. This technique has also played a key role in the production of various flavors, natural colorants, sweeteners and pharmaceutical products (Gaurav et al. 2018). Tissue culture facilitates an extensive program for genetically superior clones, germplasm conservation, disease-free plants and secondary metabolite production. Today, numerous ornamental and medicinal plants have been propagated by plant tissue culture (Chandran et al. 2020). Therefore, this review chapter aims to focus on the application of plant tissue culture for improvement of the eminent medicinal plant *Centella asiatica* as well as secondary metabolite enhancement through the use of elicitor molecules, transformation and bioreactors.

6.2 Tissue Culture in *Centella asiatica*

Micropropagation is an *in vitro* culture method of quickly growing elite plant species by the use of plant tissue culture technology (Moraes et al. 2021). Its application is popular in forestry, horticulture and agriculture (Sidhu 2010). Tissue culture technology has been applied for the production of small antidote molecules, large antidote molecules and standardized antidote extracts. Micropropagation allows mass selection with bioassay, and it may help in the assessment of elite phenotype lines (Mathe et al. 2015). For medicinal plants, these techniques facilitate hundreds to thousands of plantlets along with enhancement of secondary metabolite production (Yoshimatsu 2008; Oseni et al. 2018). There are various strategies to scale up secondary metabolite production, such as bioreactors, application of elicitor molecules and transformation (Pant 2014).

6.2.1 Source of Explants for Plant Tissue Culture in *Centella*

Materials used for micropropagation are called explants. The result of tissue culture depends on the types, age and position of explants. Plants have unequal totipotency; therefore, in *Centella asiatica*, the most used explants are nodal segment, leaf, stem and petiole. Large explants can be responsible for contamination, while small explants sometimes show less growth (Iliev et al. 2010; Chandana et al. 2018; Gaurav et al. 2018). Nodal and shoot tip explants showed multiple shoot regeneration in Murashige and Skoog (MS) medium fortified with different combinations and compositions of cytokinin and auxin (Jaheduzzaman et al. 2012). In micropropagation of *Centella asiatica*, a very high rate of contamination was reported during inoculation of explants. Leaf and stolon tip explants were used as the surface for sterilization and establishment of a reproducible and feasible protocol for somatic embryogenesis. Field-grown and collected *Centella* explants were found to contain high levels of fungal and bacterial contamination. It was found that dipping in 70% ethyl alcohol for 30 s followed by bleach treatment for 12 min and soaking for 60 min in plant preservative mixture before inoculation of explants for somatic embryogenesis was better for surface sterilization (Joshee et al. 2007).

Krishnan et al. (2018) investigated adventitious shoot culture from shoot explants with maximum shoot proliferation to determine the influence of different manipulated nutrients. They observed the effects of carbon source, potassium source, nitrogen source, phosphorus source, micronutrients and macronutrients. The development of shoots was influenced by relative ratio of nitrogen source contain highest number of shoot (47.16 ± 1.52). MS medium containing 3% sucrose showed optimum shoot multiplication; increasing the concentration of carbon source led to drastically decreased shoot multiplication. MS medium supplemented with 30% potassium also showed the maximum shoot number (31.66 ± 1.52), while 150% phosphorus showed the maximum shoot number (28 ± 1), respectively. The highest number of shoots (31 ± 1) was obtained from macronutrient magnesium concentration 1.5 mM, whereas the micronutrient manganese showed shoot number (43 ± 2) at a concentration of 200 μM, respectively.

Nodal segment and leaf explants were selected for regeneration and callus induction and regeneration in MS medium augmented with different combinations and compositions of plant growth regulators. The authors used response surface methodology to yield healthy regenerated plantlets with better height as compared with conventional method and multiple shoots (Gururajan et al. 2021). Nodal segment was used to optimize surface sterilization. Field-grown *Centella asiatica* was used, which was highly exposed to various bacterial and fungal contaminations. The authors used $HgCl_2$ and plant preservative mixture along with bavistin, cetrimide and trimethoprim followed by the addition of plant preservative mixture in culture medium. The result showed that a combination of (clean culture 90 ± 1.33%) and TS5 (decon + Bavistin 150 mg/l + cetrimide 1% + trimethoprim 50 mg/l + $HgCl_2$ 0.1% + plant preservative mixture 2% in soak and 2 ml/l in medium) was the best method for surface sterilization in *Centella* (Moghaddam et al. 2011).

6.2.2 Plant Tissue Culture Media and Combination of Plant Growth Regulators

In plant tissue culture, a medium is required for *in vitro* growth of explants. A broad range of media are available for *in vitro* culture, but among all the media, MS medium is most widely used. Growth media possess carbon source, vitamins, growth regulators, macro and micro nutrients, and other organic components. Growth hormones modulate the morphological and physiological process of plants and are therefore called plant growth regulators (Gaspar et al. 1996). They are synthesized in specific sites and translocated into different parts of the plant. Many plant species do not require plant growth regulators in the culture medium and grow successfully without the addition of external supplements. Addition of plant growth regulators to culture medium upregulates plant growth as well as enhancing secondary metabolite synthesis (Dias 2019; Martinez et al. 2021).

Shoot tips were used to investigate the effects of growth hormones on clonal propagation. MS medium was fortified with a combination of BA, NAA and Kn for microshoots. A combination of BA 4 mg/l and NAA 0.1 mg/l showed the maximum number of shoots (3.38). Developed microshoots were transferred into rooting medium containing IBA (1–3 mg/l) and NAA (0.5–2 mg/l). Maximum rooting of 46.8 per shoot with root length 19.7 cm was obtained. This micropropagation protocol may be used for establishment of genetically identical germplasm (Nath and Buragohain 2003). Ling et al. (2009) reported the effect of different plant growth regulators on adventitious root induction in leaf and petiole explants. They used indole acetic acid (IAA), IBA and NAA at concentrations of 0, 1, 3, 5 and 7 mg/l for adventitious root induction. Among the three plant growth hormones, IBA showed the best response over NAA and IAA. No adventitious root formation was observed without a plant growth regulator. In leaf explants, the highest number of roots per explant, rooting percentage and root length per explant were obtained at IBA 7 mg/l. Petiole explants showed better rooting efficiency on IBA 5 mg/l. Petiole explants showed a better response than leaf explants. The effect of sucrose on petiole explants was also observed. Without sucrose, no adventitious roots were obtained in MS medium. Sucrose 4% and IBA 5 mg/l showed the maximum number of roots per explant and root length. Micropropagation of *Centella* offers multiplication of elite clones and help in dispersal along with *ex situ* conservation. It was observed that a combination of 6-benzylaminopurine (BAP) (1 mg/l), sucrose (20 g/l) and agar (8 g/l) was best for culture initiation and also auxiliary shoot proliferation. The highest number of roots per shoot was obtained in MS medium, NAA 1 mg/l, charcoal 200 mg/l and sugar 20 g/l with maximum secondary roots (Singh and Bhati 2011).

The effects of BAP, Kn, NAA and IBA on regeneration in *Centella* were also reported by researchers. MS medium fortified with 3% sucrose, 4 mg/l BAP and 0.5 mg/l NAA showed maximum shoot bud initiation (4.68 in number) and response (92.64%). The combination of Kn and NAA obtained maximum response (58.84%) with 2.24 shoot buds. Maximum rooting was observed on MS medium augmented with IBA 0.5 mg/l (4.5/plantlet) (Kumari et al. 2018).

The effect of low concentration of plant growth regulator was also observed in mass multiplication of *Centella asiatica* from axillary meristem explants. MS medium fortified with BAP 0.5 mg/l and NAA 0.1 mg/l showed the best shoot induction (82%). Combinations of BAP 0.5 mg/l and NAA 0.5 mg/l, and BAP 0.5 mg/l and Kn 0.5 mg/l were most effective for initiated meristem culture. Lower concentrations of IBA showed a better response as compared with NAA (Siddiqui et al. 2019).

6.2.3 Callus and Suspension Culture

Callus comprises an undifferentiated and unorganized mass of cells containing three step process induction, differentiation and cell development. Callus has wide applications in industry worldwide (Rahayu et al. 2016; Zaman et al. 2020). *In vitro* culture has emerged as an alternative to produce large amounts of secondary metabolites because wild plants have limitations and are dependent on geographical, seasonal and environmental factors. Another disadvantage of *in vivo* culture is that extraction and purification are very time consuming and result in a low yield of bioactive molecules (Martinez et al. 2021). Callus and suspension culture are alternative methods to overcome these obstacles and efficiently produce bioactive molecules (Bouhouche et al. 1998; Singh et al. 2011). The *in vitro* culture offers a continuous supply of secondary metabolites with uniform quality as well as high yield. Through this method, any plants, either endangered or threatened, can be easily multiplied and maintained to produce molecules of interest without dependence on external factors.

Bouhouche et al. (1998) reported the glycosylation of 3-demethylthiocolchicine into 3-O-glucosylthiocolchicine through suspension culture. Approximately 30% 3-demethylthiocolchicine was converted into 3-O-glucosylthiocolchicine after 11 days of incubation. Other phenolic compounds were also assayed to investigate their effects on the glucosyltransferase reaction.

In vitro grown petiole explants were used to assess cell growth and asiaticoside concentration by inoculation in suspension culture. It was observed that the cell growth and asiaticoside concentration were maximal at 24 d of culture with agitation speed 150 r/m and aeration 2.5 l/min. The highest cell growth obtained was 302.45 g (dry weight 31.45 g) with growth index 3.03 (Loc and Nhat 2013).

For suspension culture, *in vitro* grown leaf explants were excised in two or three equal sections and inoculated in MS medium fortified with different concentrations and combinations of BAP, NAA and 2,4-D in liquid medium. It was found that the cells in suspension culture grew higher as compared with callus culture (Nath and Buragohain 2005).

Loc and An (2010) reported callus induction from petiole explants to determine asiaticoside concentration through suspension culture. MS media augmented with sucrose 20 g/l, BAP 1 mg/l and NAA 1 mg/l were used for callus induction. The maximum concentration of asiaticoside was found in callus inoculated in suspension culture as compared with regenerated leaf. Nodal segment, leaf and petiole explants were used for callus induction in MS fortified with NAA (0.5–5 mg/l) alone or combined with 2,4-D. This induced callus was used for proliferation into shoots (Naidu et al. 2010). The effect of auxin on callus induction was also determined. Leaves and petiole explants were treated with three kinds of auxin, 2,4-D, picloram and dicamba, at concentrations of 2, 4 and 6 mg/l. The best callus induction was observed in 2,4-D or dicamba at 4 mg/l with maximum percentage, friable texture and highest fresh weight (Rahayu et al. 2016).

A callus induction protocol was optimized by using field-grown nodal and leaf explants for the conservation of *Centella asiatica* through micropropagation. For callus induction, full-strength MS medium supplemented with 2 mg/l IAA alone or 0.5 mg/l Kn and 1 mg/l IAA was used. A combination of 0.5 mg/l Kn and 1.5 mg/l IAA was applied for callus proliferation. This callus was further inoculated for regeneration to conserve this valuable medicinal plant (Gururajan et al. 2021).

6.3 Approaches for Scaling Up Secondary Metabolite Production through Application of Plant Tissue Culture

Advances in plant tissue culture technology along with genetic engineering, especially transformation, have unlocked new methods for production of nutraceuticals, pharmaceuticals and bioactive molecules in large amounts. For large-scale production of secondary metabolites from plant cells, specific bioreactors have been used. Different type of bioreactors containing specific impellers have been used in industry for commercial production of valuable secondary metabolites. Synthetic seed technology can be applied as a promising proposal that can be applied for exchange of plant matter between public and private *in vitro* culture laboratories. This technology is used for germplasm conservation as well as propagules that are isolated from *in vitro* culture and applied in the field. Elicitors are compounds involved in the synthesis of phytoalexins and other bioactive molecules in plants. There are two kind of elicitor molecules: biotic, such as *Aspergillus niger*, and abiotic, such as mannitol and methyl jasmonate. These are used for secondary metabolite enhancement (Pant 2014; Chandana et al. 2018; Chandran et al. 2020).

6.3.1 Induction of Elicitor Molecules in *Centella*

Elicitors are molecules that stimulate secondary metabolism production. These molecules may be biotic, abiotic or physical factors that trigger a reaction in a living organism and lead to accumulation of secondary metabolite production. Elicitation can be applied to enhance secondary metabolites through *de novo* synthesis in plant tissue culture techniques. These molecules reduce production cost by increasing synthesis of bioactive molecules (Lambert et al. 2011; Siddiqui et al. 2013; Singh and Dwivedi 2018). Several research studies have been reported regarding elicitor molecules.

In *Centella asiatica*, 2,3-oxidosqualene, which is a precursor molecule of triterpene and sterol resulted in a higher content of triterpene and sterol (up to 152 times) after culturing with 100 µM methyl jasmonate. Triterpenes are directly synthesized from 2, 3-oxidosqualene (Mangas et al. 2006). An abiotic elicitor molecule (2-hydroxybenzoic acid) was more efficient as compared with a biotic elicitor (yeast extract).

The concentrations of 50–200 µM benzoic acid and yeast extract 2–5 g/l had different eliciting effects. The addition of yeast extract and 2-hydroxybenzoic acid showed powerful enhancement of asiaticoside production in *Centella asiatica*; 100 µM of 2-hydroxybenzoic acid and 4 g/l of yeast extract at day 10 of inoculation enhanced asiaticoside production 5- and 3.5-fold as compared with reference cells (Loc and Giang 2012). Hidalgo et al. (2016) reported that the plant tissue culture of *Centella* produces high amounts of centellosides. This study revealed that the combination of elicitor molecules with natural sources of amryins (centelloside precursors), copal and Manila elemi resins, increased centelloside production in *Centella* cell suspension culture.

The effect of methyl jasmonate, an elicitor molecule, was investigated for enhancing the rate of biosynthesis of asiaticoside and asiatic acid. A 100 μM concentration of methyl jasmonate was used for callus, shoot and suspension culture of *Centella asiatica*. It was observed that the concentration of asiaticoside was enhanced 69-fold in callus culture and 39-fold in shoot culture, and the concentration of asiatic acid was enhanced 1.9 -fold in cell suspension culture (Krishnan et al. 2019).

A study was conducted on the effect of methyl jasmonate on triterpenoid production in diploid and tetraploid *Centella asiatica* hairy roots. The hairy root was developed by infection with *Agrobacterium rhizogenes* strain ATCC 43057. Methyl jasmonate triggered triterpenoid production in both diploid and tetraploid hairy roots of *Centella*, while untreated roots were unable to produced triterpenoids. It was observed that treatment with 50 μM of methyl jasmonate increased the highest triterpenoid production up to 27.25 ± 0.27 μg/mg dry weight at 21 days in diploid hairy root culture. In tetraploid hairy root culture, the maximum amount of triterpenoids was 16.29 ± 6.32 μg/mg at 28 days of culture with 50 μM methyl jasmonate. At 14 day of culture, treatment with 100 μM methyl jasmonate produced the same amount of triterpenoids in both the hairy root cultures (16.31 ± 9.24 μg/mg DW) (Nguyen et al. 2019).

6.3.2 Transformation through Tissue Culture in *Centella* Species

The application of recombinant DNA technology has led to unexpectedly great improvements in human health. For secondary metabolite production, *Agrobacterium rhizogenes*-mediated transformation has been used in hairy root culture (Ramesh et al. 2011). The importance of plant tissue culture has increased in transgenic hairy root cultures for enhancement of secondary metabolites. This technique facilitates genetic stability and a rapid growth rate without the application of hormones. Infections of wounded plants result in hairy root disease, characterized by high growth and maximum branching of roots in the host plant's wounded sites (Chandra and Chandra 2011). Particles between 1 and 100 nm in size, called nanoparticles, have been used to increase germination efficiency, improve secondary metabolites and amplify plant growth (Zhao et al. 2018).

For transient analysis and genetic transformation in *Centella asiatica*, the particle bombardment transformation method was used in callus culture. Synthetic green fluorescent protein bound to CaMV 35S as a reporter was used to optimize eight parameters that affect the transformation system. It was found that the DNA delivery conditions of 9 cm target distance, 1,100 psi helium pressure, 27 mm Hg chamber vacuum pressure, 1 μm gold particle size, spermidine as a precipitation agent, two bombardments, 60 h post-bombardment incubation time and 2 μg plasmid DNA were optimal for callus transformation in *Centella*. Expression of synthetic green florescent protein was investigated by fluorescent microscope and confirmed using real-time polymerase chain reaction (RT-PCR) (Lai et al. 2011). Hairy root transformation was also conducted for enhancement of secondary metabolites. Leaf explants were inoculated into MS medium augmented with 1 mg/l IAA and 1 mg/l BAP containing 4% sucrose and 0.9% agar. Leaf-derived callus was cultured with *Agrobacterium rhizogenes* ATCC 15834 strain for 40 min. *A. rhizogenes* induced hairy root in developed callus (Ruslan et al. 2011). Muhsinin et al. (2021) reported enhanced metabolism of the secondary metabolite asiaticoside by using a semi-quantitative PCR method in *Centella asiatica*. This method was applied to the gene responsible for increasing asiaticoside formation. The gene RNA UGT73AH1 influences asiaticoside synthesis (UDP-glycosyltransferase) in *Centella* species. This study focused on expression of the gene UGT73AH 1 in treated and untreated plants. The analysis of the gene was performed on the thickness of the band on the electropherogram, which was used to produce area under the curve (AUC) values. The results were analyzed using a semi-quantitative PCR method, and it was observed that the callus of *Centella* had the highest average AUC (24,879.59), followed by plantlets (23,780.04), while the value for plants was 7802.26. This was significant at $p < 0.05$ on analysis of variance (ANOVA). This value showed significant average differences among the groups of plantlets, callus and wild *Centella asiatica* plant species.

6.3.3 Bioreactor and Synthetic Seed Technology

Bioreactors are an important biotechnological engineering system that is capable of scaling up the manufacture of valuable secondary metabolites in plants through plant tissue culture (Omar et al. 2006). The selection of the bioreactors totally depends on the type of *in vitro* culture that has to be cultured (Allan et al. 2019). When scaling up suspension culture to the pilot bioreactor level for commercial production, the parameters to be chosen are production behavior, growth, morphology and shear tolerance for better plant cell culture in the bioreactor. The bioreactor should have proper oxygen transfer, a good mixing system and low shear tension. The maximum cell growth and production of secondary metabolites requires a constant supply of appropriate nutrients, which manage the mixing and shear sensitivity of plant cells to decrease cell damage. This parameter depends on the cell line. Other parameters such as pH, oxygen concentration, temperature and substrate concentration should be optimized and regularly monitored (Gallego et al. 2014). Synthetic seed technology is mostly used for encapsulated somatic embryos or other vegetative organs (cell aggregates, auxillary buds, shoot buds or other micropropagules), which may be applied as seed that further develops into the whole plant in both *in vivo* and *in vitro* conditions (Rihan et al. 2017). Biotechnological research that focuses on plant cell tissue culture has developed a new outlook in the case of germplasm storage and conservation, mass propagation, secondary metabolite production and genetic transformation (Rihan et al. 2017; Jang et al. 2020). In this technology, alginate encapsulation of both *in vitro* and *in vivo* grown explants facilitates a competent system that is involved in the repository and interchange of seedless medicinal plants as well as their multiplication. These plants contain desirable characteristics that are complicated to propagate by conventional methods. However, the optimization of production, repository and interchange of synthetic seed is impacted by several other factors. The success of synthetic seed technology in medicinal plants is determined by the selection of explants, encapsulating agent and matrix (Gantait et al. 2015).

Prasad et al. (2014) reported synthetic seed germination from axillary buds/nodal segments derived from multiple shoot culture. Excised segments were encapsulated in 4% (w/v) sodium alginate beads followed by complexation in 75 mM calcium chloride solution. The seed were kept on moist filter paper in sealed petri plates and stored at 25 ± 3 °C for 200 days. The germination rate was up to 85%, and plantlets were transferred on to hormone-free MS medium.

6.4 Conclusion

Medicinal plants have been used for curing diseases since ancient times. Plant tissue culture is an approach for mass propagation along with production and enhancement of secondary metabolites. Through this technique, secondary metabolites can be produced in a short time as compared with conventional methods. By the application of plant tissue culture, the biosynthetic activity of cultured cells or tissues may be increased by managing environmental factors and artificial synthetic seed technology as well as transformation. Because of the many applications of plant tissue culture technology in medicinal plants, it has been adopted by the phytopharmaceutical industry. With time, more advances have been made in the technology to yield high amounts of secondary metabolites. Large-scale bioreactors and commercial production of secondary metabolites derived from plant cells and tissues are expected to increase in the forthcoming future. Thus, plant tissue culture facilitates improved phytopharmaceutical production along with more exploration of medicinal plant physiology.

REFERENCES

Abdulla, M. A; ALbayaty, F. H; Younis, L. T; Abu, H. M. I; 2010. Antiulcer activity of *Centella asiatica* leaf extract against ethanol-induced gastric mucosal injury in rats. *Journal of Medicinal Plants Research*. 4, 1253–1259.

Allan, S. J; De Bank, P. A; Ellis, M. J; 2019. Bioprocess design considerations for cultured meat production with a focus on the expansion bioreactor. *Frontiers in Sustainable Food Systems*. DOI 10.3389/fsufs.2019.00044.

Aziz, M. A; Adnan, M; Khan, A. H; Shahat, A. A; AL-Said, M. S; Ullah, R; 2018. Traditional uses of medicinal plants practiced by the indigenous communities at Mohmand Agency, FATA, Pakistan. *Journal of Ethnobiology and Ethnomedicine*. 14,2. DOI 10.1186/s13002-017-0204-5.

Aziz, Z. A; Davey, M. R; Power, J. B; Anthony, P; Smith, R. M; Lowe, K. C; 2007. Production of asiaticoside and madecassoside in *Centella asiatica in vitro* and *in vivo*. *Biologia Plantarum*. 51, 34–42.

Bajaj, Y. P. S; Furmanowa, M; Olszowska, O; 1988. Biotechnology of the micropropagation of medicinal and aromatic plants. In: Y. P. S Bajaj (ed.). *Biotechnology in agriculture and forestry, medicinal and aromatic plants*. Springer Verlag, Berlin Heidelberg.

Bhavana, D; Jyoti, K; 2011. *Centella asiatica*: The elixir of life. *International Journal of Research in Aurveda and Pharmacy*. 2, 431–438.

Bouhouche, N; Solet, J. M; Ramisa, A. S; Bonaly, J; Cosson, L; 1998. Conversion of 3-demethylthiocolchicine into thiocolchicoside by *Centella asiatica* suspension cultures. *Phytochemistry*. 47 (5), 743–747.

Bunpo, P; Kataoka, K; Arimochi, H; Nakayama, H; Kuahara, T; Bando, Y; Izumi, K; Viniketkumneun, U; Ohnishi, Y; 1997. Inhibitory effects of *Centella asiatica* on azoxymethane induced aberrant crypt focus formation and carcinogenesis in the intestines of F344 rats. *Food Chem. Toxicol*. 42, 1987–1997.

Chandana, B. C; Nagaveni, H. C; Kumari; Lakshmana, D; Kolakar, S. S; Heena, M. S; 2018. Role of plant tissue culture in micropropagation, secondary metabolites production and conservation of some endangered medicinal crops. *Journal of Pharmacognosy and Phytochemistry*. SP3: 246–251.

Chandra, S; Chandra, R; 2011. Engineering secondary metabolite production in hairy roots. *Phytochem Rev*. 10, 371–395.

Chandran, H; Meena, M; Barupal, T; Sharma, K; 2020. Plant tissue culture as a perpetual source for production of industrially important bioactive compounds. *Biotechnological Reports*. 26, DOI 10.1016/j.btre.2020.e00450.

Chong, N. J; Aziz, Z; 2011. A systematic review on chemical constituents of *Centella asiatica*. *Research Journal of Pharmaceutical Biological and Chemical Sciences*. 2, 445–459.

Dias, J. P. T; 2019. Plant growth regulators in horticulture practices and perspectives. *Biotechnologia Vegetal*. 19, 3–14.

Gallego, A; Estrada, K. R; Limon-Vidal, H. R; Hidalgo, D; Lalaleo, L; Kayani, W. K; Cusido, R. M; Palazon, J; 2014. Biotechnological production of centellosides in cell cultures of *Centella asiatica* (L) urban. *Eng. Life Sci*. 00, 1–10.

Gantait, S; Kundu, S; Ali, N; Sahu, N. C; 2015. Synthetic seed production of medicinal plants: A review on influence of explants, encapsulation agent and matrix. *Acta. Physiol Plant*. 37, 98.

Gaspar, T; Kevers, C; Penel, C; Greppin, H; Reid, D. M; Thorpe, T. A; 1996. Plant hormones and plant growth regulators in plant tissue culture. *In Vitro Cell. Dev. Biol-Plant*. 32, 272–289.

Gaurav, N; Komal, Juyal, P; Tyagi, M; Chauhan, N; Kumar, A; 2018. A review on in vitro propagation of medicinal plants. *Journal of Pharmacognosy and Phytochemistry*. 7 (6), 2228–2231.

Ghareeb, Z. F; Taha, L. S; 2018. Micropropagation protocol for *Antigonon leptopus*, an important ornamental and medicinal plant. *Journal of Genetic Engineering and Biotechnology*. 16, 669–675.

Gnanapragasam, A; Ebenezar, K. K; Sathish, V; Govindaraju, P; Devaki, T; 2004. Protective effect of Centella asiatica on antioxidant tissue defence system against adriamycin induced cardiomyopathy in rats. *Life Science*. 76, 585–597.

Gray, N. E; Magana, A. A; Lak, P; Wright, K. M; Joseph, Q; Stevens, J. F; Maier, C. S; Soumyanath, A; 2018. *Centella asiatica*: Phytochemistry and mechanisms of neuroprotection and cognitive enhancement. *Phytochem Rev*. 17, 161–194.

Gururajan, K; Hedge, D; Udupa, S; Shetty, S; Ujwal, P; 2021. Optimization of shoot initiation using response surface methodology and callus media and standardization for micropropagation of *Centella asiatica*. *J Microbiol Biotech Food Sci*. 10 (4), 685–690.

Hidalgo, D; Steinmetz, V; Brossat, M; Couturier, L. T; Cusido, R. M; Corchete, P; Palazon, J; 2016. An optimized biotechnological system for the production of centellosides based on elicitation and bioconversion of *Centella asiatica* cell cultures. *Eng Life Sci*. 17 (4), 413–419. DOI 10.1002/elsc.201600167.

Hosseinzadeh, S; Jafarikukhdan, A; Hosseini, A; Armand, R; 2015. The application of medicinal plants in traditional and modern medicine: A review of *Thymus vulgaris*. *International Journal of Clinical Medicine*. 6, 632–645.

Jagtap, N. S; Khadabadi, S. S; Ghorpade, D. S; Banarase, N. B; Naphade, S. S; 2009. Antimicrobial and antifungal activity of *Centella asiatica* (L.) Urban Umbeliferae. *Research J. Pharm. and Tech*. 2, 328–330.

Jaheduzzaman; Habib, M. A; Akter, S; Banu, N. A; Rahman, R. B; Banu, T. A; Khan, S; Islam, S; 2012. In vitro regeneration of an important medicinal plant *Centella asiatica* L. Urban. *Bangladesh J. Sci. Ind. Res*. 47(3), 269–272.

Jamil, S. S; Nizami, Q; Salam, M; 2007. *Centella asiatica* (L) Urban: A review. *Natural Products Radiance*. 6, 158–170.

Jang, B. K; Cho, J. S; Lee, C. H; 2020. Synthetic seed technology development and production studies for storage, transport and industrialization of bracken spores. *Plants*. 9. DOI 10.3390/plants9091079.

Joshee, N; Biswas, B. K; Yadav, A. K; 2007. Somatic embryogenesis and plant development in *Centella asiatica* L. a highly prized medicinal plant of the tropics. *Hortscience.* 42 (3), 663–637.

Khan, H; 2014. Medicinal plants in light of history recognized therapeutic modality. *Journal of Evidence Based Complementary & Alternative Medicine.* 19, 216–219.

Kia, F. J; Lorigooini, Z; Khoei, H. A; 2018. Medicinal plants past history and future perspective. *J. Herbmed Pharmacol.* 7, 1–7.

Krishnan, M. L; Roy, A; Bharadvaja, N; 2019. Elicitation effect on the production of asiaticoside and asiatic acid in shoot, callus and cell suspension culture of *Centella asiatica. Journal of Applied Pharmaceutical Science.* 9 (06), 067–074.

Krishnan, M. L; Roy, A; Bharadvaja, N; 2018. Influence of different macro and micro nutrients on the shoot multiplication of *Centella asiatica. J. Plant Biochem Physiol.* 6, 2.

Kumari, R; Abbas, S. A; Singh, A; 2018. Micropropagation of *Centella asiatica* through tissue culture: An attempt to develop an efficient protocol for large scale production. *Indian J. Sci. Res.* 08 (2), 41–45.

Lai, K. S; Abdullah, P; Yusoff, K; Mahmood, M; 2011. An efficient protocol for particle bombardment mediated transformation of *Centella asiatica* callus. *Acta Physiol Plant.* 33, 2547–2552.

Lambert, E; Faizal, A; Geelen, D; 2011. Modulation of triterpene saponin production in vitro cultures, elicitation and metabolic engineering. *Appl Biochem Biotechnol.* 164, 220–237.

Lee, J; Jung, E; Kim, Y; Park, J; Park, J; Hong, S; Kim, J; Hyun, C; Kim, Y. S; Park, D; 2004. Asiaticoside induces human collagen I synthesis through TGF beta receptor I kinase (TbetaRI kinase) independent smad signaling. *Planta Med.* 72, 324–328.

Lemma, D. T; Banjaw, D. T; Megersa, H. G; 2020. Micropropagation of medicinal plants: Review. *International Journal of Plant Breeding and Crop Science.* 7 (2), 796–802.

Ling, A. P. K; Chin, M. F; Hussein, S; 2009. Adventitious root production of *Centella asiatica* in response to plant growth regulators and sucrose concentration. *Medicinal and Aromatic Plant Sciences and Biotechnology.* 3, 36–41.

Loc, N. H; Nhat, N. T. D; 2013. Production of asiaticoside from *Centella asiatica* L. urban cells in bioreactor. *Asian Pac J Trop Biomed.* 3 (10), 806–810.

Loc, N. H; Giang, N. T; 2012. Effects of elicitors on the enhancement of asiaticoside biosynthesis in cell cultures of *Centella* (L. urban). *Chemical Papers.* 66 (7), 642–648.

Loc, N. H; An, N. T. T; 2010. Asiaticoside production from *Centella* (*Centella asiatica* L. Urban) cell culture. *Biotechnology and Bioprocess Engineering.* 15, 1065–1070.

Mangas, S; Bonfill, M; Osuna, L; Moyano, E; Tortoriello, J; Cusido, R. M; Pinol, M. T; Palazon, J; 2006. The effect of methyl jasmonate on triterpene and sterol metabolisms of *Centella asiatica, Ruscus aculeatus and Galphimia glauca* cultured plants. *Phytochemistry.* 67, 2041–2049.

Martinez, M. E; Jorquera, L; Poirrier, P; Diaz, K; Chamy, R; 2021. Effect of the carbon source and plant growth regulators in the induction and maintenance of an in vitro callus culture of *Taraxacum officinale* (L) weber ex F. H. wig. *Agronomy.* 11, 1181. DOI 10.3390/agronomy11061181.

Mathe, A; Hassan, F; Kader, A. A; 2015. In vitro micropropagation of medicinal and aromatic plants. In: A. Mathe (ed.) *Medicinal and aromatic plants of the world.* DOI 10. 1007/ 978-94-017-9810-5-15.

Moghaddam, S. S; Jaafar, H. W; Aziz, M. A; Rusli, I; Rahmat, A. Bt; Philip, E; 2011. Optimization of an efficient semi-solid culture protocol for sterilization and plant regeneration of *Centella asiatica* (L.) as a medicinal herb. *Molecules*. 16, 8981–8991.

Moraes, R. M; Cerdeira, A. L; Lourenco, M. V; 2021. *Using micropropagation to develop* medicinal plants into crops. *Molecules*. 26 (6), 1752.

Muhsinin, S; Ahmad, F. S. T; Marliani, L; 2021. Study of UGT73AH1 gene expression in *Centella asiatica* L. from tissue culture using semiquantitative PCR method. *International Journal of Pharmaceutical and Phytopharmacological Research*. 11 (1), 167–171.

Naidu, T, B; Rao, S. N; Mani, N. S; Mohan, Y. S. Y. V. J; Pola, S; 2010. Conservation of an endangered medicinal plant *Centella asiatica* through plant tissue culture. *Drug Invention Today*. 2, 17–21.

Nath, S; Buragohain, A. K; 2005. Establishment of callus and cell suspension cultures of *Centella asiatica*. *Biologia Plantarum*. 49 (3), 411–413.

Nath, S; Buragohain, A. K; 2003. In vitro method for propagation of *Centella asiatica* (L) urban by shoot tip culture. *Journal of Plant Biochemistry and Biotechnology*. 12, 2. DOI 10.1007/BF03263181.

Nguyen, K. V; Pongkitwitoon, B; Pathomwichaiwat, T; Viboonjun, U; Prathanturarug, S; 2019. Effects of methyl jasmonate on the growth and triterpenoid production of diploid and tetraploid *Centella asiatica* (L.) Urb. Hairy root cultures. 9, 18666. DOI 10.1038/541598-019-54460-z.

Omar, R; Abdullah, M. A; Hasan, M. A; Rosfarizan, M; Marziah, M; 2006. Kinetics and modelling of cell growth and substrate uptake in *Centella asiatica* cell culture. *Biotechnology and Bioprocess Engineering*. 11, 223–229.

Oseni, O. M; Pande, V; Nailwal, T. K; 2018. A review on plant tissue culture, a technique for propagation and conservation of endangered plant species. *Int. J. Curr. Microbiol. App. Sci*. 7 (7), 3778–3786.

Pant, B; 2014. Application of plant cell and tissue culture for the production of phytochemicals in medicinal plants. In: R. Adhikari and S. Thapa (eds.) *Infectious diseases and nanomedicine II, advances in experimental medicine and biology Vol. 808*, Springer India. DOI 10.1007/978-81322-1774-9-3.

Prasad, A; Singh, M; Yadav, N. P; Mathur, A. K; Mathur, A; 2014. Molecular chemical and biological stability of plants derived from artificial seeds of *Centella asiatica* (L.) Urban, an industrially important medicinal herb. *Industrial Crops and Products*. 60, 205–211.

Rahayu, S; Roostika, I; Bermawie, N; 2016. The effect of types and concentrations of auxins on callus induction of *Centella asiatica*. *Nusantara Bioscience*. 8 (2), 283–287.

Ramesh, M; Karthikeyan, A; Vijayakumar, K; Largia, M. J. V; Pandian, S. K; 2011. *Agrobacterium* mediated transformation of pharmaceutically important Indian medicinal herb *Bacopa monnieri* (L.). *Journal of Medicinal Plants Research*. 5 (11), 2316–2321.

Rao, S. B; Chetana, M; Uma, D. P; 2005. *Centella asiatica* treatment during postnatal period enhances learning and memory in mice. *Physiol. Behav*. 86, 449–457.

Rihan, H. Z; Kareem, F; El-Mahrouk, M. E; Fuller, M. P; 2017. Artificial seeds (principle, aspects and applications). *Agronomy*. 7, 71. DOI 10.3390/agronomy7040071.

Roy, A; Bharadvaja, N; 2017. *Centella asiatica*: A pharmaceutically important medicinal plant. *Current Trends in Biomedical Engineering and Biosciences*. 5 (3). DOI 10.19080/CTBEB.2017.05.555661.

Ruksiriwanich, W; Khantham, C; Sringarm, K; Sommano, S; Jantrawut, P; 2020. Depigmented Centella asiatica extraction by pretreated with supercritical carbon dioxide fluid for wound healing application. Processes. 8, 277. DOI 10.3390/pr8030277.

Ruslan, K; Selfitri, A. D; Bulan, S. A; Rukayadi, Y; Elfahmi; 2011. Effect of *Agrobacterium rhizogenes* and elicitation on the asiaticoside production in cell cultures of *Centella asiatica*. *Pharmacognosy Magazine*. 8 (30), 111–115.

Seevaratnam, V; Banumathi, P; Premalatha, M. R; Sundaram, S. P; Arumugam, T; 2012. Functional properties of *Centella asiatica* (L.): A Review. *International Journal of Pharmacy and Pharmaceutical Sciences*. 4, 8–14.

Siddiqui, S; Thomas, T; Khan, S; 2019. Standardization of an efficient protocol for in vitro micropropagation of *Centella asiatica* L., an important medicinal plant. *Pharmaceutical and Biosciences Journal*. 7 (4), 22–66.

Siddiqui, Z. H; Mujib, A; Mahmooduzzafar; Aslam, J; Hakeem, K. R; 2013. In vitro production of secondary metabolites using elicitor in Catharanthus roseus: A case study. In: K. R. Hakeem et al. (eds.). *Crop Improvement*. Springer. DOI 10.1007/978-1-7028-1_14.

Sidhu, Y; 2010. In vitro micropropagation of medicinal plants by tissue culture. *The Plymouth Student Scientist*. 4 (1), 432–449.

Singh, A; Dwivedi, P; 2018. Methyl jasmonate and salicylic acid as potent elicitors for secondary metabolite production in medicinal plants: A review. *Journal of Pharmacognosy and Phytochemistry*. 7 (1), 750–757.

Singh, D; Bhati, S; 2011. Study of new protocol for in vitro micropropagation of *Centella asiatica*, an important medicinal plant. *Pharm Biotech & Microbiol*. 3, 1–5.

Singh, N; Meena, M. K; Patni, V; 2011. Effect of plant growth regulators, explants type and efficient plantlet regeneration protocol through callus induction in *Naringi crenulata* (Roxb.) Nicolson and its biochemical investigation. *African Journal of Biotechnology*. 10 (77), 17769–17777.

Sofowora, A; Ogunbodede, E; Onayade, A; 2013. The role and place of medicinal plants in the strategies for disease prevention. *Afr J Tradit Complement Altern Med*. 10 (5), 210–229.

Yoshimatsu, K; 2008. Tissue culture of medicinal plants: Micropropagation, transformation and production of useful secondary metabolites. In: Atta-ur-Rahman (ed.) *Studies in natural products chemistry*. Elsevier B. V.

Zaidan, M. R. S; Noor, R. A; Badrul, A. R; Adlin, A; Norazah, A; Zakiah, I; 2005. *In vitro* screening of five local medicinal plants for antibacterial activity using diffusion method. *Trop. Biomed*. 22, 165–170.

Zainol, N. A; Voo, S. C; Sarmidi, M. R; Aziz, R. A; 2008. Profiling of *Centella asiatica* (L.) urban extract. *The Malaysian Journal of Analytical Sciences*. 12, 322–327.

Zaman, M. A. K; Azzeme, A. M; Ramle, L. K; Normanshah, N; Ramli, S. N; Shaharuddin, N. A; Ahmad, S; Abdullah, S. N. A; 2020. Induction, multiplication and evaluation of antioxidant activity of *Polyalthia bullata* callus, a woody medicinal plant. *Plants*. 9, 1772. DOI 10.3390/plants9121772.

Zhao, Y; Zhang, Y; Su, P; Yang, J; Huang, L; Gao, W; 2018. Genetic transformation system for woody plant *Tripterygium wilfordii* and its application to product natural celastrol. *Frontiers in Plant Science*. 8, 2221. DOI 10.3389/fpls.2017.02221.

7

Improvement of Seed Protein Quality in Some Important Food Crops Using Genetic Engineering Approaches

Jitendra Kumar Sharma
Maharshi Dayanand University
Rohtak
Haryana, India

Anita Rani Santal
Maharshi Dayanand University
Rohtak
Haryana, India

Nater Pal Singh
Maharshi Dayanand University
Rohtak
Haryana, India

CONTENTS

7.1 Introduction

Protein is one of the seven major nutrients: carbohydrates, fats, fiber, minerals, protein, vitamins, and water. Proteins are a fundamental component in the nutrition of organisms, and because of their nutritious and health values, they are crucial for human and animal rations. Studies show that human protein intake should account for 10–30% of total daily calorie intake, or 0.8 g of protein per kilogram of body weight (de Carvalho et al., 2020; Wolfe et al., 2017). Plants like legumes, grains, and nuts, and animal items like meat, egg, and milk, provide dietary proteins. Indeed, the source of these dietary proteins greatly impacts their health and nutritional value (Bernstein et al., 2012; Ohanenye et al., 2020). However, humans and livestock consume the

DOI: 10.1201/9781003239932-7

majority of their dietary protein from plants. This is because plant proteins are significantly less expensive to produce than meat. However, due to a lack of certain essential amino acids, most plant proteins are nutritionally imbalanced. Therefore, plant proteins are usually partially complete when used in the diet, and this is because only some essential amino acids are present in a particular plant.

Some plants, such as cereals, have low lysine and tryptophan content, while legumes are low in methionine and cysteine (Leinonen et al., 2019). Essential amino acids such as methionine (Met), lysine (Lys), and tryptophan (Trp) cannot be synthesized by humans or animals (Trp). As a result, these amino acids must be obtained through food. In addition, several essential amino acids are inadequate or entirely absent in human food and animal feed crops. Soybeans, for example, are low in Met, while maize is low in Lys and Trp (Le et al., 2016). Grains and grain legumes are important sources of protein for both humans and animals. In many parts of the world, grain legumes such as soybean and soy products, beans, chickpeas, groundnuts, lentils, and peas are staple foods. The global value of legume crops is estimated to be around $200 billion per year. Many of the world's poorest countries get around 10–20% of their total dietary energy from beans (Akibode and Maredia, 2012). Cereals also provide 68% of the world's nutritional calories. However, certain important amino acids are in short supply in legumes and cereals. Pulse storage proteins, for example, are high in Lys but low in sulfur-containing amino acids, primarily Met.

On the other hand, cereal crops are nearly devoid of Lys and Trp (Apostolatos, 1984; Galili et al., 2005; Wenefrida et al., 2009). Thus, Met is considered the first limiting amino acid in legumes. Because of the low Met concentration, even soybean protein, regarded as the greatest plant protein, is not a complete protein (Hanafy et al., 2013; Pfarr et al., 2018). Stable foods, such as legumes, grains, and nuts, have much lower Lys, Trp, and Met levels than animal-derived proteins, according to a report by Le et al. (2016). Largely, proteins contain less methionine, but genetic modification of gene encoding proteins can increase the methionine content in proteins. For example, the transformation of tobacco with a chimeric gene encoding the methionine-rich Brazil nut protein increases the methionine content by 30% in tobacco seeds and potato tubers (Altenbach et al., 1989; Tu et al., 1998).

The nature of the biosynthetic pathway and the distance between sources and sink organs affect the abundance of amino acids. In plants, asparagine and glutamine amino acids in developing seeds are transported from leaf tissues, where they are synthesized, and in developing seeds, their conversion into lysine occurs (Le et al., 2016). Our bodies require food for energy, and crops are an essential source of that. The demand for food crops is on the rise today. When it comes to maintaining the balance of food, we need to protect crops from harmful pathogens such as ectoparasites, bacteria, viruses, etc. and improve their nutritional value.

Traditional plant breeding programs rely on the phenotype-based selection of breeding progenies, which is a labor-intensive and time-consuming process; also, the long generation time of many crop plants limited their outcome (Yu and Tian, 2018). Moreover, crop breeding to improve nutritional quality in the context of essential amino acids is not satisfactory. Biotechnological approaches may solve these problems to enhance cereals and pulses with essential amino acids. For many years, various transgenic strategies have been performed to change the amino acid composition of plant proteins, especially with essential amino acids. Various strategies have been developed and tested for improving the nutritional value of plants over the last decade.

The transgenic approach includes synthetic proteins, protein sequence modification, heterologous or homologous protein overexpression, and metabolic engineering of the free essential amino acid pool and protein sink (Sun and Liu, 2004).

This chapter describes the progress of and potential approaches to genetic engineering for the improvement of seed protein quality. Dietary protein consumed by humans and animals is usually obtained from plant sources, which are comparatively less expensive to produce than meat. So, improving the quality of plant protein will help us meet our future food needs. In this area of interest, research and development are making promising progress toward this goal.

7.2 Seed Protein Improvement in Cereals

Cereal crops are central food crops worldwide for food products; they can be consumed in various ways, such as whole grains of rice, barley, maize, oats, millet, and sorghum, or as flour of wheat, maize, and rye, or as flakes of oats, barley, and maize (Guerrieri and Cavaletto, 2018). Protein quality and quantity have an influence on seed quality. Seed protein contents are variable, as different varieties show a difference in protein content. Also, environmental and genetic factors play a significant role in the protein content of seeds. The variation in protein content of seeds occurs after every harvest. In contrast, it is important to obtain seeds and flour with uniform properties to produce end-use products such as baked foods and many others. This problem can be solved by better understanding the physicochemical properties of seed components such as lipids, proteins, polysaccharides, and fibers and their interactions with minor components like enzymes and micronutrients.

The world's population mostly relies on cereals for energy, and cereals are mostly grown for conventional uses such as making bread, breakfast cereals, etc. Among cereal crops, maize, rice, and wheat are the most popular. However, other minor cereals are also grown (Reeves et al., 2015). An organism requires essential amino acids for its growth and development, and tryptophan is an essential amino acid and has an important role in its growth and development. However, most cereal crops have a low content of Trp. In the biosynthetic pathway of Trp, anthranilate synthase catalyzes the formation of anthranilate from chorismate (Tozawa et al., 2001). Rice is a major staple food and consumed by more than 40% of the world's population. Among transgenic cereal crops, a large number of transgenic rice varieties have been developed with enhanced efficiencies (Sindhu et al., 1997). Rice grains contain 6.6–8.4% of protein but lack the amino acids lysine and threonine. When the protein content of rice glutelin protein Gt1 and pea legumin protein LegA is compared, pea legumin has a higher content of lysine residue. Sindhu et al. (1997) developed transgenic rice (*Oryza sativa* L.) cultivar Nipponbare, which contains pea legumin gene LegA, and reported improved nutritional value of rice grains. In higher plants, two key enzymes, aspartate kinase (AK) and dihydrodipicolinate synthase (DHDPS), catalyze the rate-limiting steps in lysine biosynthesis, and both are highly sensitive to feedback inhibition by lysine (Yang et al., 2021) (Figure 7.1).

Kafirins, seed storage proteins of sorghum, have low digestibility and influence the grain's poor nutritional value (Elkonin et al., 2018). Therefore, it is important to change the kafirin composition to increase digestibility and improve the nutritional value of sorghum grains. Chiquito-Almanza et al. (2016) studied the γ-kafirin gene in 12 Mexican tannin-free white sorghum genotypes and characterized its relationship

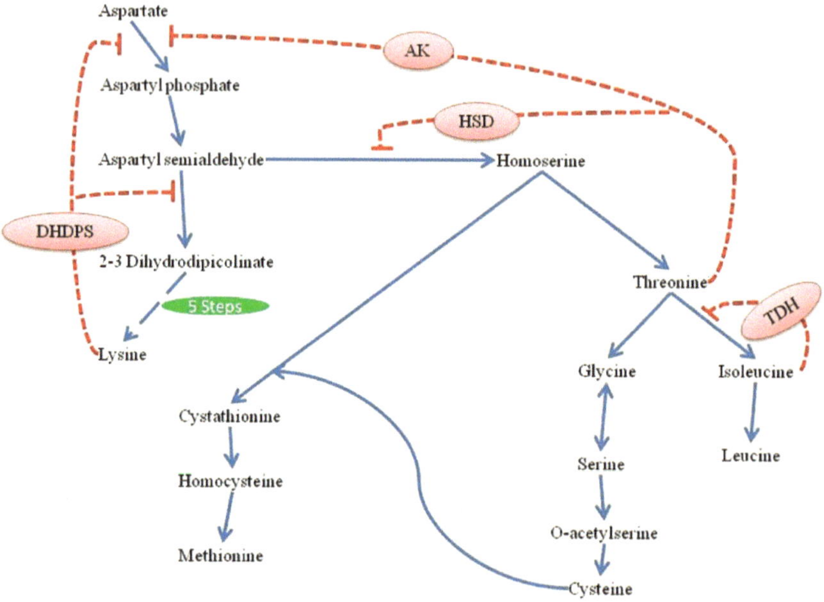

FIGURE 7.1 Biosynthetic pathway of aspartate-derived amino acids and end-product feedback inhibition. AK, aspartate kinase; DHDPS, dihydrodipicolinic acid synthase; TDH, threonine dehydratase; HSD, homoserine dehydrogenase. (Falco et al., 1995; Hanafy et al., 2013; Yang et al., 2021).

with low digestibility of kafirin protein and lysine content. Two alleles, namely allele 1 and allele 7, code for γ-kafirin, and a missense (C235G) mutation on allele 7 increases sorghum grains' lysine content. Another group of researchers studied the α-kafirin protein coded by k1C family genes. They created variants of sorghum with reduced kafirin level by targeting k1C genes using the (CRISPR)/CRISPR-associated protein 9 (Cas9) gene-editing approach (Li et al., 2018).

In maize, zein is the major seed storage protein, and assimilation of sulfur may reduce the incorporation of cysteine and methionine in zein by limiting the availability of cysteine and methionine. Transgenic maize has been developed with change in sulfate reduction capacity by expression of *Escherichia coli* gene cysH for enzyme 3′-phosphoadenosine-5′-phosphosulfate reductase exclusively in leaf, which results in enhanced methionine accumulation in seedlings. Transgenic kernels exhibited an elevated expression of 10 kDa δ-zein, which is rich in methionine and sulfur in total protein. However, other zeins are unchanged. The increase in the expression of sulfur-rich zeins explained one aspect of these proteins' regulation under enhanced sulfur assimilation. Accumulation of methionine in the kernel was 57.6% greater in transgenic line PE5 than in the inbred line. Experimentally, transgenic kernels promote significant weight gain in chicks. As a result, increasing the source strength of maize can improve its nutritional value while causing no marked loss in yield and may reduce the cost of feed supplementation (Planta et al., 2017). Developments in the past few years in the genetic engineering approach to improve the seed protein quality of different cereal crops are presented in Table 7.1.

TABLE 7.1

Seed Protein Quality Improvement in Some Important Cereal Crops Using Genetic Engineering

Gene	Gene source	Genetically modified crop	Encoding protein/ amino acids	Improvement	Reference
Albumin gene (*ama*1)	A. *hypochondriacus*	*Triticum aestivum* L. cv Cadenza	35-kDa AmA1 protein	AmA1 protein increased by 1.78–2.41% in seeds	(Tamás et al., 2009)
r9-LOX1 gene	*Oryza sativa* L. ssp. *indica* cv. Swarna	*O. sativa* L. ssp *indica* cv. PSII	LOX enzyme	Reduced LOX activity enhanced grain nutritional quality	(Gayen et al., 2015, 2014)
Gene SBgLR (Potato), TSRF1(Tomato)	*Potato*	Maize (*Zea mays* L.) cv X178	Lysine-rich protein, transcription factor	Increase in protein content from 7.7% to 24.38%, and lysine content from 8.70% to 30.43%	(Wang et al., 2013)
TKTKK1 and *TKTKK2*	Rice	Rice	Lysine and threonine	Lysine content increased by 33.87%, threonine content by 21.21%, total amino acids by 19.43%, and crude protein content by 20.45%	(Jiang et al., 2016)
Overexpression of *AK* and *DHPS* and silencing of *LKR/SDH* by RNAi	*E. coli* strain TOC R21	*Oryza sativa* ssp. *Japonica* cv.Wuxiangjing 9	Free Lys	About 12-fold increase in free Lys in leaves and about 60-fold in seeds	(Long et al., 2013)
GhLRP	Cotton	Maize	High-lysine protein	Increased lysine content up to 65.0%	(Yue et al., 2014)
AtMAP18 gene	*Arabidopsis thaliana*	Maize	Microtubule-associated protein with high lysine content	Increased level of zein and non-zein protein	(Chang et al., 2015)
SBgLR	Potato	Maize	Zeins and non-zein proteins	High-protein and high-lysine maize seeds	(Liu et al., 2015)
γ-KAFIRIN-1 gene	RNAi-silencing	Sorghum cv. Avans	Kafirin protein	Digestibility of kafirin protein enhanced in transgenic kernels from 57% to 93% *in vitro*	(Elkonin et al., 2021)

7.3 Seed Protein Improvement in Pulses

Pulses are consumed in various ways around the world, as a vegetable (e.g., green pea, green beans, and cluster beans) and for oil extraction (e.g., peanut and soybean) (Iriti and Varoni, 2017). Regardless of their various uses, one common factor is pulses providing nutrition. Pulses have immense benefits in human nutrition and have been recognized and added to diets for a long time. They are a good source of protein and are known for their nutritional value. Still, in many parts of the world, their consumption and cultivation are also neglected, scientifically understudied areas (Robinson et al., 2019). Nevertheless, pulses are emerging as an important crop to fulfill protein requirements in the future and provide an environmentally sustainable source of dietary protein and micronutrients when climate changes occur due to the anthropogenic activities of humans. Furthermore, enhancement in the nutritional qualities of pulse crops might help reduce the risk of chronic diseases such as type 2 diabetes (Robinson et al., 2019).

It is assumed that the consumption of pulses fulfills the nutritional requirements of the vegetarian diet. Still, methionine (Met) essential amino acid is a limiting amino acid in legumes (Le et al., 2016). Legume storage proteins are generally low in sulfur-containing amino acids such as Met and Cys. Genetic engineering can play an important role in improving the nutritional quality of storage proteins that lack important amino acids (Torio et al., 2011). Various groups of researchers have attempted to improve the nutritional quality of soybean and mungbean. Torio et al. (2011) engineered the 8Sα globulin or vicilin (major storage protein) of mungbean using site-directed mutagenesis and increased the Met residues in the protein molecule. The engineered proteins were thermally stable, had greater solubility, and also showed no allergenic potential.

The lysine content in canola and soybean seeds is increased by bypassing the normal feedback regulation of two biosynthetic pathway enzymes, AK and dihydrodipicolinic acid synthase (DHDPS) (Figure 7.1). Falco et al. (1995) developed transgenic canola and soybean by transforming them with lysine-feedback-insensitive bacterial DHDPS (*Corynebacterium* dapA gene) and AK enzymes (mutant *E. coli* lysC gene). Accumulation of free lysine in canola seeds was more than 100-fold greater upon expression of DHDPS gene under seed-specific promotor, while in transgenic soybean, expression of *Corynebacterium* DHDPS and lysine-insensitive *E. coli* AK increased the free lysine in the seeds several hundred fold.

Lactostatin (Ile-Ile-Ala-Glu-Lys) was reported to lower cholesterol absorption in mice. It suppresses the expression of cholesterol 7 α-hydroxylase (CYP7A1) mRNA and induces the expression of intestinal ABCA-1 mRNA (Nagaoka et al., 2006). Genetic engineering of 8Sα globulin of mungbean with the cholesterol-lowering bioactive peptide lactostatin using site-directed mutagenesis enhanced the protein level and cholesterol-lowering activity (Gamis et al., 2020; Medina et al., 2020). The Brazil nut 2S protein (BN2S) is high in Met content and a suitable candidate for enhancing the nutritional value of plants that are deficient in essential sulfur amino acids, such as leguminous and root/tuber crops (Tu et al., 1998). Aragão et al. (1999) genetically engineered bean (*Phaseolus vulgaris*) cv Olathe plants to improve their nutritional value. Transgenic plants expressed the 2S-albumin gene from the Brazil nut, and methionine content was 10–23% higher than in non-transgenic plants. Pickardt et al. (1995) evaluated the effect of transferring the 2S albumin gene from Brazil nut to *Vicia narbonensis*. They reported an increase in sulfur amino acids in grain legumes (total sodium dodecyl sulfate (SDS)-soluble seed protein increased from 1% to 4.8%).

Multi-gene families encode pea seed storage proteins, and their composition is quite complex. Encoded proteins are assembled in trimers such as vicilins and con-vicilins and as hexamers such as legumins with a different post-translational processing pattern (Bourgeois et al., 2011). Since multi-genes encode pea storage protein, a mutation in a single gene has a negligible effect on total protein concentration. However, the quality of the storage protein may be improved by disrupting the production of poor-quality proteins (Robinson et al., 2019). Lectin and con-vicilin are two storage proteins of pea seeds that showed potential for altering and improving the seed protein composition (Domoney et al., 2013).

Mungbean contains 24–28% of proteins on a dry basis and could be an excellent dietary protein source; the seeds are also rich in iron and folate. Humans can consume mungbean as sprouts, flour, soups, porridge, and noodles. In addition, it can be used as feed or forage for cattle or as haulms (Das et al., 2018). The common bean is one of the most widely consumed grain legumes in the world. Although beans are high in some essential amino acids, such as Lys, Thr, Val, Ile, and Leu, their nutritional value is limited due to low levels of methionine and cysteine, which are essential amino acids. The expression of methionine-rich storage albumin from Brazil nuts increased the methionine content of common beans (Aragão et al., 1999). The lupine is a major grain legume. The sulfur-containing amino acids methionine and cysteine are deficient in lupine seed protein, as in most other grain legume proteins. Lupine seeds were transformed with the sunflower seed albumin gene, and its expression increased the methionine content in lupine seeds by 94%, with a 12% reduction in cysteine content (Molvig et al., 1997). Improvement of seed proteins of different pulses using genetic modification in recent years is summarized in Table 7.2.

7.4 Protein Improvement in Other Important Crops

Potato is another important crop, with around 3–6% protein of dried weight, and two major storage proteins, patatin and 11S globulin, are rich in lysine. However, the concentration of sulfur-containing amino acids is low in tuber protein. A PrLeg polypeptide of Perilla seeds contains a high level of sulfur-containing amino acids. PrLeg cDNA was transformed into a potato plant and overexpressed under the control of the tuber-specific promoter patatin (Goo et al., 2013). The first enzyme specific to methionine biosynthesis in higher plants is cystathionine synthase (CgS). A mutated gene from Arabidopsis encoding CgS (CgS90, not regulated by methionine) when co-transformed with the methionine-rich 15-kD zein increases the methionine content in storage protein of potato. Transgenic potato exhibits two- to six-fold increased free methionine content and also soluble isoleucine and serine content (Dancs et al., 2008). Nutritional quality and the iconic pleasant aroma in baked and fried potatoes are associated with methionine content, which improves these qualities. Methionine content in potato tuber can be increased by up-regulating a rate-limiting step in methionine biosynthesis and silencing methionine lyase (StMGL). Overexpression of *A. thaliana* cystathionine synthase (AtCGS) in potato is achieved when up-regulating a rate-limiting step in the biosynthesis of methionine, increasing the methionine content in potato tuber.

TABLE 7.2

Seed Protein Quality Improvement in Some Important Pulses by Genetic Engineering Approach

Gene	Gene source	Genetically modified crop	Encoding protein/ amino acids	Improvement	Reference
Gy1 (A1aB1b), hpt and V3-1 genes	--	Soybean	Glycinin	Seeds enriched with glycinin	(El-Shemy et al., 2007)
β-zein gene	Maize	Soybean cultivar Jack	β-zein protein	Accumulation of β-zein protein and improved total Met content of soybean	(Guo et al., 2020)
EAA-rich artificial storage protein (ASPx)	Artificially synthesized	Peanut plant cultivar Georgia Green	Methionine, lysine, threonine, isoleucine, tryptophan, valine, phenylalanine, and leucine	Levels of essential amino acids Val, Iso, Leu, Met, and Threonine were increased	(Diby et al., 2020)
MB16	A synthetic gene	Soybean	MB-16 protein	Boosted seed methionine content	(Zhang et al., 2014)
Cystathionine γ-synthase	*Arabidopsis*	Soybean cultivars, Zigongdongdou and Jilinxiaoli 1	Essential amino acid methionine	Up to 7.3-fold increase in the levels of soluble methionine	(Yu et al., 2018)
OASA1D	Rice	Soybean [*Glycine max* (L.) Merrill]	Tryptophan (Trp)	Increased level of free Trp in soybean seeds	(Ishimoto et al., 2010)
β-amyrin synthase genes	RNAi-silencing	Soybean cultivar Jack	β-amyrin synthase	Suppressed saponin biosynthesis resulted in enhanced taste and foam in tofu	(Takagi et al., 2011)
OASA1D	Rice	Soybean cultivar Jack	Arginine (Arg) and asparagine (Asn)	Increased level of Arg and Asn in seed	(Kita et al., 2010)
18 kDa δ-zein and 27 kDa γ-zein	Maize	Soybeans cultivar 'Williams 82'	δ-zein and γ-zein protein	Enhanced methionine content	(Kim and Krishnan, 2019)

On the other hand, silencing of methionine lyase (StMGL) in potatoes causes less methionine degradation into 2-ketobutyrate, raising methionine levels. A high ratio of biosynthesis and degradation can cause increases in tuber methionine content in potato. Potatoes cv. Désirée, with AtCGS overexpression and StMGL silenced by RNA interference, have normal morphology and accumulate more free methionine (Kumar and Jander, 2017). Chakraborty et al. (2010) produced transgenic potatoes with increased nutritive value through the tuber-specific expression of a seed protein, AmA1 (Amaranth Albumin 1). Total protein content in transgenic tubers increased by up to 60%. In addition, the concentrations of several essential amino acids, normally limited in potatoes, significantly increased in transgenic tubers. Mustard is an economically important crop that is widely grown for oil production around the world. Therefore, it is desirable to boost the nutritional value of unsaturated fatty acids. This was accomplished by expressing the enzyme Δ6 fatty acid desaturase in transgenic mustard, which resulted in the production of gamma-linoleic acid (Hong et al., 2002).

REFERENCES

Akibode, C.S., Maredia, M.K. (Eds.), 2012. *Global and Regional Trends in Production, Trade and Consumption of Food Legume Crops*. https://doi.org/10.22004/ag.econ.136293

Altenbach, S.B., Pearson, K.W., Meeker, G., Staraci, L.C., Sun, S.S., 1989. Enhancement of the methionine content of seed proteins by the expression of a chimeric gene encoding a methionine-rich protein in transgenic plants. *Plant Molecular Biology* 13, 513–522.

Apostolatos, G., 1984. Isolation and characterization of a methionine-rich protein fraction from edible dry bean (*Phaseolus vulgaris* L.). *Plant Science Letters* 33, 39–46.

Aragão, F.J.L., Barros, L.M.G., De Sousa, M.V., Grossi de Sá, M.F., Almeida, E.R.P., Gander, E.S., Rech, E.L., 1999. Expression of a methionine-rich storage albumin from the Brazil nut (*Bertholletia excelsa* HBK, Lecythidaceae) in transgenic bean plants (*Phaseolus vulgaris* L., Fabaceae). *Genetics and Molecular Biology* 22, 445–449.

Bernstein, A.M., Pan, A., Rexrode, K.M., Stampfer, M., Hu, F.B., Mozaffarian, D., Willett, W.C., 2012. Dietary protein sources and the risk of stroke in men and women. *Stroke* 43, 637–644.

Bourgeois, M., Jacquin, F., Cassecuelle, F., Savois, V., Belghazi, M., Aubert, G., Quillien, L., Huart, M., Marget, P., Burstin, J., 2011. A PQL (protein quantity loci) analysis of mature pea seed proteins identifies loci determining seed protein composition. *Proteomics* 11, 1581–1594.

Chakraborty, S., Chakraborty, N., Agrawal, L., Ghosh, S., Narula, K., Shekhar, S., et al. 2010. Next-generation protein-rich potato expressing the seed protein gene AmA1 is a result of proteome rebalancing in transgenic tuber. *Proc. Natl. Acad. Sci. USA* 107 (41), 17533–17538. DOI: 10.1073/pnas.1006265107

Chang, Y., Shen, E., Wen, L., Yu, J., Zhu, D., Zhao, Q., 2015. Seed-specific expression of the Arabidopsis AtMAP18 gene increases both lysine and total protein content in maize. *PLoS ONE* 10, e0142952. https://doi.org/10.1371/journal.pone.0142952

Chiquito-Almanza, E., Ochoa-Zarzosa, A., López-Meza, J.E., Pecina-Quintero, V., Nuñez-Colín, C.A., Anaya-López, J.L., 2016. A new allele of γ-kafirin gene coding for a protein with high lysine content in Mexican white sorghum germplasm. *Journal of the Science of Food and Agriculture* 96, 3342–3350.

Dancs, G., Kondrák, M., Bánfalvi, Z., 2008. The effects of enhanced methionine synthesis on amino acid and anthocyanin content of potato tubers. *BMC Plant Biol* 8, 65. https://doi.org/10.1186/1471-2229-8-65

Das, A., Singh, P., Kushwah, N.S., Thakur, S., Rathore, M., Pratap, A., Singh, N.P., 2018. Genetic and genomic approaches for improvement in mungbean (*Vigna radiata* L.). In: Wani, S.H., Jain, M. (Eds.), *Pulse Improvement*. Springer, pp. 175–189.

de Carvalho, K.M., Pizato, N., Botelho, P.B., Dutra, E.S., Gonçalves, V.S., 2020. Dietary protein and appetite sensations in individuals with overweight and obesity: A systematic review. *European Journal of Nutrition* 1–16. https://doi.org/10.1007/s00394-020-02321-1

Diby, N'N. A. S., Konan, K. N'D., Ananga, A.O., Dodo, H., 2020. Improving peanut protein quality: Expression of a synthetic storage protein. *African Journal of Biotechnology* 19, 265–275.

Domoney, C., Knox, M., Moreau, C., Ambrose, M., Palmer, S., Smith, P., Christodoulou, V., Isaac, P.G., Hegarty, M., Blackmore, T., 2013. Exploiting a fast neutron mutant genetic resource in *Pisum sativum* (pea) for functional genomics. *Functional Plant Biology* 40, 1261–1270.

Elkonin, L., Italyanskaya, J., Panin, V., 2018. Genetic modification of sorghum for improved nutritional value: State of the problem and current approaches. *J Investig Genomics* 5, 39–48.

Elkonin, L.A., Panin, V.M., Kenzhegulov, O.A., Sarsenova, S.K., 2021. RNAi-mutants of *Sorghum bicolor* (L.) Moench with improved digestibility of seed storage proteins. In: Jimenez-Lopez, J.C. (Ed.), *Grain and Seed Proteins Functionality*. IntechOpen.

El-Shemy, H.A., Khalafalla, M.M., Fujita, K., Ishimoto, M., 2007. Improvement of protein quality in transgenic soybean plants. *Biologia Plantarum* 51, 277–284.

Falco, S.C., Guida, T., Locke, M., Mauvais, J., Sanders, C., Ward, R.T., Webber, P., 1995. Transgenic canola and soybean seeds with increased lysine. *Nat Biotechnol* 13, 577–582. https://doi.org/10.1038/nbt0695-577

Galili, G., Amir, R., Hoefgen, R., Hesse, H., 2005. Improving the levels of essential amino acids and sulfur metabolites in plants. *Biological Chemistry* 386, 817–831.

Gamis, M., Uy, L.Y., Laurena, A., Hurtada, W., Torio, M.A., 2020. Protein engineering of mung bean (*Vigna radiata* (L.) Wilczek) 8Sα globulin with lactostatin. *Applied Sciences* 10, 8787. https://doi.org/10.3390/app10248787

Gayen, D., Ali, N., Ganguly, M., Paul, S., Datta, K., Datta, S.K., 2014. RNAi mediated silencing of lipoxygenase gene to maintain rice grain quality and viability during storage. *Plant Cell, Tissue and Organ Culture (PCTOC)* 118, 229–243.

Gayen, D., Ali, N., Sarkar, S.N., Datta, S.K., Datta, K., 2015. Down-regulation of lipoxygenase gene reduces degradation of carotenoids of golden rice during storage. *Planta* 242, 353–363. https://doi.org/10.1007/s00425-015-2314-4

Goo, Y.-M., Kim, T.-W., Lee, M.-K., Lee, S.-W., 2013. Accumulation of PrLeg, a Perilla legumin protein in potato tuber results in enhanced level of sulphur-containing amino acids. *C R Biol* 336, 433–439. https://doi.org/10.1016/j.crvi.2013.09.002

Guerrieri, N., Cavaletto, M., 2018. Cereals proteins. In: Yada, R.Y. (Ed.), *Proteins in Food Processing*. Elsevier, pp. 223–244.

Guo, C., Liu, X., Chen, L., Cai, Y., Yao, W., Yuan, S., Wu, C., Han, T., Sun, S., Hou, W., 2020. Elevated methionine content in soybean seed by overexpressing maize β-zein protein. *Oil Crop Science, Special issue for soybean* 5, 11–16. https://doi.org/10.1016/j.ocsci.2020.03.004

Hanafy, M.S., Rahman, S.M., Nakamoto, Y., Fujiwara, T., Naito, S., Wakasa, K., Ishimoto, M., 2013. Differential response of methionine metabolism in two grain legumes, soybean and azuki bean, expressing a mutated form of Arabidopsis cystathionine γ-synthase. *Journal of plant physiology* 170, 338–345. https://doi.org/10.1016/j.jplph.2012.10.018

Hong, H., Datla, N., Reed, D.W., Covello, P.S., MacKenzie, S.L., Qiu, X., 2002. High-level production of γ-linolenic acid in *Brassica juncea* using a Δ6 desaturase from *Pythium irregulare*. *Plant Physiology* 129, 354–362.

Iriti, M., Varoni, E.M., 2017. *Pulses, Healthy and Sustainable Food Sources for Feeding the Planet.*

Ishimoto, M., Rahman, S.M., Hanafy, M.S., Khalafalla, M.M., El-Shemy, H.A., Nakamoto, Y., et al., 2010. Evaluation of amino acid content and nutritional quality of transgenic soybean seeds with high-level tryptophan accumulation. *Mol Breeding* 25, 313–326. https://doi.org/10.1007/s11032-009-9334-3

Jiang, S.-Y., Ma, A., Xie, L., Ramachandran, S., 2016. Improving protein content and quality by over-expressing artificially synthetic fusion proteins with high lysine and threonine constituent in rice plants. *Sci Rep* 6, 34427. https://doi.org/10.1038/srep34427

Kim, W.-S., Krishnan, H.B., 2019. Impact of co-expression of maize 11 and 18 kDa δ-zeins and 27 kDa γ-zein in transgenic soybeans on protein body structure and sulfur amino acid content. *Plant Science* 280, 340–347. https://doi.org/10.1016/j.plantsci.2018.12.016

Kita, Y., Nakamoto, Y., Takahashi, M., Kitamura, K., Wakasa, K., Ishimoto, M., 2010. Manipulation of amino acid composition in soybean seeds by the combination of deregulated tryptophan biosynthesis and storage protein deficiency. *Plant Cell Rep* 29, 87–95. https://doi.org/10.1007/s00299-009-0800-5

Kumar, P., Jander, G., 2017. Concurrent overexpression of *Arabidopsis thaliana* cystathionine γ-synthase and silencing of endogenous methionine γ-lyase enhance tuber methionine content in *Solanum tuberosum*. *J. Agric. Food Chem.* 65, 2737–2742. https://doi.org/10.1021/acs.jafc.7b00272

Le, D.T., Chu, H.D., Le, N.Q., 2016. Improving nutritional quality of plant proteins through genetic engineering. *Curr Genomics* 17, 220–229. https://doi.org/10.2174/1389202917666160202215934

Leinonen, I., Iannetta, P.P.M., Rees, R.M., Russell, W., Watson, C., Barnes, A.P., 2019. Lysine supply is a critical factor in achieving sustainable global protein economy. *Frontiers in Sustainable Food Systems* 3, 27. https://doi.org/10.3389/fsufs.2019.00027

Li, A., Jia, S., Yobi, A., Ge, Z., Sato, S.J., Zhang, C., et al., 2018. Editing of an alpha-kafirin gene family increases digestibility and protein quality in sorghum. *Plant Physiology* 177, 1425–1438.

Liu, C., Li, S., Yue, J., Xiao, W., Zhao, Q., Zhu, D., Yu, J., 2015. Microtubule-associated protein SBgLR facilitates storage protein deposition and its expression leads to lysine content increase in transgenic maize endosperm. *International Journal of Molecular Sciences* 16, 29772–29786. https://doi.org/10.3390/ijms161226199

Long, X., Liu, Q., Chan, M., Wang, Q., Sun, S.S.M., 2013. Metabolic engineering and profiling of rice with increased lysine. *Plant Biotechnology Journal* 11, 490–501. https://doi.org/10.1111/pbi.12037

Medina, M., Uy, L., Manuel, M., Recuenco, M., Torio, M., 2020. Hypocholesterolemic activity of mungbean 8Sα globulin engineered with lactostatin. *PJCS* 45, 13–26.

Molvig, L., Tabe, L.M., Eggum, B.O., Moore, A.E., Craig, S., Spencer, D., Higgins, T.J., 1997. Enhanced methionine levels and increased nutritive value of seeds of transgenic lupins (*Lupinus angustifolius* L.) expressing a sunflower seed albumin gene. *Proceedings of the National Academy of Sciences* 94, 8393–8398. https://doi.org/10.1073/pnas.94.16.8393

Nagaoka, S., Fujimura, W., Morikawa, K., Nakamura, A., Kanamaru, Y., Hori, G., et al., 2006. Lactostatin (IIAEK) and CSPHP: New cholesterol-lowering peptides derived from food proteins. In: Huang, Y. S., Yanagita, T., Knapp, H. R. (Eds), *Dietary Fats and Risk of Chronic Disease*, 168–185.

Ohanenye, I.C., Tsopmo, A., Ejike, C.E., Udenigwe, C.C., 2020. Germination as a bioprocess for enhancing the quality and nutritional prospects of legume proteins. *Trends in Food Science & Technology* 101, 213–222.

Pfarr, M.D., Kazula, M.J., Miller-Garvin, J.E., Naeve, S.L., 2018. Amino acid balance is affected by protein concentration in soybean. *Crop Science* 58, 2050–2062. https://doi.org/10.2135/cropsci2017.11.0703

Pickardt, T., Saalbach, I., Waddell, D., Meixner, M., Müntz, K., Schieder, O., 1995. Seed specific expression of the 2S albumin gene from Brazil nut (*Bertholletia excelsa*) in transgenic *Vicia narbonensis*. *Molecular Breeding* 1, 295–301.

Planta, J., Xiang, X., Leustek, T., Messing, J., 2017. Engineering sulfur storage in maize seed proteins without apparent yield loss. *Proceedings of the National Academy of Sciences* 114, 11386–11391. https://doi.org/10.1073/pnas.1714805114

Reeves, T.G., Thomas, G., Ramsay, G., 2015. *Save and grow in practice: Maize, rice, wheat*. FAO, Rome, Italy.

Robinson, G.H.J., Balk, J., Domoney, C., 2019. Improving pulse crops as a source of protein, starch and micronutrients. *Nutrition Bulletin* 44, 202–215.

Sindhu, A.S., Zheng, Z., Murai, N., 1997. The pea seed storage protein legumin was synthesized, processed, and accumulated stably in transgenic rice endosperm. *Plant Science* 130, 189–196.

Sun, S.S., Liu, Q., 2004. Transgenic approaches to improve the nutritional quality of plant proteins. *In Vitro Cellular & Developmental Biology-Plant* 40, 155–162.

Takagi, K., Nishizawa, K., Hirose, A., Kita, A., Ishimoto, M., 2011. Manipulation of saponin biosynthesis by RNA interference-mediated silencing of β-amyrin synthase gene expression in soybean. *Plant Cell Rep* 30, 1835. https://doi.org/10.1007/s00299-011-1091-1

Tamás, C., Kisgyörgy, B.N., Rakszegi, M., Wilkinson, M.D., Yang, M.-S., Láng, L., Tamás, L., Bed\Ho, Z., 2009. Transgenic approach to improve wheat (*Triticum aestivum* L.) nutritional quality. *Plant Cell Reports* 28, 1085–1094.

Torio, M.A.O., Adachi, M., Garcia, R.N., Prak, K., Maruyama, N., Utsumi, S., Tecson-Mendoza, E.M., 2011. Effects of engineered methionine in the 8Sα globulin of mungbean on its physicochemical and functional properties and potential nutritional quality. *Food Research International* 44, 2984–2990.

Tozawa, Y., Hasegawa, H., Terakawa, T., Wakasa, K., 2001. Characterization of rice anthranilate synthase α-subunit genes OASA1 and OASA2. Tryptophan accumulation in transgenic rice expressing a feedback-insensitive mutant of OASA1. *Plant Physiology* 126, 1493–1506. https://doi.org/10.1104/pp.126.4.1493

Tu, H.M., Godfrey, L.W., Sun, S.S., 1998. Expression of the Brazil nut methionine-rich protein and mutants with increased methionine in transgenic potato. *Plant Molecular Biology* 37, 829–838.

Wang, M., Liu, C., Li, S., Zhu, D., Zhao, Q., Yu, J., 2013. Improved nutritive quality and salt resistance in transgenic maize by simultaneously overexpression of a natural lysine-rich protein gene, SBgLR, and an ERF transcription factor gene, TSRF1. *International Journal of Molecular Sciences* 14, 9459–9474. https://doi.org/10.3390/ijms14059459

Wenefrida, I., Utomo, H.S., Blanche, S.B., Linscombe, S.D., 2009. Enhancing essential amino acids and health benefit components in grain crops for improved nutritional values. *Recent Patents on DNA & Gene Sequences (Discontinued)* 3, 219–225.

Wolfe, R.R., Cifelli, A.M., Kostas, G., Kim, I.-Y., 2017. Optimizing protein intake in adults: Interpretation and application of the recommended dietary allowance compared with the acceptable macronutrient distribution range. *Advances in Nutrition* 8, 266–275. https://doi.org/10.3945/an.116.013821

Yang, Q.-Q., Yu, W.-H., Wu, H.-Y., Zhang, C.-Q., Sun, S.S.-M., Liu, Q.-Q., 2021. Lysine biofortification in rice by modulating feedback inhibition of aspartate kinase and dihydrodipicolinate synthase. *Plant Biotechnology Journal* 19, 490–501. https://doi.org/10.1111/pbi.13478

Yu, S., Tian, L., 2018. Breeding major cereal grains through the lens of nutrition sensitivity. *Molecular Plant* 11, 23–30.

Yu, Y., Hou, W., Hacham, Y., Sun, S., Wu, C., Matityahu, I., Song, S., Amir, R., Han, T., 2018. Constitutive expression of feedback-insensitive cystathionine γ-synthase increases methionine levels in soybean leaves and seeds. *J. Integr. Agric* 17, 54–62.

Yue, J., Li, C., Zhao, Q., Zhu, D., Yu, J., 2014. Seed-specific expression of a lysine-rich protein gene, GhLRP, from cotton significantly increases the lysine content in maize seeds. *International Journal of Molecular Sciences* 15, 5350–5365. https://doi.org/10.3390/ijms15045350

Zhang, Y., Schernthaner, J., Labbé, N., Hefford, M.A., Zhao, J., Simmonds, D.H., 2014. Improved protein quality in transgenic soybean expressing a de novo synthetic protein, MB-16. *Transgenic Res* 23, 455–467. https://doi.org/10.1007/s11248-013-9777-5

8

Somatic Embryogenesis and Transformation Studies in Ginger

Valiyaparambath Musfir Mehaboob
MES Ponnani College
Ponnani
Kerala, India

Kunnampalli Faizal
Jamal Mohamed College
Tiruchirappalli
Tamil Nadu, India

Palusamy Raja
Jamal Mohamed College
Tiruchirappalli
Tamil Nadu, India

Ganesan Thiagu
Jamal Mohamed College
Tiruchirappalli
Tamil Nadu, India

Kizhakke Modongal Shamsudheen
Jamal Mohamed College
Tiruchirappalli
Tamil Nadu, India

Abubakker Aslam
Jamal Mohamed College
Tiruchirappalli
Tamil Nadu, India

Appakan Shajahan
Jamal Mohamed College
Tiruchirappalli
Tamil Nadu, India

DOI: 10.1201/9781003239932-8

CONTENTS

8.1 Introduction

Ginger (*Zingiber officinale* Rosc.) is an important spice crop belonging to the family Zingiberaceae. The commercial production of ginger is limited by several factors. Plant diseases like bacterial wilt (*Pseudomonas solanacearum*) and soft rot (*Pythium aphanidermatum*) are causing heavy yield losses of ginger (Sharma and Singh 1997). Conventional breeding methods have limited success due to its obligatory asexual nature, stigmatic incompatibility and lack of genetic variability. These problems point to the necessity of biotechnological approaches for ginger improvement. *Agrobacterium tumefaciens*-mediated transformation can be a resourceful alternative for integrating a single copy of a transgene into the plant genome (Sood et al. 2011). Somatic embryogenesis and regeneration is considered the most suitable plant propagation method for genetic transformation. In this chapter, we describe an efficient *Agrobacterium*-mediated transient transformation protocol for ginger *via* the somatic embryogenesis system.

8.2 Uses of Ginger

Ginger is a unique spice crop used in many countries for medicinal and culinary preparations. Ginger is used as a common condiment in many foods and beverages as it gives them a special flavor. It is used in the preparation of gingerbread, soups, biscuits, puddings, ginger jams, cakes, pickles and drinks like ginger beer, ginger tea and ginger wine. Fresh ginger paste is used in curries, and dried ginger powder is used in curry powder, syrup, candy and sauces (Pruthi 1993; Vasala 2012). India is the largest producer and consumer of ginger in the world, accounting for 50% of total production (Sundararaj et al. 2010).

The rhizome of ginger is a widely used in the Chinese, Japanese and Indian traditional medicine systems. It possesses several medicinal properties, such as a stimulant of the gastrointestinal tract, a carminative and a diuretic, and has antioxidant, anti-inflammatory and diaphoretic effects (Nirmal Babu et al. 2016). Ginger has also been shown to have potential action against stomach discomfort, tumors, asthma, cough, rheumatism and osteoporosis (Zheng et al. 2008). In the Chinese system of medicine, ginger is used for the treatment of diarrhea, blurred vision, vomiting, light-headedness, decrease in blood pressure, high blood pressure and dyspepsia (Ravindran and Babu 2005).

8.3 Somatic Embryogenesis

Somatic embryogenesis is an important plant regeneration method that resembles zygotic embryogenesis. At the same time, somatic embryogenesis enables non-zygotic plant cells to form embryos and a whole plant (Rose et al. 2010). Standardization of a somatic embryogenesis protocol facilitates the commercial production of plants (Loyola-Vargas and Vazquez-Flota, 2006). It is also considered to be a highly desirable plant regeneration system for genetic transformation, with few or no somaclonal variations and higher genetic uniformity (Gaj 2001; Zhao et al. 2012).

8.3.1 Somatic Embryogenesis in Ginger Family

A somatic embryogenesis protocol has been successfully established in many species of the Zingiberaceae family (Table 8.1). Different tissues such as leaf sheath, leaf base, young inflorescence, shoot buds and the inner core region of the rhizome have been used to produce somatic embryos in different species of Zingiberaceae. Growth regulators also play an important role in induction of embryogenic tissues. Murashige and Skoog (MS) medium (Murashige and Skoog 1962) supplemented with auxin 2,4-dichlorophenoxyacetic acid (2,4-D) is found to be the most successful growth regulator in somatic embryo induction. Auxin in combination with cytokinin exhibits better results in the majority of species. Most studies on somatic embryogenesis of the Zingiberaceae species have reported indirect somatic embryogenesis (Rahman et al. 2004; Manohari et al. 2008; Wong et al. 2013; Zuraida et al. 2014). Direct somatic embryogenesis has been reported in *Curcuma longa* and *Curcuma amada* (Raju et al. 2015; Shajahan et al. 2016).

8.4 *Agrobacterium*-mediated Transformation

A. tumefaciens-mediated genetic transformation is the most widely used transformation technique in dicotyledonous plants. Earlier, transformation of monocots was considered a difficult technique, because they are not natural hosts for *Agrobacterium* (Wu et al. 2014). However, *Agrobacterium*-mediated transformation systems have been successfully developed in agronomically important monocot crops, including rice (Raineri et al. 1990; Ozawa 2009), wheat (Cheng et al. 1997; Wu et al. 2003), barley (Shrawat et al. 2006), maize (Ishida et al. 1996), sugarcane (Arencibia et al. 1998) and turmeric (He and Gang 2013) using the optimized transformation system. An efficient plant regeneration system, the *Agrobacterium* strain, binary vector, phenolic substances and selection markers are important factors influencing the monocot transformation (Cheng et al. 2004).

TABLE 8.1
Somatic Embryogenesis in Zingiberaceae Family

Species	Explant	Embryogenic callus/somatic embryo induction	Somatic embryo germination	Reference
Boesenbergia rotunda	Shoot buds	MS medium, 1 mg/l 2,4-D and 0.5 mg/l BA	1 mg/l NAA and 3 mg/l BA	Wong et al. 2013
Curcuma amada	Leaf sheath	MS medium, 2.0 mg/l 2,4-D and 0.5 mg/l BA	0.25 mg/l GA_3	Raju et al. 2016
Curcuma caesia	Sprouted buds	MS medium, 2 mg/l 2,4-D and 5 mg/l BAP	5 mg/l BAP and 0.2 mg/l 2,4-D	Zuraida et al. 2014
Curcuma longa	Young inflorescence	Gamborg B5 Medium, 5.0 g/l NAA, 1.0 g/l BAP	2 mg/l KT, 0.2 mg/l NAA	He and Gang 2013
Curcuma longa	Leaf base	MS medium, 4.49 μM 2,4-D	1.44 μM GA_3	Raju et al. 2015
Elettaria cardamomum	Inner core region of rhizome	MS medium, 4.4 μM BAP and 0.5 μM NAA	13.2 μM BAP and 0.5 μM NAA	Manohari et al. 2008
Kaempferia galanga	Leaf base	MS medium, 1.5 mg/l 2,4-D and 1 mg/l BA	2 mg/l BA and 0.1 mg/l NAA	Rahman et al. 2004
Zingiber officinale	Leaf	MS medium, 2.7 μM dicamba	8.9 μM BA	Kackar et al. 1993
Zingiber officinale	Shoot tips	MS medium, 1.0 mg/l 2,4-D and 0.2 mg/l Kn	3.0 mg/l BA and 0.1 mg/l NAA	Guo and Zhang 2005
Zingiber officinale	Leaf sheath	MS medium, 9.06 μM 2,4-D, 2.27 μM TDZ	2.22 μM BAP and 2.6 μM NAA	Mehaboob et al. 2019

2,4-D: 2,4-dichlorophenoxyacetic acid; BA: 6-benzyladenine; BAP: 6-benzylaminopurine; NAA: naphthaleneacetic acid; TDZ: thidiazuron.

8.5 *Agrobacterium*-mediated Transformation via Somatic Embryogenesis in Ginger

The number of somatic embryogenesis and transformation protocols reported for ginger is still considered low. So far, only indirect somatic embryogenesis and plant regeneration have been described in ginger using young leaf segments (Kackar et al. 1993) and shoot tips (Guo and Zhang 2005). Kackar et al. (1993) obtained embryogenic culture from leaf explants of ginger on MS medium containing 2.7 μM dicamba alone. But, Guo and Zhang (2005) reported the induction of embryogenic calli from ginger shoot tip explant on MS medium containing 1.0 mg/l 2,4-D and 0.2 mg/1 kinetin. Later, Lincy et al. (2009) reported indirect somatic embryogenesis in ginger using *in planta* leaf explant cultured on MS medium supplemented with 2,4-D and 6-benzylaminopurine (BAP). They also described the induction of direct somatic embryos from the ginger *in planta* leaf explant by using thidiazuron alone or in combination with indole, 3-butyric acid. However, no plantlet regeneration was observed for direct somatic embryos in ginger.

A transformation protocol in ginger was attempted previously using young bud derived callus as explant (Suma et al. 2008). *A. tumefaciens* EHA 105 containing binary vector p35SGUS INT was used for the transformation. A vector having hygromycin phosphotransferase (*hpt*II) and gusA genes driven by the cauliflower mosaic virus (CaMV) 35S promoter was successfully introduced into the ginger genome.

Later, in ginger, leaf sheath explants were infected with *Agrobacterium* strains (EHA105 and LBA4404) binary vector harboring pGFPGUS*Plus* containing *hpt*II selection marker and *gus* reporter gene (Mehaboob et al. 2019). The transformation and embryo regeneration protocol is described in the following (Figure 8.1).

8.5.1 *Agrobacterium*-mediated Genetic Transformation

Bacterial strains were cultured on LB medium (Table 8.2) in an orbital shaker (28 °C, 180 rpm) overnight. After centrifugation of the culture at 10,000 rpm for 5 min, the resultant pellet was suspended in liquid MS medium (basal) containing acetosyringone. Then, 1–2-cm long leaf sheath explants obtained from 6–8-week-old *in vitro* grown plantlets were infected with bacterial culture with gentle shaking (20 min, 80 rpm). Explants were then blotted dry on filter paper for 5 min and placed on co-cultivation medium (MSC) (Table 8.2) in the dark (25 °C, 2 days). Infected explants were subcultured at regular intervals of 3 to 4 days. To remove excess growth of bacteria, transformants were washed with distilled water, blot dried on sterile paper and transferred to resting medium (MSR) (Table 8.2) containing antibiotics. After 4–6 days in dark conditions, the cultures were placed on selection medium (MSS) (Table 8.2) for 4 weeks.

8.5.2 Selection and Somatic Embryo Regeneration

Hyg-resistant embryogenic callus developed from leaf explant was selected and shifted onto MSM medium (Table 8.2) containing BAP alone. Somatic embryos were produced after regular subculturing on the same medium. Fully grown somatic

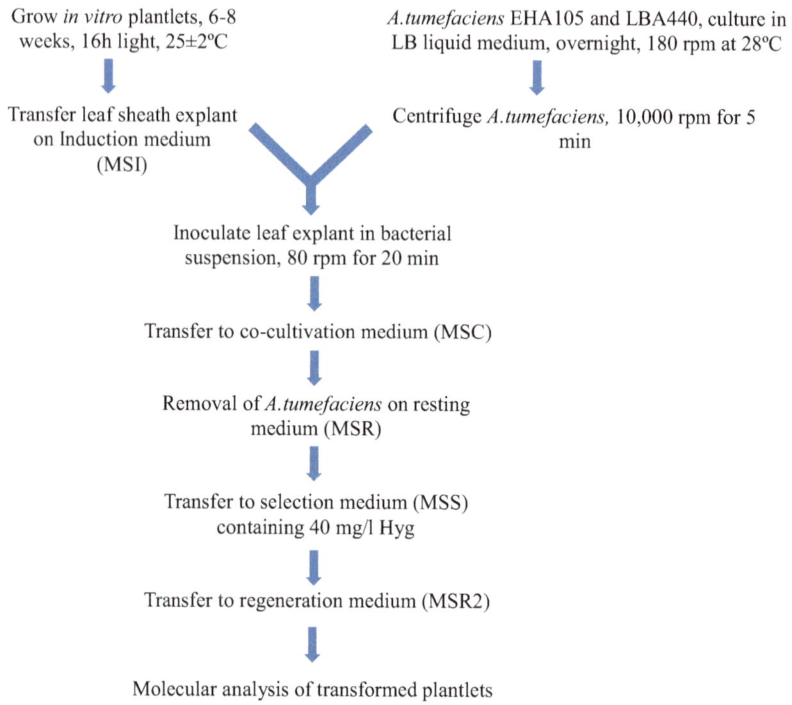

FIGURE 8.1 Flow chart of transformation protocol of ginger.

TABLE 8.2

Composition of Media Used for *Agrobacterium*-mediated Transformation in Ginger

Media	Composition
Induction medium (MSI)	MS, 30 g/l sucrose, 8 g/l agar, PGRs (9.06 µM 2,4-D + 2.27 µM TDZ), pH 5.8
Maturation medium (MSM)	MS, 30 g/l sucrose, 1.33 µM BAP, pH 5.8
Bacterial culture medium (LB)	5 g/l yeast extract, 10 g/l tryptone, 10 g/l Nacl, pH 7.2, kanamycin added just before use
Co- Cultivation medium (MSC)	MS, 30 g/l sucrose, PGRs (9.06 µM 2,4-D + 2.27 µM TDZ), pH 5.2, 100 µM acetosyringone added just before use
Resting medium (MSR)	MS, 30 g/l sucrose, 8 g/l agar, PGRs (9.06 µM 2,4-D + 2.27 µM TDZ) pH 5.8, cefotoxime and timentin added just before use
Selection medium (MSS)	MS, 30 g/l sucrose, 8 g/l agar, PGRs (9.06 µM 2,4-D + 2.27 µM TDZ), pH 5.8, hygromycin added just before use
Regeneration medium (MSR2)	MS, 30 g/l sucrose, 8 g/l agar, PGRs (2.22 µM BAP + 2.6 µM NAA), pH 5.8

FIGURE 8.2 Developmental stages of somatic embryogenesis in ginger.

embryos were placed on regeneration medium (MSR2) (Table 8.2) for regeneration (Figure 8.2). Different factors improving transformation efficiency were examined in our study. Bacterial cell density of 0.6 OD_{600}, 150 µM acetosyringone and a period of 2 days were optimum for co-cultivation of bacterial culture and leaf explants. High transformants were obtained using the selection regime of 40 mg/l hygromycin. Transient expression of *gus* gene and *hpt*II gene were confirmed by histochemical β-glucuronidase (GUS) assay and polymerase chain reaction (PCR) analysis.

8.6 Conclusion

Agrobacterium-mediated transformation via somatic embryos has become one of the most important biotechnological tools for genetic improvement of monocot species. Plant regeneration methods through somatic embryogenesis have been achieved in several species of the Zingiberaceae family. This chapter describes the *Agrobacterium*-mediated transformation protocol for ginger by infecting the leaf explant. The effects of *Agrobacterium* strains, bacterial cell density, doses of acetosyringone, co-cultivation period and hygromycin are important determinants for the efficient transformation of somatic embryos. This transformation system helps with a quick expression of marker and reporter genes in transformed ginger plants.

Acknowledgments

Dr. A. Shajahan and the authors thank the Department of Science & Technology, Govt. of India for providing facilities through the DST-FIST program and the Department of Biotechnology, Govt. of India for their support through Star college scheme.

REFERENCES

Arencibia AD, Carmona ER, Tellez P, Chan MT, Yu SM and Trujillo LE (1998) An efficient protocol for sugarcane (*Saccharum spp.* L.) transformation mediated by Agrobacterium tumefaciens. *Transgenic Res* 7:213–222

Cheng M, Fry JE, Pang SZ, Zhou HP, Hironaka CM, Duncan DR, Conner TW and Wan YC (1997) Genetic transformation of wheat mediated by Agrobacterium tumefaciens. *Plant Physiol* 115(3):971–980

Cheng M, Lowe BA, Spencer TM, Ye X and Armstrong CL (2004). Factors influencing Agrobacterium-mediated transformation of monocotyledonous species. *In Vitro Cellular & Developmental Biology-Plant* 40:31–45.

Gaj MD (2001) Direct somatic embryogenesis as a rapid and efficient system for *in vitro* regeneration of *Arabidopsis thaliana. Plant Cell Tiss Organ Cult* 64:39–46

Guo YH and Zhang ZX (2005) Establishment and plant regeneration of somatic embryogenic cell suspension cultures of the *Zingiber officinale* Rosc. *Sci Hortic* 107:90–96

He R and Gang DR (2013) Somatic embryogenesis and *Agrobacterium*-mediated transformation of turmeric (*Curcuma longa*). *Plant Cell, Tissue and Organ Culture* 116(3):333–342

Ishida Y, Saito H, Ohta S, Hiei Y, Komari T and Kumashiro T (1996) High efficiency transformation of maize (*Zea mays* L.) mediated by Agrobacterium tumefaciens. *Nat Biotechnol* 14:745–750

Kackar A, Bhat SR, Chandel KPS and Malik SK (1993) Plant regeneration via somatic embryogenesis in ginger. *Plant Cell Tiss Organ Cult* 32:289–292

Lincy AK, Remashree AB and Sasikumar B (2009) Indirect and direct somatic embryogenesis from aerial stem explants of ginger (*Zingiber officinale* Rosc.). *Acta Bot Croat* 68(1):93–103

Loyola-Vargas VM and Vazquez-Flota F (2006) Methods in Molecular Biology: Plant Cell Culture Protocols. Humana Press, Totowa, NJ

Manohari C, Backiyarani S, Jebasingh T, Somanath A and Usha R (2008) Efficient plant regeneration in small cardamom (*Elettaria cardamomum* Maton.) through somatic embryogenesis. *Ind J Biotechnol* 7:407–409

Mehaboob VM, Faizal K, Thilip C, Raja P, Thiagu G, Aslam A and Shajahan A (2019) Indirect somatic embryogenesis and Agrobacterium-mediated transient transformation of ginger (*Zingiber officinale* Rosc.) using leaf sheath explants. *The Journal of Horticultural Science and Biotechnology* 94(6):753–760.

Murashige T and Skoog F (1962) A revised medium for rapid growth and bioassays with tobacco tissue cultures. *Physiol Plant* 15:473–497

Nirmal Babu K, Samsudeen K, Divakaran M, Pillai GS, Sumathi V, Praveen K, Ravindran PN and Peter KV (2016) Protocols for *in vitro* propagation, conservation, synthetic seed production, embryo rescue, microrhizome production, molecular profiling and genetic transformation in ginger (*Zingiber officinale* Roscoe.). In: Mohan Jain S (ed.), Protocols for in vitro *cultures and secondary metabolite analysis of aromatic and medicinal plants, second edition*, Springer, pp. 403–426

Ozawa K (2009) Establishment of a high efficiency *Agrobacterium*-mediated transformation system of rice (*Oryza sativa* L.). *Plant Sci* 176:522–527

Pruthi JS (1993) *Major spices in India: Crop management and postharvest technology*. Publication and Information Division, Indian Council of Agricultural Research: Krishi Anusandhan Bhavan, Pusa, New Delhi

Rahman MM, Amin MN, Ahamed T, Ali MR, Habib A (2004)Efficient plant regeneration through somatic embryogenesis from leaf base derived callus of *Kaempferia galanga* L. *Asian J Plant Sci* 3:675–678

Raineri DM, Bottino P, Gordon MP and Nester EW (1990) *Agrobacterium*-mediated transformation of rice (*Oryza sativa* L.). *Nat Biotechnol* 8(1):33–38

Raju CS, Aslam A and Shajahan A (2015) High-efficiency direct somatic embryogenesis and plant regeneration from leaf base explants of turmeric (*Curcuma longa* L.). *Plant Cell Tiss Organ Cult* 122(1):79–87

Raju CS, Aslam A and Shajahan A (2016) Germination and storability of calcium-alginate coated somatic embryos of mango ginger (*Curcuma amada* Roxb.). *Horticulture, Environment, and Biotechnology* 57(1):88–96

Ravindran PN and Babu NK (2005) *Ginger, the genus Zingiber, medicinal and aromatic plants—industrial profiles*. CRC Press, Boca Raton

Rose RJ, Mantiri FR, Kurdyukov S, Chen SK, Wang XD, Nolan KE, Sheahan MB (2010) Developmental biology of somatic embryogenesis. In: *Plant developmental biology: Biotechnological perspectives*, pp. 3–26, Springer, Berlin, Heidelberg

Shajahan A, Raju CS, Thilip C, Varutharaju K, Faizal K, Mehaboob VM and Aslam A (2016) Direct and indirect somatic embryogenesis in mango ginger (*Curcuma amada* Roxb.). In: Loyola-Vargas VM, Ochoa-Alejo N (eds), *Somatic embryogenesis: Fundamental aspects and applications*, pp.367–379, Springer, Cham

Sharma TR and Singh BM (1997) High frequency *in vitro* multiplication of disease-free *Zingiber officinale* Rosc. *Plant Cell Rep* 17:68–72

Shrawat KA, Becker D and Lorz H (2006) Agrobacterium tumefaciens-mediated genetic transformation of barley (Hordeum vulgare L.). *Plant Science* 172:281–290

Sood P, Bhattacharya A and Sood A (2011) Problems and possibilities of monocot transformation. *Biologia Plantarum* 55(1):1–15

Suma B, Keshavachandran R and Nybe EV (2008) *Agrobacterium tumefaciens* transformation and regeneration of ginger (*Zingiber officinale* Rosc.). *Journal of Tropical Agriculture* 46:38–44

Sundararaj SG, Agrawal A and Tyagi RK (2010) Encapsulation for in vitro short-term storage and exchange of ginger (*Zingiber officinale* Rosc.) germplasm. *Sci Hortic* 125:761–766

Vasala PA (2012) Ginger. In: Peter KV (ed.), *Handbook of Herbs and Spices*, vol. 1, pp. 195–206, Woodhead Publishing, Cambridge

Wong SM, Salim N, Harikrishna JA and Khalid N (2013) Highly efficient plant regeneration via somatic embryogenesis from cell suspension cultures of *Boesenbergia rotunda. In Vitro Cellular & Developmental Biology-Plant* 49(6):665–673

Wu H, Sparks C, Amoah B and Jones HD (2003) Factors influencing successful *Agrobacterium*-mediated genetic transformation of wheat. *Plant Cell Rep* 21(7):659–668

Wu HY, Liu KH, Wang YC, Wu JF, Chiu WL, Chen CY, Wu SH, Sheen J and Lai EM (2014) AGROBEST: an efficient Agrobacterium-mediated transient expression method for versatile gene function analyses in Arabidopsis seedlings. Plant Methods 2014:19

Zhao J, Cui J, Liu J, Liao F, Henny RJ and Chen J (2012) Direct somatic embryogenesis from leaf and petiole explants of *Spathiphyllum* 'Supreme' and analysis of regenerants using flow cytometry. *Plant Cell Tiss Org Cult* 110:239–249

Zheng Y, Liu Y, Ma M and Xu K (2008) Increasing *in vitro* microrhizome production of ginger (*Zingiber officinale* Roscoe). *Acta Physiol Plant* 30: 513–519

Zuraida AR, Izzati KFL, Nazreena OA, Radziah CMZC, Asyikin SGSN and Sreeraman S (2014) *In vitro* regeneration of *Curcuma caesia* plantlets from basal part and via somatic embryogenesis. *Adv Biosci Biotechnol* 5:363–372

9

Role of Biotechnology in Genetic Improvement of Clitoria ternatea*: A Rare Medicinal Plant*

Ambika Gupta
Central University of South Bihar
Gaya
Bihar, India

Nitish Kumar
Central University of South Bihar
Gaya
Bihar, India

CONTENTS

9.1 Introduction

Due to growing modernization and changing life style, some traditional ways are being given up. One of these is the traditional medicinal system of India—Ayurveda, "Science of life"—which is also mentioned in the ancient Vedas and other scriptures. Ayurveda teaches us how to rejuvenate our body through diet and nutrition. Due to the

DOI: 10.1201/9781003239932-9

high cost of Western drugs and their side effects, microbial resistance scientists are focusing on medicinal plants to pave the way towards affordable medicine. India is very rich in a great variety of medicinal plants used in the traditional medicinal system. About 20,000 medicinal plants are reported in India, of which 7000 plants have so far been used in the medicinal field (Mukherjee et al., 2008).

One of the most important and well known is *Clitoria ternatea* L. of the Fabaceae family, which originated from tropical Asia and is commonly known as Asian pigeonwing. The medicinal properties of the plant have been investigated scientifically in considerable detail. In India, *C. ternatea* is traditionally known as Aprajita (Bengali), Aprajit (Hindi) and Kakktam (Tamil Naidu). Due to the presence of primary and secondary metabolites as medicinal components, the root, leaves and flowers have been used in the Ayurveda medicinal system for a long time.

9.1.1 Plant Description

The plant *C. ternatea* is a perennial twining climber that reaches 2–3 m in height. It grows in the wild and also in gardens. It bears imparipinnately compound, alternate, stipulate in leaf showing reticulate venation. It possesses five to seven leaflets, 6–13 cm long. Their stomata are sub-coriaceous, rubiaceous with wavy cell wall, and present on both upper and lower epidermis of the leaflets. The leaf shows a dorsiventral structure on transverse section. The plant shows solitary, axillary inflorescence, having a blue- or white-colored flower resembling a conch shell. The flowers are pentamerous, zygomorphic and pea shaped. The pods of *C. ternatea* are sharply beaked, flat and 5–10 cm long, having 6–11 seeds. Initially, the pods are green in color, and after maturing or ripping, they look brownish. The seeds are non-endospermous and kidney shaped, yellowish brown or blackish in color. *C. ternatea* is also a very nutritious plant; its seed contains around 500 cal/100 g and also some natural acids, including palmitic acid (19%), oleic acid (52%), stearic acid (10%), linoleic acid (17%) and linolenic acid (4%) (Joshi et al., 1981).

C. ternatea has an extensive deep root system, which is adapted to drought conditions and enables the plant to survive up to 7–8 months. The root system consists of few branches and many slender lateral roots, which grow more than 2 m long. The root is woody and produces large nodules for nitrogen fixation. The transverse section study of the root shows that the phloem is composed of 12–15 rows of thin-walled, longitudally elongated cells, some of which are compressed and some exfoliating in nature.

Although *C. ternatea* can withstand arid conditions, its germination and establishment are most favorable when the temperature is between 24 and 32 °C, and when seeds are sown in moist soil (Oguis et al., 2019).

9.1.2 Geographical Distribution

C. ternatea originated in tropical Asia, but it is now neutrally distributed and widely grown as an ornamental, fodder or medicinal plant. In India, it is widely cultivated as fodder grass in Punjab, Gujrat, Tamil Naidu, Karnataka, Uttar Pradesh and Andhra Pradesh because of the key characteristics of this plant, that is, tolerance to drought conditions, non-reliance on specific pollinators (self-pollination) and nitrogen fixation capability. It is distributed pantropically, including in Africa (Kenya, Tanzania,

Nigeria, Gambia). In America, this species is cultivated from Florida to Texas and from New Jersey to Kentucky and Arkansas. It is also widely distributed in Mexico, in the Southwestern Pacific (Fiji, Solomon Islands, New Caledonia), and in South America in Paraguay and Argentina.

9.2 Genetic Diversity in *C. ternatea*

Molecular markers are widely used in genetic improvement and analysis of interspecific or intra-specific genetic diversity. They are also employed in hybrid development to select the desired genotype and to identify identical species and the percentage of evolution from wild type. Information based on molecular markers can also be utilized in efficient management of germplasm, breeding, and conservation of medicinal and economically important plant diversity.

Ali et al. (2013) studied intra-specific genetic diversity in 11 accessions of *C. ternatea* collected from different geographical regions of India. Out of 25, only 7 random amplified polymorphic DNA (RAPD) primers amplified 72 clear bands, of which 32 bands showed phenotypic nature with 45.07%. The maximum polymorphic band (75%) was generated from primer OPN-4, followed by 60% (OPN-09) and 57% (OPN-01). OPN-10 generated the lowest polymorphism at 14.28%. The polymorphism and genetic distance in different populations were 0–0.75 and 0.02–0.28, respectively. From primer OPN-01, eight polymorphism bands were obtained, and two of them had unique bands (1.5 kb) detected in the Haryana and Uttar-Pradesh *C. ternatea* samples. The authors grouped 14 genotypes of *C. ternatea* into two clusters, cluster I and cluster II. There is less divergence value within a cluster than between the clusters. Yeotkar et al. (2011) observed polymorphism and genetic diversity between four variants: A, B, C of *C. ternatea* and D of *C. biflora*. RAPD analysis revealed 202 polymorphic fragments from 100 random primers. Primer OPF-10 generated the maximum number of bands (11). Primers OPC11 and OPB11 generated 150-bp and 1500-bp DNA fragments, respectively. The genotypes A and C of *C. ternatea* showed the highest similarity index (0.57). The dendrogram from RAPD data indicated that *C. biflora* is a distinct species from *C. ternatea*, with a number of well-defined and consistent characters. Similarly, Bishoyi et al. (2014) investigated genetic diversity in the *C. ternatea* population by using RAPD and ISSR (inter simple sequence repeat) marker. They worked on 17 accessions of *C. ternatea* population from 9 different states of India. They found 137 and 105 DNA fragments of sizes ranging from 150 to 3000 bp via 23 RAPD primer and 18 ISSR primer, respectively. RAPD analysis showed 81–97% Jaccard's co-efficient similarity, whereas this was 80–98% by ISSR analysis. There is <80% genetic similarity between *C. ternatea* populations, indicating low genetic variation.

9.3 Tissue Culture

Plant tissue culture is an applied plant biotechnology tool that involves in vitro aseptic culture of cells, tissue, organ or whole plant under controlled nutritional and environmental conditions. It has been widely employed in the areas of agriculture, forestry, horticulture and plant breeding. This technique attracts interest from researchers because of its key features, including mass propagation, virus elimination, independence from the seasons, less time and labor need, in vitro cloning of plants, etc. Improvement in regeneration methods is a prerequisite for the development of transgenic plants, enhancing the phytoconstituents of medicinally and economically important plants. In vitro generation and genetic manipulation approaches give a potential opportunity for mass propagation and genetic enhancement of plants in a limited time.

9.3.1 Direct Plant Regeneration in *C. ternatea* Explant

Regeneration of plants without going to the callus stage is known as direct regeneration. Due to the generation of genetically similar plants, it is a valuable approach for the micropropagation of economically important and endangered plant species. A number of in vitro culture studies on *C. ternatea* have been carried out to investigate its regeneration potential using direct organogenesis of shoot buds from leaf, axillary nodal part, root and cotyledonary explants (Mohamed and Taha, 2011; Rout, 2005; Shahzad et al., 2007). Mohamed and Taha (2011) studied in vitro culture of *C. ternatea* from leaf explant on Driver and Kuniyukai walnut (DKW) medium. The medium was supplemented with different plant hormones in different combinations and concentrations. Their results showed that medium containing 1 mg/l 6-benzylaminopurine (BAP) only shows good shoot proliferation with the highest number per explant, i.e. 4.03 ± 0.39, compared with the other three BAP concentrations: 0.5, 1.5 and 2 mg/l. The lowest shoot generation was observed in 1.5 mg/l BAP. For the root formation, 2.0 mg/l naphthaleneacetic acid (NAA) showed the highest root proliferation (3.17 ±_0.92). It was also proved that the combination of BAP and NAA in DKW did not show any shoot or root initiation; instead, callus was formed.

DKW with 1.0 mg/l BAP showed the best shoot induction in cotyledonary nodal explant, and for Murashige and Skoog (MS) medium, the best result was found at 1.5 mg/l BAP. In both types of medium, the addition of 0.5 mg/l indole acetic acid (IAA) did not show any results (Roy et al., 2017). Rout (2005) studied the effect of growth regulators on shoot and root multiplication on nodal explants of *C. ternatea*. They observed that buds started to elongate within 6–7 days of inoculation in MS media supplemented with varying concentrations of BAP or kinetin as compared with control containing no growth regulator. They reported that good shoot elongation occurred in MS media containing 6.6–11.1 μM BAP alone. They further reported that shoots were stunned at higher concentrations of BAP (above 17.8 μM) or kinetin (above 18.6 μM), resulting in basal callusing of the shoot. NAA showed better enhancement of shoot multiplication than IAA when supplemented with BAP-containing media. They found that the maximum percentage of multiple shoots (89%) occurred within 4 weeks of culture in medium containing 8.9 μM BAP, 1.43 μM NAA and 3% sucrose. They also observed that the rate of multiplication increased as the number of subcultures increased. A combination of NAA (1.34 μM) and IBA (0.49 μM) showed the highest rooting percentage (72.4%) in newly generated shoots. Ismail et al. (2011) generated multiple shoots via culturing the cotyledonary node of *C. ternatea* in MS media. They reported that after 1 week, a superior response of shoot initiation occurred in MS medium fortified with 5.0 μM BAP alone as compared with kinetin alone. Indole-3-butyric acid (IBA) (2.0 μM) showed a strong in vitro rooting response. Mohan et al. (2014) found that MS + BAP (2.0 mg/l) showed a high percentage of shoot regeneration as well as multiple shoot response of 4.80 ± 1.81, followed by 4.23 ± 2.01 in MS + BAP (1.5 mg/l). IBA (1.5 mg/l) showed better rooting efficiency (12.13 ± 0.116). Shahzad et al. (2007) studied in vitro micropropagation of *C. ternatea* using root segment as explant. They found that after 3 weeks of incubation, seven or eight well developed shoots (6–7 cm) appeared in medium containing 20 μM BAP. Further, the addition of 2.0 μM NAA to BAP (20 μM)-containing MS medium showed enhancement in regeneration potential. Regular removal of elongated shoots (5–8 cm) and sub-culturing them on medium fortified with 5 μM 6-benzyladenine (BA) shows better shoot elongation. For continuous differentiation and elongation of shoots, a low concentration of BA is required. Roots are achieved only in auxin-containing medium. After 4 weeks, 80% regenerated shoots showed the best rooting on MS medium containing 5.0 μM IBA. The authors also observed that more than 70% of rooted plantlets survived in soil. Kumar et al. (1993) cultured *C. ternatea* shoot tip as explant in B_5 medium containing 0.1–5 mg/l BA and induced multiple shoots. As per their report, low concentrations of BA (0.1–2 mg/l) did not show any effect on shoot induction. After 20 days, medium containing 4 mg/l BA showed the highest shoot induction as compared with 3 and 5 mg/l BA concentrations. Friable callus was formed in B_5 medium augmented with both BA and IAA. Significant root induction was shown by 3 mg/l NAA . Mukhtar et al. (2012) did a comparative study on influencing micropropagation in *C. ternatea* using Thidiazuron (TDZ), a potent plant growth regulator having cytokine-like activity. Among various levels of TDZ tested, 0.1 μM proved effective in inducing shoot regeneration (49.6 ± 3.69%), shoot length (5.8 ± 0.23 cm) and maximum number of shoots (5.2 ± 0.37) in cotyledonary node (CN) explant. The same concentration of TDZ shows less response in nodal explant: only 31.6 ± 2.54% shoot regeneration rate, 3.2 ± 0.25 cm shoot length and 2.6 ± 0.50 number of shoots. However, after 4 weeks, nodal explant shows the highest shoot regeneration frequency (45.8 ± 4.14%) with shoot length (5.0 ± 1.01) having (3.6 ± 0.24) shoot number in 1.0 μM TDZ supplemented with MS medium. They observed that irrespective of the explant type, an increase in TDZ concentration (2.5 μM) results in low shoot regeneration potential, with a low number of shoots and also a decrease in shoot length. As per their report, prolonged culture in TDZ, even at the optimum concentration, also negatively affected shoot regeneration. To overcome this, they transferred the TDZ-exposed explant after 4 weeks to MS medium lacking TDZ. This type of sub-culturing provided a good response in regeneration. After the fourth sub-culturing on control MS medium, there were 7.4 ± 0.60 shoots per CN explant and 6.8 ± 0.58 shoots per nodal explant. MS medium containing 1.0 μM IBA showed the best rooting percentage (88.4 ± 1.20%) with the highest root number (5.2 ± 0.37) per shoot. Singh and Tiwari (2011) observed that all cytokinins (BAP, kinetin, TDZ) support shoot regeneration and elongation from decapitated embryonic axes explant. Varying concentration except 1.0 mg/l of BAP, kinetin and TDZ induces shoot. Among the three cytokines, BAP shows the most efficient shoot regeneration: 2 mg/l BAP, 2 mg/l kinetin, 0.02 mg/l TDZ are the optimum concentrations, with 100%, 60% and 45% shoot regeneration frequency, respectively. Above this, the frequency and number of shoot generation both decrease. A varying range of GA_3 (0.05–1.5 mg/l) was used to elicit the shoot length. GA_3 at 1 mg/l resulted in a good effect, with 90% elongation frequency having 4.50 ± 0.22 shoots with 5.80 ± 0.27cm length. MS medium with 1.0 mg/l IBA showed 90% rooting frequency with 4.80 ± 0.24 cm length.

9.3.2 Indirect Plant Regeneration in *C. ternatea*

In plant tissue culture, organogenesis is frequently used to generate a number of de novo plants in lower numbers, although it has the drawback of genetic instability. It is a process of dedifferentiation, forming callus from explant, and then redifferentiation into plant organs like shoots, roots, leaves, buds, etc. Plant regeneration via organogenesis is a mono-polar structure.

Lakshmanan and Dhanalakshmi (1990) reported callus induction on the root of a *C. ternatea* cultured in vitro seedling on MS medium lacking growth regulator. MS + kinetin (0.1 mg/l) and MS + kinetin (0.5 mg/l) differentiate the callus on lateral root and hypocotyl after 25 days and 40 days, respectively. They found that unsplit hypocotyl produced buds in MS medium supplemented with kinetin (0.5 mg/l). Split root shows high root generation frequency and also the maximum number of shoots as compared with hypocotyl. The combination of IAA and kinetin in the same concentrations promotes shoot buds, but the rate and number are lower than with kinetin alone. Efficient rooting is shown with 0.1 or 0.5 mg/l IBA as compared with IAA at the same concentration. Kumar et al. (1993) developed callus on Gamborg's medium with stem and leaf of *C. ternatea* as explant. Both explants developed into friable white callus in 2,4-D (0.01–1 mg/l) and kinetin (0.01–2 mg/l) or 1–3 mg/l BAP alone in medium. After 40 days of culture on B_5 + 0.1 mg/l IAA and 0.01 mg/l BAP, friable callus became nodular and green. Medium supplemented with 3–4 mg/l IAA alone or in combination with 1–3 mg/l BAP was responsible for shoot bud appearance. Callus derived from different medium shows significant shoot but in different medium having different growth regulator in varying concentration. Callus of 2,4-D and kinetin medium shows shoot bud appearance in medium containing 1–2 mg/l BAP and 2–4 mg/l IAA, while callus of B_5 + (1–3 mg/l) BAP shows shoot bud appearance only in medium with 2–4 mg/l IAA alone. The frequency of shoot development is high in the callus of B_5 + 2,4-D + kinetin medium as compared with other combinations, and also, the number of shoots regenerated is maximum in 2 mg/l IAA and 2 mg/l BAP-containing medium. Along with the regeneration medium, the initial hormonal combination and the concentration at which the callus initiates plays a vital role in shoot regeneration. Jaiswal and Singh (2016) observed rapid callus growth from CN appearing in MS medium fortified with 0.5 mg/l BAP + 1 mg/l 2,4-D. Small callus was reported in MS + 1 mg/l BAP + 1 mg/l IAA. Similarly, Shahzad et al. (2007) found that MS medium containing 10 μM 2,4-D and 5 μM BAP was efficient for better callus initiation at 4 weeks. Mohamed and Taha (2011) stated that 2 mg/l BAP and 1 mg/l NAA with DKW was the best combination for callus formation, with 79.33 ± 1.51 percentage per explant; 10 μM BAP and 2.0 μM NAA was the optimum combination, on which 90% of callus formed shoots, with the highest number (8.7 ± 0.10) of shoots per culture, while 80% of shoots showed the best rooting response in MS medium augmented with 5.0 μM IBA. Rout (2005) observed that 73.33% explants produce callus in DKW + 1.0 mg/l BAP + 1.0 mg/l NAA, the best result from different combinations of the same hormone, and 80% response was found in MS medium supplemented with the same hormones at the same concentrations. The maximum shoot induction (90%) from callus was found in MS + 0.5 mg/l IAA, followed by 80% in MS + 0.25 mg/l IBA and DKW + 2.0 mg/l shows its best result with 80% among other concentrations. The best rooting (90%) was observed in DKW supplemented with 1.0 mg/l BAP+ 0.5 mg/l NAA, and 80% in MS medium fortified with 1.0 mg/l BAP + 0.5–1.0 mg/l NAA.

9.3.3 Embryogenesis in *C. ternatea*

Somatic embryogenesis is a process of in vitro embryo formation from a single somatic cell. There is no endosperm or seed coat around the somatic embryo. Firstly, the source tissue is cultured to form an undifferentiated mass of cell called callus. Plant growth regulators in different concentrations are used to induce embryo from callus. Auxins are commonly used plant growth regulators (PGRs) for this purpose. This method reduces the culturing on multiple types of medium for shoot and root formation, as somatic embryos are bipolar in nature, allowing them to form a whole plant in a single type of medium. Dhanalakshmi and Lakshmanan (1992) reported in vitro plant regeneration occurring in the callus stage in *C. ternatea*. They observed that the somatic embryo developed indirectly in basal medium via callus of mature embryo as explant.

Nair and Reghunath (2009) found that leaf-derived callus from MS + 0.5 mg/l 2,4-D + BA 0.1 mg/l exhibited embryogenic potential after 4 weeks of culture. Within 45 days, MS medium augmented with 0.5–1.0 mg/l BAP promoted the development of globular, heart and torpedo stages of embryo. BAP (1.0 mg/l) induced 91.67% somatic embryo, followed by 83.33 at 0.5 mg/l BAP. Combination of BAP with coconut milk or NAA produced globular-shaped embryo only. A high concentration of BAP (2.0–3.0 mg/l) did not show any somatic embryo even with an increase in coconut milk level. A similar result was obtained when BAP was combined with other supplements, including gibberellic acid (GA), glutamine, activated charcoal and abscisic acid. GA (0.5–1.0 mg/l) did not promote embryogenesis, but in combination with NAA (1.0 mg/l), induced the development of globular and heart-shaped embryo. Shoot and root initials were developed after inoculating the torpedo-shaped embryo on germination medium. Kumar and Thomas (2012) observed that 75% cultured cotyledon showed callus on MS + 2.0 mg/l 2,4-D. MS medium supplemented with varying concentrations (1.0–4.0 mg/l) of BAP or kinetin along with 0.5 mg/l NAA was used as transfer medium for somatic embryo formation. BAP (2.0 mg/l) with 0.5 mg/l NAA showed the optimum response for somatic embryo (60%) with 22 embryos per 1.0 g of callus. Similarly, kinetin at 3.0 mg/l with 0.5 mg/l NAA showed 33% response with 19 embryos. Within 2 weeks, easily separable small, green, globular embryos developed from calli. The torpedo stage was observed within 1 week and then bipolar structure. After 4 weeks, individual embryos reached a size of 2 cm and developed with a root and shoot system. Along with auxin and cytokinin, sucrose also plays a vital role in improving the shape and number of somatic embryos: 4% sucrose in MS medium with 2.0 mg/l BAP and 0.5 mg/l NAA produces a high response in globular embryos. Abscisic acid at a level of 3.0 mg/l increases the somatic embryo response from 37.3% to 80% per gram of callus.

9.4 Genetic Transformation in *C. ternatea*

9.4.1 *Agrobacterium*-mediated Genetic Transformation in *C. ternatea*

In plant biotechnology, *Agrobacterium tumefaciens* is used as a potential tool to transfer a desired gene into a host. Genetic engineering in plants relies heavily on this approach because of low re-arrangement of transgenes, cost-effectiveness, easy culturability and manageability in the lab, transfer of relatively large DNA fragments, etc. Few studies are reported on genetic transformation in *C. ternatea* using the *Agrobacterium*-mediated approach (Buddharak and Chundet, 2009; Swain et al., 2012a and Swain et al., 2012b).

Buddharak and Chundet (2009) developed a protocol for flavonoid 3′ hydroxylase (F3′H) gene transformation in leaf explant of *C. ternatea* via *A. tumefaciens*. Successful transformation was obtained via the AGLO strain of *A. tumefaciens* containing the pBI121/C_uDFR plasmid. MS + 10 mg/l BAP and 0.1 mg/l IAA were used for a shoot regeneration medium for leaf explant. Luria-Bertani medium containing 100 mg/l kanamycin and 50 mg/l rifampicin was used for culturing *A. tumefaciens* at 28 °C. $OD_{600\ nm}$ = 0.6–0.8 of *A. tumefaciens* was optimum for efficient transformation. To enhance the transformation efficiency, MS20 medium containing 20% sucrose and supplemented with 100 μM acetosyringone and 0.1 M betain hydrochloride was used. They found that 2 days co-culturing of leaf explant and *A. tumefaciens* was best for transformation. On the basis of β-glucuronidase (GUS) assay, *C. ternatea* shows 20% transformation efficiency with C_uDFR gene.

Similarly, Swain et al. (2012) developed an efficient protocol for a transformed *C. ternatea* rhizoclone for rapid and large-scale production of hairy root using a virulent strain of *A. rhizogenes*. They also optimized the key factors influencing transformation efficiency. They observed that explants from outside plants were more susceptible and required less time than in vitro *C. ternatea*. Out of two types of explant nodal segment, internode and leaf, internode shows more response. Strain A4T of *A. rhizogenes* shows better emergence of hairy roots at the infection site as well as high transformation frequency as compared with other strains used, including A4, 8196 and LBA9402. Strain 8196 gave zero response in hairy root induction. Bacterial growth phase and cell density play a vital role in influencing the transformation efficiency. The authors found that late log phase at an optical density (OD) of 0.6 at 660 nm with cell density of 10^9 cells/ml was most effective for transformation in *C. ternatea*. At this stage transformation occurred with A4T (85.8%) followed by A4 (78.3%) and LBA9402 (72.4%). Acetosyringone-treated explant showed early emergence of hairy root from the infected site escaping the nodular intermediate. Irrespective of *A. rhizogenes* strain, 100 μM was the optimum concentration of acetosyringone for the most effective result. Of the three different methods used for infecting the explant with *A. rhizogenes*, the method of creating a wound on the explant and dipping in solution was most productive and gave a high percentage of transformants (85.8%) with A4T strain. This was followed by 80.3% via the direct inoculation method. No transformants were observed with the unwound method. Along with the infection method, the co-cultivation period also affected the efficiency. Five days of co-cultivation was optimum for maximum transformation with 85% via A4T strain, followed byA4 and LBA9402 strains with 78.3% and 72.4%, respectively. To establish the transformed root, MS0 medium solidified with 0.6% agar was optimized. Transformed root shows a slow growth rate in liquid MS0 medium compared with the solidified version. MS0 supplemented with 0.25 mg/l IBA increased the biomass of transformed root clone. Exceeding the IBA concentration triggered callus formation.

Swain et al. (2012) again worked on gene transformation of *C. ternatea* to enhance the production of the anti-cancer compound texaxerol using *Agrobacterium*-

transformed root culture. Southern blotting was done for individual selected transformed rhizoclones for molecular confirmation of T_LDNA and T_RDNA genes. They observed that texaxerol yield was maximum (13.3 mg/g) in transformed rhizoclone HRL1A3induced by A4T strain as compare with the wild type containing 3.33 mg/g.

9.5 Omic Technologies in *C. ternatea*

To understand the pharmaceutical potential of phytocomponents and their regulation, it is important to identify and characterize the responsible genes, mRNA and proteins of *C. ternatea*. An increase in the level of gene expression is essential to enhance the medicinally important phytoconstituents. It will also be economical for pharmaceutical companies to fulfill the increasing demands of a rapidly growing population.

9.5.1 Identified Genes in *C. ternatea*

Togami et al. (2006) performed molecular characterization of F3′5′H gene of *C. ternatea* and analyzed its function in transgenic *Verbena*. They demonstrated that F3′5′H is flavonoid 3′, 5′-hydroxylase, a key gene determining flower (Violet). As compared with wild-type *Verbena* F3′5 ′H gene, *C. ternatea* F3′5′H gene yielded more delphinid compound under the control of CMV 35S promoter. They found that *C. ternatea* F3′5′H gene had strong phenotype-changing potential and altered the color of transgenic *Verbena* towards violet. Wunnakup et al. (2018) studied the effect of homogentisate phytyltransferase gene from *C. ternatea* (*Ct*HPT) on alpha-tocopherol (vitamin E) in tomato leaves. The result showed that the transient expression of *Ct*HPT gene increases the alpha-tocopherol level along with phytol and various fatty acids. *Ct*HPT expression also led to chlorophyll deficiency in tomato leaves. With this study they also confirmed that HPT is a key enzyme in tocopherol biosynthesis, which has anti-cancer properties. Poth et al. (2011) obtained a *Cter*M precursor gene that encodes cyclotides having a diverse range of pharmaceutically important activities. They also sequenced the gene using tandem mass spectrometry.

9.5.2 Identified Proteins in *C. ternatea*

Hiromoto et al. (2015) found that anthocyanidin-3-*O*-glucosyltransferase (UGT78K6) of *C. ternatea* acts as a catalyst and transfers glucose from UDP-glucose to anthocyanidin (delphinidin), a natural color pigment. Kelemu et al. (2004), using ultrafiltration with a centricon-3 membrane tube and granulated bed isoelectric focusing technique, identified an antimicrobial and insecticidal protein named ‘finotin’ from the seed source. They observed that finotin has a broad and potent inhibitory effect on the growth of fungal pathogens (*Fusarium solani, Bipolaris oryzae*) and a bacterium (*Bacillus subtilis*). Ajesh and Sreejith (2014) isolated a novel 14 KDa protein from seeds from *C. ternatea* and designated it as *Ct* protein. The protein shows antifungal properties against various species, including *Candida albicans*, *Cryptococcus albidus*, *Aspergillus niger*, *A. flavus*, etc. Similarly, Naeem et al. (2007) purified and characterized another novel protein, ‘lectin’, specific to β-D-galactosides. They also studied the structure of lectin, composed of two identical sub-units having a molecular weight of 34.7 KDa, obtained a complete sequence of it, and found it to be homologous to S-64 protein of *Glycine max.*

9.6 Conclusion and Future Outlook

In an era of rapidly growing modernization and changing life style, there is a transformation from a traditional, rural, agrarian society to a secular, urban, industrial society, which results in a complex trap of a range of diseases, from common (cough) to lethal (cancer). *Clitoria ternatea* is a promising medicinal plant, as it has a number of potential pharmacological properties, including nootopic, anti-stress, anti-Alzheimer's, anti-diabetic, etc. These features are only present because of its secondary metabolites, such as cyclotides, flavonoids, triterpenoids, ternatin, etc. Not only its medicinal features but also its nutritive value, nitrogen fixing capacity and negligible toxicity have made it a universally accepted important medicinal plant. A number of scientific studies tending to confirm these medicinal properties have expanded its use from traditional to modern-day. However, work is still needed in respect to its genetic transformation, molecular markers, genes responsible for secondary metabolites, etc. to develop novel and value-added variety. These studies would also help in enhancing pharmaceutical plant extraction and its use in the medicinal field at the commercial level.

REFERENCES

Ajesh, K., & Sreejith, K. (2014). A novel antifungal protein with lysozyme-like activity from seeds of *Clitoria ternatea. Applied Biochemistry and Biotechnology*, *173*(3), 682–693.

Ali, Z., Ganie, S. H., Narula, A., Sharma, M. P., & Srivastava, P. S. (2013). Intra-specific genetic diversity and chemical profiling of different accessions of *Clitoria ternatea* L. *Industrial Crops and Products*, *43*, 768–773.

Bishoyi, A. K., Pillai, V. V., Geetha, K. A., & Maiti, S. (2014). Assessment of genetic diversity in *Clitoria ternatea* populations from different parts of India by RAPD and ISSR markers. *Genetic Resources and Crop Evolution*, *61*(8), 1597–1609.

Buddharak, P., & Chundet, R. (2009). Isolation and characterization of flavonoid 3' hydroxylase (F3'H) gene and genetic transformation in butterfly pea (*Clitoria ternatea* Linn.) via *Agrobacterium tumefaciens. Acta Hortic*, *836*, 247–253. doi: 10.17660/ActaHortic.2009.836.35

Dhanalakshmi, S., & Lakshmanan, K. K. (1992). In vitro somatic embryogenesis and plant regeneration in *Clitoria ternatea. Journal of Experimental Botany*, *43*(2), 213–219.

Hiromoto, T., Honjo, E., Noda, N., Tamada, T., Kazuma, K., Suzuki, M., Blaber, M., & Kuroki, R. (2015). Structural basis for acceptor-substrate recognition of UDP-glucose: Anthocyanidin 3-O-glucosyltransferase from *Clitoria ternatea. Protein Sci*, *24*, 395–407. doi:10.1002/pro.2630

Ismail, N., Rani, U., Sharma, M. M., & Batra, A. (2011). Influence of plant growth regulators on in vitro shoot regeneration via cotyledonary node in *Clitoria ternatea* L. *International Journal of Pharmaceutical Sciences and Research*, *2*(3), 552.

Jaiswal, A., & Singh, S. P. (2016) In-vitro micropropagation of a medicinal plant: Aparajita (*Clitoria ternatea* L.). *International Journal of Agricultural Science and Research*, *6*(5), 117–122.

Joshi, S. S., Shrivastava, R. K., & Shrivastava, D. K. (1981). Chemical examination of *Clitoria ternatea* seeds. *Journal of the American Oil Chemists' Society*, *58*(6), 714.

Kelemu, S., Cardona, C., & Segura, G. (2004). Antimicrobial and insecticidal protein isolated from seeds of *Clitoria ternatea*, a tropical forage legume. *Plant Physiology and Biochemistry*, *42*(11), 867–873.

Kumar, G. K., & Thomas, T. D. (2012). High frequency somatic embryogenesis and synthetic seed production in *Clitoria ternatea* Linn. *Plant Cell, Tissue and Organ Culture (PCTOC)*, *110*(1), 141–151.

Kumar, K. P., Soniya, E. V., Lawrence, B., & Nair, G. M. (1993). Micropropagation of *Clitoria ternatea* L. (Papilionaceae) through callus regeneration and shoot tip multiplication. *Journal of Spices and Aromatic Crops*, *2*(1), 41–46.

Lakshmanan, K. K., & Dhanalakshmi, S. (1990). Callus, organogenesis and plantlet formation in tissue cultures of *Clitoria ternatea*. *Annals of Botany*, *66*(4), 451–455.

Mohamed, N., & Taha, R. M. (2011). Plant regeneration of *Clitoria ternatea* from leaf explants cultured in vitro. *Journal of Food, Agriculture & Environment*, *9*(3&4), 268–270.

Mohan, C., Manjula, P., Kumar, B. K., Naresh, B., Ramadevi, B., & Devi, B. P. (2014). In vitro propagation of *Clitoria ternatea* L., a valuable medicinal plant. *Trends in Biosciences*, *7*(18), 2698–2701.

Mukherjee, P.K., Kumar, V., Kumar, N. S., & Heinrich, M. (2008). The Ayurvedic medicine *Clitoria ternatea*—from traditional use to scientific assessment. *Journal of ethnopharmacology*, 120(3), 291–301.

Mukhtar, S., Ahmad, N., Khan, M. I., Anis, M., & Aref, I. M. (2012). RETRACTED ARTICLE: Influencing micropropagation in *Clitoria ternatea* L. through the manipulation of TDZ levels and use of different explant types. *Physiology and Molecular Biology of Plants*, *18*(4), 381–386.

Naeem, A., Haque, S., & Khan, R. H. (2007). Purification and characterization of a novel β-D-galactosides-specific lectin from *Clitoria ternatea*. *The Protein Journal*, *26*(6), 403–413.

Nair, D. S., & Reghunath, B. R. (2009) Cryoconservation and regeneration of axillary shoot meristems of *Indigofera tinctoria* (L.) by encapsulation–dehydration technique. In Vitro Cellular & Developmental Biology-Plant, *45*(5), 565–573.

Oguis, G. K., Gilding, E. K., Jackson, M. A., & Craik, D. J. (2019). Butterfly pea (*Clitoria ternatea*), a cyclotide-bearing plant with applications in agriculture and medicine. *Frontiers in Plant Science*, *10*, 645. doi: 10.3389/fpls.2019.00645

Poth, A. G., Colgrave, M. L., Lyons, R. E., Daly, N. L., & Craik, D. J. (2011). Discovery of an unusual biosynthetic origin for circular proteins in legumes. *Proceedings of the National Academy of Sciences*, *108*(25), 10127–10132.

Rout, G. R. (2005). Micropropagation of *Clitoria ternatea* Linn. (Fabaceae)—An important medicinal plant. *In Vitro Cellular & Developmental Biology-Plant*, *41*(4), 516–519.

Roy, C. K., Mohiuddin, T. M., & Rahman, S. S. (2017). Establishment of in vitro micropropagation and callus formation protocols for large scale production of *Clitoria ternatea* L. *Zeitschrift für Arznei-und Gewürzpflanzen (Journal of Medicinal and Spice Plants)*, *22*(3), 136–144.

Shahzad, A., Faisal, M., & Anis, M. (2007). Micropropagation through excised root culture of *Clitoria ternatea* and comparison between in vitro–regenerated plants and seedlings. *Annals of Applied Biology*, *150*(3), 341–349.

Singh, J., & Tiwari, K. N. (2011). In vitro plant regeneration from decapitated embryonic axes of *Clitoria ternatea* L.—an important medicinal plant. *Industrial Crops and Products*, *35*(1), 224–229.

Swain, S. S., Rout, K. K., & Chand, P. K. (2012a). Production of triterpenoid anti-cancer compound taraxerol in Agrobacterium-transformed root cultures of butterfly pea (*Clitoria ternatea* L.). *Applied Biochemistry and Biotechnology*, *168*(3), 487–503.

Swain, S. S., Sahu, L., Pal, A., Barik, D. P., Pradhan, C., & Chand, P. K. (2012b). Hairy root cultures of butterfly pea (*Clitoria ternatea* L.): Agrobacterium× plant factors influencing transformation. *World Journal of Microbiology and Biotechnology*, *28*(2), 729–739.

Togami, J., Tamura, M., Ishiguro, K., Hirose, C., Okuhara, H., Ueyama, Y., ... & Fukui, Y. (2006). Molecular characterization of the flavonoid biosynthesis of *Verbena hybrida* and the functional analysis of verbena and *Clitoria ternatea* F3′ 5′ H genes in transgenic verbena. *Plant Biotechnology*, *23*(1), 5–11.

Wunnakup, T., Vimolmangkang, S., & De-Eknamkul, W. (2018). Transient expression of the homogentisate phytyltransferase gene from *Clitoria ternatea* causes metabolic enhancement of α-tocopherol biosynthesis and chlorophyll degradation in tomato leaves. *Journal of Plant Biochemistry and Biotechnology*, *27*(1), 55–67.

Yeotkar, S. D., Malode, S. N., Waghmare, V. N., & Thakre, P. (2011). Genetic relationship and diversity analysis of *Clitoria ternatea* variants and *Clitoria biflora* using random amplified polymorphic DNA (RAPD) markers. *African Journal of Biotechnology*, *10*(79), 18065–18070.

10

Molecular Clonal Fidelity Assessment of Micropropagated Orchids Using DNA Markers

Paonam Sonia
Manipur University
Canchipur
Manipur, India

Nandeibam Apana
Manipur University
Canchipur
Manipur, India

Leimapokpam Tikendra
Manipur University
Canchipur
Manipur, India

Abhijit Dey
Presidency University
Kolkata, India

Imlitoshi Jamir
Nagaland University
Dimapur, India

Potshangbam Nongdam
Manipur University
Canchipur
Manipur, India

CONTENTS

DOI: 10.1201/9781003239932-10

10.1 Introduction

Orchids represent one of the most advanced and largest families of angiosperms with over 25,000 species and innumerable hybrids and varieties (Arditti, 1992). They have incredible floral appeal due to their stunning flowers with a wide range of floral shapes, sizes, coloration, and fragrances (Chongtham et al. 2006). Orchids are cultivated as cash crops in Thailand, Malaysia, Japan, and the USA, as they are regarded as one of the most expensive ornamentals. Orchid hybrids are in great demand in the international floriculture trade due to their variedly colored attractive flowers with long shelf life (Hsiao et al. 2011). In addition to their high ornamental values, orchids are also primarily applied in traditional medicines due to their rich contents of alkaloids, glycerides, and other valuable phytochemicals (Gutiérrez, 2010). The therapeutic values of orchids have been extensively utilized in the indigenous medicine system to treat many ailments. Despite rich orchid natural resources, their populations are dwindling at an alarming rate, mainly due to excessive unregulated commercial collection and mass habitat destruction. In the last few years, an increasing number of orchid species have been threatened with the danger of extermination. Many of them are prominently featured in the Red Data Book of the International Union of Conservation of Nature and Natural Resources (IUCN). The entire family is now listed in Appendix-II of the Convention on International Trade in Endangered Species of Wild Fauna and Flora (CITES) (Chugh et al. 2009). Efficient conservation strategies must be devised to save these valuable plants from the brink of extinction and harness their wide range of economic potential. Orchid micropropagation through tissue culture techniques provides an excellent opportunity for rapidly propagating them at a larger scale without sacrificing the mother plants (Nongdam and Chongtham, 2011). However, the micropropagated orchids must be adequately screened using sophisticated DNA markers for their clonal fidelity, as they are often associated with molecular and phenotypic defects. The goal of successful micropropagation is to rapidly mass propagate plants by preserving the genetic integrity of the natural mother plants. DNA markers have been employed successfully for clonal fidelity assessment of several in vitro propagated orchids (Antony et al. 2015; Bhattacharyya et al. 2017; Khor et al. 2020; Tikendra et al. 2021a). This chapter discusses orchid micropropagation using different explants and the application of varied DNA markers in determining their clonal fidelity.

10.2 Micropropagation of Orchids

10.2.1 Factors Influencing Orchid Propagation

The success of orchid micropropagation depends on several factors that influence the in vitro culture growth and response. Culture media and plant growth regulators (PGRs) are critical factors for effective in vitro orchid propagation. The media provide the required nutrients for the growing plant tissues. MS (Murashige and Skoog, 1962), M (Mitra et al. 1976), KC (Knudson, 1946), VC (Vacin and Went, 1949), and B5 (Gamborg et al. 1968) are some of the primary media, which can be modified to suit the requirement of a particular species. Several organic additives can also be incorporated into standard culture media to improve the nutritional content. Media can be supplemented with different combinations and concentrations of PGRs to enhance culture growth and development. Auxins are mainly responsible for root induction, expansion, and multiplication, while cytokinins function in shoot initiation, proliferation, and plant regeneration (Agbadje et al. 2021). Though growth hormones act primarily as enhancers of tissue development, they can also adversely affect some species. Auxins retard the embryo formation on leaf explants of *Oncidium* 'Gower Ramsey' (Chen and Chang, 2001), while cytokinins in high concentration induced somaclonal variation in *Dendrobium fimbriatum* (Tikendra et al. 2021a).

Choosing the proper explant to initiate culture is crucial to effective orchid propagation. Seeds, shoot and root tips, leaf, nodal segments, rhizome, pseudobulb, and inflorescences are important plant parts employed as explants for orchid micropropagation. The responsiveness of the explants depends largely on juvenility (Kaur and Bhutani, 2009), their location, position, size, and orientation (Wu et al. 2012), and season of collection (Do et al. 2019). Juvenile tissues are preferable to matured ones as they have more regeneration capacity than the differentiated tissues (Kaur and Bhutani, 2009). In vitro raised plant-derived explants are favorable because the chances of contamination are lower than in wild-grown plants. Microbial contamination of the culture can be controlled by proper surface sterilization of explants with chemical sterilants (Teixeira da Silva et al. 2016). However, as the sterilants are toxic to the plant tissues, there must be a balance between reducing infection and the survivability of the explant (Tikendra et al. 2021b). Selecting a proper sterilant concentration and treatment duration is vital for effective contamination control leading to successful orchid culture. At varying concentrations and treatment periods, sodium hypochlorite, mercuric chloride, calcium hypochlorite, and bromine water are commonly used for explant surface sterilization. Antimicrobial agents can also be incorporated into the media to treat endogenous microbes, which surface sterilization cannot remove.

10.2.2 Orchid Propagation from Different Explants

Several orchids have been micropropagated using different explants. Orchid seeds, which are very small, dust-like, and non-endospermic, germinate poorly in nature. Since the formulation of KC medium for asymbiotic orchid seed germination, several orchids have been successfully propagated in vitro using seeds (Nongdam et al. 2006; Wu et al. 2014; Srivastava et al. 2015; Utami and Hariyanto, 2019; Kim et al. 2021). Orchid seeds, after germination, formed protocorms, which subsequently developed into complete seedlings with leaves and roots (Figure 10.1). However, the extent of seed germination varied with the media types and different concentrations of PGRs employed. Paul et al. (2012) found MS medium to be the most effective for seed germination of *Dendrobium hookerianum*. Nongdam and Tikendra (2014) observed the highest seed germination of *Dendrobium chrysotoxum* in Mitra medium supplemented with 2.0 mg/L benzylaminopurine (BAP) and 2.0 mg/L indole-3-acetic acid (IAA). Not all the germinated seeds developed into seedlings, as some protocorms morphogenetically differentiated into callus tissues. David et al. (2015)

FIGURE 10.1 Micropropagation of *Dendrobium* orchids: (A) In vitro seed germination; (B) protocorm formation; (C) shooting initiation from protocorms; (D) in vitro shoot and leaf development; (E) shoot multiplication upon sub-culturing onto fresh media; (F) pseudobulb formation (red arrow); (G) in vitro root proliferation; (H) mature in vitro propagated orchids with well-developed leaves and roots; (I) hardening of *Dendrobium* orchids with potting mixture containing coconut husk, bricks and charcoal pieces (1:1:1).

used three types of basal media, KC, MS, and VW, to investigate superiority in seed germination of *Vanda helvola*. KC medium proved to be most favorable, but supplementation with organic additives like tomato juice, coconut water, peptone, and yeast extract at different concentrations affected the seed germination percentage. The germination was enhanced with 15% tomato juice, while a reduction took place with yeast extract and coconut water in the medium.

Ever since the production of protocorm-like bodies (PLBs) from *Cymbidium* leaf culture by Wimber (1965), several orchids have been successfully propagated in vitro using leaf explants (Arditti and Krikorian, 1996; Khoddamzadeh et al. 2011; Chookoh et al. 2019; Bhowmik and Rahman, 2020a). Propagation using leaves is highly advantageous, especially to monopodial orchids, as the sacrifice of the mother plant is not required (Hardjo and Savitri, 2017). Also, the availability of leaf explant is not restricted to any season. Naing et al. (2011) observed early and better responses to regeneration by younger leaves of *Coelogyne cristata* compared with matured ones. Devi et al. (2013) noticed only the leaf base of the explant producing calli successfully, while other parts failed to respond, in half-strength MS medium supplemented with various concentrations of BAP (0.5–8.0 mg/L) and 0.1–4.0 mg/L thidiazuron (TDZ). Balilashaki and Ghehsareh (2016) found the combination of 15.0 mg/L BAP and 3.0 mg/L napthaleneacetic acid (NAA) favorable for maximum development of PLBs in *Phalenopsis amabilis* var. 'Manila'. BAP at a lower concentration (1.5 mg/L) inhibited the regeneration response in *Rhynchostylis gigantea* but promoted shoot bud formation at a higher concentration (2.0 mg/L) (Pathak et al. 2017). The leaf explants did not show any proliferation and died in the absence of PGRs in *Coelogyne flaccida* (De and Sil, 2015).

Morel (1960), after propagating *Cymbidium* using apical meristem (part of the shoot tip) for the first time, stated that 'very often the PLBs divided into a clump of four to five identical structures, each of them producing a new plant' (Yam and Arditti, 2017). Since then, several workers have applied shoot tip culture to micropropagate different orchids (Pant and Thapa, 2012; Borah et al. 2015; Winarto and Samijan, 2018; Ma et al. 2020). Do et al. (2019) stressed the shoot tip collection season of *Paphiopedilum* as one of the crucial factors for explant survival. Preeta et al. (2017) examined the effect of explant structure (a longitudinally bisected and an intact shoot tip) of *Phaleanopsis* hybrid culture on a medium supplemented with different TDZ concentrations. The bisected shoot tips proved to be most favorable for PLB and shoot induction and their proliferation. Devi et al. (2013) used shoot tip explants to study the influence of PGRs on in vitro culture development of *Aerides odorata*. A high percentage of callus formation was noticed in medium supplemented with 2.0 mg/L NAA. In comparison, medium enriched with 2.0 mg/L NAA and 4.0 mg/L BAP gave the most elevated axillary shoot formation. The synergistic effect of BAP and NAA (2.0 mg/L BAP + 0.5 mg/L NAA) on shoot growth and multiplication was also reported in *Dendrobium densiflorum* when shoot tip explants were employed (Pradhan et al. 2013).

Kerbauy (1984), for the first time, reported the generation of PLBs directly from root tips of *Catasetum* without intervening callus formation. MS medium was the most favorable for PLB regeneration in *Catasetum pileatum*. But supplementation with additives like bacto-peptone and activated charcoal (AC) further enhanced the frequency of PLB regeneration. Kraus and Kerbauy (1992) demonstrated the importance of explant size, as PLB formation was reduced with increased root explant dimension. Location of isolated explants, maturity of the donor tissues, and PGR types significantly influenced the regenerative potential of the root explants. Sharma (2012) reported proximal root explants of *Cattleya* 'Almakee' responding more positively to PGRs, with PLB initiation noticed in Mitra medium incorporating 1.0 mg/L kinetin (KN) and 1.0 mg/L NAA. Sharma (2019a) further demonstrated the proximal region of root explants in *Rhynchostylis gigantea* reacting more positively to regeneration depending on the concentration of (BAP/KN) incorporated. Mitra medium augmented with 1.0 mg/L each of KN and NAA initiated PLB production in 3 weeks. However, indirect somatic embryogenesis was mediated in 12.5% of explants when KN was replaced with 1.0 mg/L BAP. The increasing concentration of BAP showed a marginal regenerative response, but KN impaired it, leading to the formation of fewer plantlets.

Rotor (1949) first initiated the orchid inflorescence segment as an explant in *Phalaenopsis* culture. As with leaf segment culture, the donor mother plant is not sacrificed, but there is the limitation of being able to assess the explant (flower stalk) only in the flowering season. PLB production from the flower stalk of *Epidendrum radicans* was observed only in medium enriched with 0.5 μM TDZ. The lowest response to PLB proliferation was noticed in the absence of sucrose. Chen and Chang (2000) also reported the enhanced production of somatic embryos and shoot buds in *Oncidium* Sweet Sugar with an increased concentration of TDZ. Martin and Madassery (2006) observed the highest percentage of shoot induction in both hybrid *Dendrobium* sonia 17 and *Dendrobium* sonia 28 with 6.97 μM KN in the medium. AC also helped in increasing root production in both the hybrids, but not in a dose-dependent manner. Nuraini and Shaib (1992) employed the scape nodes of *Oncidium*, *Dendrobium*, and *Phaleanopsis* to investigate the effect of explant age on shoot induction. The percentage of shoot induction was higher with younger scapes (fully formed flower bud stage) than with scapes from whole-bloomed flower stalks or scapes with intact flower buds. Brief notes on the various investigations of orchid micropropagation using different explants are given in Table 10.1.

10.3 Clonal Fidelity Assessment of Micropropagated Orchids

10.3.1 Somaclonal Variation

Somaclonal variation is the genetic variation noticed among plants regenerated from in vitro culture of any plant tissues (Larkin and Scowcroft, 1981). The genetic variability in the in vitro clones might be due to epigenetic effects or alterations in the genome of developing cells, as they are unceasingly confronted with different stresses during tissue culture (Tikendra et al. 2021b). Somaclonal variation may also appear among micropropagated plantlets due to long-term maintenance of cells and tissues in the micro-environment, prolonged exposure to high concentrations of synthetic hormones, and stress inflicted on the cells during frequent sub-culturing (Rani and Raina, 2000; Jain, 2001; Tikendra et al. 2021c). The explant types used in the tissue culture also have a vital role in inducing somaclonal variations in plants propagated in vitro (Salvi et al. 2001). The involvement of the callus phase in plant regeneration is more prone to genetic variation than plants developed through the organized meristematic tissues like axillary buds and shoot tips (Rani and Raina, 2000; Bhatia et al. 2009). Duncan (1997) also reported more variation from differentiated tissues of roots, leaves, and stems than from explants with pre-existing meristems due to the existence of the intermediate callus phase. Plants propagated via somatic embryogenesis (SE) are expected to be more stable than those regenerated through the organogenesis pathway, as DNA methylation is lower in the early stages of SE development (Vázquez, 2001). The stresses encountered by the cultured cells and tissues, which disrupted the normal cell development, were considered important factors for induction of somaclonal variation (Cassels and Curry, 2001). PGR-induced stress is regarded as one of the most important contributory factors for causing genetic variations among the regenerants (Stover, 1987). The growth regulators generate a somaclonal variation by affecting cell division (Gao et al. 2010), inducing rapid, disorganized growth (Karp, 1995), and changing the DNA methylation level (Vanyushin, 1984). The plant hormones at a particular concentration may be toxic, and in combination with other nutrient components, may become mutagens, producing variability among the regenerants (D'Amato, 1985). Tikendra et al. (2021a) demonstrated the appearance of somaclonal variation under high BAP concentration in micropropagated *Dendrobium fimbriatum*. Trujillo and Garcia (1996) also showed the induction of high variability in banana

TABLE 10.1

Micropropagation of Orchids Using Different Explants

Name of the species	Explant type	Culture media with PGR combinations producing optimal growth response	References
Dendrobium officinale	Seeds	Half-strength MS medium with 2.0 mg/L BA (Benzyl adenine) + 0.1 mg/L NAA + 100 g/L potato extract provided the optimal proliferation medium for *D. officinale*, and half-strength MS with (0.2 mg/L BA + 1.0 mg/L NAA) gave 100% rooting.	Chen et al. (2014)
Renanthera imschootiana	Seeds	Seed germination reached 93.1% on quarter-strength MS medium containing 0.5 mg/L NAA, 20% coconut water (CW), 1.0 g/L peptone, 10 g/L sucrose, and 1.0 g/L activated charcoal (AC). Quarter-strength MS medium with 1.0 mg/L NAA, 1.0 g/L peptone, 100 g/L banana homogenate (BH), and 1.0 g/L AC was suitable for plantlet formation, and 95.67% of plantlets developed from PLBs within 60 days of culture.	Wu et al. (2014)
Caladenia latifolia	Seeds	Half-strength MS with 5% CW (half MS E) showed >90% germination compared with symbiotic germination of *C. latifolia* seeds in modified oatmeal agar (OMA). However, the addition of PGRs to half MS E produced a negative effect on germination.	Bustam et al. (2014)
Cymbidium dayanum	Seeds	High seed germination observed in 0.2% AC incorporated Mitra and MS medium without PGRs. NAA at 0.7 mg/L in Mitra medium produced maximum seed germination (94.58%). Seed germination improved greatly when AC was present. MS with 1.2 mg/L of NAA and 0.2% AC recorded highest rooting (5.32 ± 0.35). Best shoot formation (5.91 ± 0.96) was noticed in Mitra medium supplemented with 1.2 mg/L of BAP and 0.2% AC.	Nongdam and Chongtham (2012)
Dactylorhiza hatagirea	Seeds	Maximum germination was attained on Lindeman orchid medium (37.12%) within 17 days of culture. The maximum number of shoots (18.12 ± 0.3), highest shoot length (17.80 cm ± 2.16), maximum root number (8.25 ± 0.69), and most extended root length (8.02 cm ± 1.45) were observed on MS medium with 3 mg/L IBA and 1 mg/L KN.	Warghat et al. (2014)
Dendrobium chrysotoxum	Seeds	Maximum seed germination (98%) was found on M medium with 2 mg/L BAP + 2 mg/L IAA + 0.4% AC in 2 weeks of culture. In 10 weeks, the highest of 4 leaves was observed in M + 2 mg/L KN + 0.5 mg/L NAA, and a maximum root number of 6 per plantlet was noticed with MS + 3.0 mg/L KN + 1.5 mg/L NAA.	Nongdam and Tikendra (2014)

(continued)

TABLE 10.1 (Continued)

Micropropagation of Orchids Using Different Explants

Name of the species	Explant type	Culture media with PGR combinations producing optimal growth response	References
Aerides ringens	Seeds	KC supplemented with 4.44 µM BAP and 500 mg/L peptone exhibited the best seed germination (89.28 ± 3.42%), producing a compact protocorm of 1.89 ± 0.38 mm size. KC with 9.3 µM KN made a maximum shoot (4.40 ± 2.20) per segment of 3.05 ± 0.46 cm length. KC with 5.71 µM IAA generated 4.44 ± 1.61 strong and stout roots of 3.22 ± 0.40 cm length per plantlet.	Srivastava et al. (2015)
Cyrtopodium saintlegerianum	Seeds	100% of seeds germinated in a KC medium with 3 g/L AC, while only 30% germinated without AC. KC with 2 mg/L BA produced the maximum shoot number (15.20) with a shoot length of 1.09 cm. The highest root number (8.30) with a root length of 1.69 cm was noticed in KC medium supplemented with 0.5 mg/L BA + 1.0 mg/L NAA.	Rodrigues et al. (2015)
Dendrobium thyrsiflorum	Seeds	Maximum shoot production (3.83 ± 0.48) was achieved in M medium fortified with 1.0 mg/L KN + 2.5 mg/L IAA, and the highest root formation (6.52 ± 0.37) was witnessed in medium supplemented with 2.0 mg/L IAA.	Tikendra et al. (2018)
Thrixspermum japonicum	Seeds	The highest seed germination percentage (88.1%) was observed on MSB (MS basal salt without vitamins) medium enriched with 0.2% CW + 0.2 mg/L NAA + 0.5 mg/L GA3. Optimized shoot induction (94.3%) and maximum shoot numbers (8.3) were obtained on MS medium supplemented with 0.1% AC, 3.0% banana pulp, 2.0% potato homogenate, 0.3 mg/L KN, 0.2 mg/L NAA, and 0.5 mg/L GA_3. Rooting of the shoots was best achieved on MS medium supplemented with 0.1% AC, 3.0% banana pulp and 2.0% potato homogenate, and 0.8 mg/L IBA.	Seon et al. (2018)
Epipactis flava	Seeds	Seeds of 6-week-old capsules showed the highest seed germination rates at 70.2% and 70.4% in semi-solid and liquid VW medium, respectively. The highest rate of protocorm developmental stage 5 (54%) was found on semi-solid BM (BM-1 Terrestrial Orchid Medium) medium. Maximum shoot number and fresh weight were obtained on liquid MS medium.	Kunakhonnuruk et al. (2018)
Phalaenopsis amboinensis	Seeds	Optimum seed germination of 90.7% was achieved on VW medium. Medium supplemented with 15% CW and 10 g/L banana homogenate (BH) grew plantlets to the highest length (62.1 mm) and highest dry weight (15.5 g). The maximum number of roots and leaves was also found in the same combination.	Utami and Hariyanto (2019)

Gastrochilus matsuran	Seeds	Maximum seed germination (93.3%) was achieved on half-strength MS medium (without vitamins) supplemented with (1 µM NAA + 1.5 µMGA_3) + 5% CW. Secondary protocorm formation was best observed on the same medium but augmented with 2 µM TDZ. Maximum conversion of protocorms into seedlings was noticed with medium containing 2 µM IBA or 1 µM NAA.	Kang et al. (2020)
Paphiopedilum insigne	Seeds	The best seed germination was achieved on MS medium augmented with 2 µM NAA + 6 µM BA + 3% sucrose. The differentiation of PLBs to plantlets and culture proliferation was maximum on MS + 4 µM NAA + 4 µM BA + 3% sucrose.	Deb and Jakha (2020)
Stanhopea tigrina	Seeds	At 120 days of culture, the seed germination reached 98% on MS basal medium containing 1% AC. MS medium with 10 g/L apple extract or 10 g/L banana extract or 30 ml/L CW or 5.0 mg/L BAP developed 1.25 ± 0.35 shoots with no significant difference. The highest root production was obtained by adding 5.0 mg/L IAA with 100 mL/L CW, with an average production of 9.00 ± 0.68 roots.	Castillo-Perez et al. (2021)
Pelatantheria scolopendrifolia	Seeds	OBTSG (Orchid BM1 Terrestrial Seed Germination and Stem Propagation Medium) gave the highest seed germination (94.1%), followed by POM (Phytamax Orchid Maintenance Medium) (90.4%), OBTSCM (Orchid BM2 Terrestrial Stem propagation and Callus Medium) (90.1%), and half-strength MS medium (87.5%).	Kim et al. (2021)
Phalaenopsis amabilis	Leaf	Half-strength MS medium with 2 mg/L BA + 0.5 mg/L NAA + 2% sucrose + 10% CW + 2 g/L peptone + 1 g/L AC was the best medium for PLB development. High-frequency PLB proliferation (250 PLBs per explant) was obtained in half-strength MS medium with 2% sucrose + 2 g/L peptone + 1 g/L AC + 10% CW + 150 mg/L L-glutamine. The addition of 1 g/L banana powder to the same medium gave the maximum root formation.	Sinha and Jahan (1970)
Aranda Wan Chark Kuan'Blue' × *Vanda coerulea*	Leaf	The optimum PLB induction (94.8%) occurred on MS medium fortified with 1.5 mg/L TDZ, producing an average of 25 PLBs from 1 cm^2 leaf segment. Well-developed roots and shoots were observed in MS medium with 1 mg/L BA + 0.5 mg/L IBA + 60 mg/L adenine sulfate.	Gantait and Sinniah (2012)
Phalaenopsis gigantea	Leaf	The highest number of PLBs (353 PLBs) was obtained on NDM (New Dogashima Medium) containing 10 mg/L chitosan and 0.1 mg/L TDZ. However, the best responses for shoot regeneration were observed in VW medium with 10 mg/L chitosan and 0.5 mg/L TDZ.	Samarfard et al. (2014)

(*continued*)

TABLE 10.1 (Continued)

Micropropagation of Orchids Using Different Explants

Name of the species	Explant type	Culture media with PGR combinations producing optimal growth response	References
Vanda tricolor	Leaf	The half-strength MS medium with 0.05 mg/L NAA and 0.01 mg/L BAP was the most favorable for the embryogenic callus formation and proliferation. The generation of somatic embryos occurred 30 days after culturing of callus onto half-strength MS without the addition of any PGRs.	Hardjo and Savitri (2017)
Rhynchostylis gigantea	Leaf	Maximum regeneration potential from the whole leaf segments (0.5–1 cm long) was observed in Mitra medium supplemented with 1.5 mg/L KN. A maximum of 25 plantlets was obtained per explant.	Pathak et al. (2017)
Smithsonia maculata	Leaf	In vitro seedling-derived leaves started producing PLBs in 30–40 days. The maximum number of shoots (10–11.25) per explant was obtained on the medium fortified with 10 mg/L BAP and 1 mg/L IAA. WPM (Woody Plant Medium) containing 5% banana pulp induced 2–3 healthy roots in 2–3 months.	Decruse and Gangaprasad (2018)
Tolumnia Louise Elmore 'Elsa'	Leaf	At 60 days under darkness, 65% of the leaf explants formed somatic proembryos in half-strength MS + 1 mg/L zeatin. After 90 days, TDZ at 1 and 3 mg/L and zeatin at 0.3, 1, and 3 mg/L induced a significantly higher percentage of somatic proembryo formation, and TDZ at 3 mg/L gave the highest percentage of somatic globular embryo formation.	Shen et al. (2018)
Tolumnia Snow Fairy	Leaf	Leaf explants taken from 1- to 2-cm high in vitro grown plantlets showed the highest percentage of PLB formation in MS medium containing 4 mg/L BA and 0.5 mg/L NAA, with the production of an average of 24.0 PLBs. No PLBs formed from the outer leaves. Only the inner expanding leaves cultured on MS basal medium supplemented with 4 mg/L BA and 0.5 mg/L NAA resulted in PLB induction at an average of 25.5 PLBs per explant. The shoot germination rate for the secondary PLBs was 33.3%, with an average of 3.5 shoots per whole PLB, with corresponding rates of 40% with 3.1 shoots for the upper PLB halves.	Chookoh et al. (2019)

Dendrobium nobile	Axillary bud	The maximum number of shoots (4.33), as well as fresh and dry weights (752.5 and 52.99 mg, respectively), was obtained on phytotechnology medium (O753) with 2 mg/L BAP. At the same time, 1.5 mg/L of KN exhibited the highest shoot length (4.18 cm). Modified MS with 2 mg/L IBA produced the maximum rooting percentage (97.5%), root number (4.70), and root length (3.47 cm).	Asghar et al. (2011)
Dendrobium longicornu	Axillary bud	The most excellent explant response (86.6%) was obtained in MS medium supplemented with 30 µM NAA, while the highest shoot number (4.42) was recorded with MS + 15 µM BAP + 5 µM NAA. The maximum number of explants forming PLBs (41.48%) was noticed in the medium containing 15 µM BAP and 15 µM 2,4-D.	Dohling et al. (2012)
Doritis pulcherrima	Axillary shoot tip	High axillary shoot formation was observed in KC medium with 0.1% peptone and 2 mg/L BAP. Maximum PLB production was witnessed in medium enriched with 2 mg/L NAA. The highest number of roots (2.41 ± 0.4) per explant was recorded in KC medium with 1 mg/L NAA.	Mondal et al. (2013)
Grammatophyllum speciosum	Shoot tip	The highest PLB (93%) production frequency was recorded on a half-strength MS liquid medium containing 2% sucrose. Plantlet regeneration with optimum shoot and root formation was found in a half-strength MS solid medium with 2.0 mg/L NAA and 1.0 mg/L BA.	Sopalun et al. (2010)
Esmeralda clarkei	Shoot tip	The best response for the shoot multiplication was achieved on MS medium supplements with 1 or 2 mg/L BAP producing around 11 shoots per culture. NAA at 0.5–1.0 mg/L gave the maximum rooting percentage (75%) with high root number (3.0) and root length (2.93 cm).	Paudel and Pant (2012)
Dendrobium primulinum	Shoot tip	The maximum number of rootless healthy shoots was witnessed on MS medium fortified with 1.5 mg/L BAP, with an average value of 4.5 shoots per culture. MS medium supplemented with various concentrations of NAA, IAA, and IBA showed a positive response in root development except for medium with 0.5 mg/L NAA incorporated. The best rooting response was observed on MS + 5 mg/L IAA.	Pant and Thapa (2012)
Dendrobium Sonia 'Earsakul'	Shoot tip	Half-strength MS medium supplemented with 4 mg/L BA was able to give an early bud break. Early (11 days) shoot multiplication with maximum numbers (4.66) of healthy shoots was observed in the medium augmented with 2.0 mg/L KN and 0.1 mg/L NAA. Medium with 0.5 mg/L NAA gave the earliest rooting (19.6 days).	Priyakumari et al. (2013)

(continued)

TABLE 10.1 (Continued)

Micropropagation of Orchids Using Different Explants

Name of the species	Explant type	Culture media with PGR combinations producing optimal growth response	References
Dimorphorchis lowii	Shoot tip	Half-strength MS medium supplemented with 3.0 mg/L TDZ and 0.046 mg/L NAA was most favorable for culture survival and callus formation. Maximum shoot proliferation from PLBs was noticed in KC medium enriched with 15% CW. In this treatment, 10.2 ± 6.2 shoots were produced from one callus explant.	Jainol and Gansau (2017)
Anoectochilus formosanus	Shoot tip	Maximum axillary shoot proliferation was found in MS medium containing 1.5 mg/L BAP and 0.25 mg/L NAA (7.0 shoots per explant, 1.0 cm shoot height, and 9.8 leaves per explant). Higher root formation of 2.4 roots per shoot and root length of 1.0 was observed on Hyponex medium containing 150 ml/L CW.	Winarto and Samijan (2018)
Dendrobium Red Bull	Shoot tip	MS medium supplemented with 3.0 mg/L BAP and 1.5 mg/L NAA showed the best response for shoot length (21.19 mm). On the other hand, the maximum shoot number (7.66) was obtained in the medium enriched with 3.0 mg/L BAP and 1.0 mg/L NAA. Maximum survivability of 92% was attained on the substrate containing cocodust.	Mamun et al. (2018)
Vanilla planifolia	Nodal	A maximum of 35% explants formed callus on MS medium supplemented with 2.0 mg/L NAA and 1.0 mg/L BA. MS with 1.0 mg/L BA and 0.5 mg/L NAA produced the highest mean number of 4.2 shoots per callus with a mean length of 3.8 cm. A mean of 3.29 roots per explant with 4.42 cm root length was obtained on MS supplemented with 1.0 mg/L NAA.	Tan et al. (2011)
Arundina graminifolia	Nodal	MS medium appended with 1.0 mg/L NAA and 2.5 mg/L KN showed the highest shoot proliferation rate with 5.33 ± 0.26 shoot number per explant and a length of 3.50 cm. MS medium, when supplemented with 3.0 mg/L IAA, induced maximum roots (6.0 ± 0.47) with the highest root length of 4.7 cm.	Das et al. (2013)

Dendrobium chrysanthum	Nodal	The highest frequency of explants forming buds (100%), the maximum shoot number/explant (14.33 ± 0.14), the bud forming capacity (BFC) index of 14.33, and the maximum shoot length (1.97 ± 0.04) were obtained in MS medium augmented with 5 μM TDZ and 5 μM BAP. 100% rooting of regenerated shoots with an average number of 11.26 roots/shoot having an average root length of around 2.45 cm was obtained in MS medium supplemented with 10 μM NAA.	Hajong et al. (2013)
Dendrobium nobile	Nodal	MS + 1.0 mg/L mT (meta-topolin) + 0.5 mg/L NAA+ 3% sucrose + 0.8% agar gave the maximum shoot number (9.2 ± 0.3) and 92% PLB formation. The highest root number (13.2 ± 0.1) and root length (5.3 ± 0.3) were witnessed in half-strength MS medium enriched with 2 mg/L IBA and 0.5 mg/L phloroglucinol.	Bhattacharyya et al. (2016a)
Dendrobium bensoniae	Nodal	Shoot regeneration was best in MS medium supplemented with 2.0 mg/L BA. The highest shoot and leaf numbers were observed in MS +1.0 mg/L BA + 1.5 mg/L IBA. MS + 0.5 mg/L BA + 1.0 mg/L IBA was the most effective for rooting.	Riva et al. (2017)
Dendrobium chrysotoxum	Nodal	The regeneration response of the explant was maximum (100%) on liquid MS media augmented with 5.37 μM NAA, generating the highest number of regenerants per explant (20).	Kaur (2017)
Ludisia discolor	Nodal	Nodal segments of *L. discolor* when cultured on half-strength MS + 1.0 mg/L NAA + 0.1 mg/L TDZ + 0.2% AC + 8% banana cultivar homogenate + 3% sucrose + 3.5 g/L Gelrite produced the best in vitro growth response.	Poobathy et al. (2019)
Aerides multiflora	Nodal	The highest average number of multiple shoot buds (8.83 ± 0.45/segment) via organogenesis was recorded on MS medium with 1.0 mg/L NAA and 2.0 mg/L BAP. The most extended shoot bud length was achieved on agar solidified MS medium with 1.0 mg/L NAA and 1.0 mg/L BAP. The best response in root length extension and root number was observed on agar solidified MS medium with 1.0 mg/L IBA.	Bhowmik and Rahman (2020a)
Coelogyne ovalis	Nodal bud	The highest regeneration frequency (80%), PLB number (22.73 ± 0.47), and shoot number per explant (14.53 ± 0.27) were achieved in KC medium augmented with 10 μM mT and 0.5 μM NAA. The best rooting response was noticed in the medium supplemented with 10 μM IAA.	Singh and Kumaria (2020)
Malaxis acuminata	Pseudobulb	A maximum of 65% explants responded best in culture initiation, regeneration frequency, PLB proliferation, and plantlet development on Mitra medium appended with 1.0 mg/L BAP, 1.0 mg/L NAA, and 2 g/L AC.	Kaur and Bhutani (2010)

(*continued*)

TABLE 10.1 (Continued)

Micropropagation of Orchids Using Different Explants

Name of the species	Explant type	Culture media with PGR combinations producing optimal growth response	References
Malaxis acuminata	Pseudobulb	98% of the explants responded positively on MS + 6 µM NAA + 6 µM BA+ 3% sucrose + 100 mg/L casein hydrolysate (CH), forming as many as 11 shoot buds per explant. The shoot buds were converted into rooted plantlets with 4–5 roots and about 18 shoots on MS + 3 µM NAA + 3 µM BA + 0.3% AC + 3% sucrose.	Deb and Arenmongla (2014)
Dendrobium palpebrae	Pseudobulb	The highest multiple shoot buds (8.21 ± 0.44) per segment in the lower part and maximum shoot buds (6.43 ± 0.40) per segment in the upper part of the explant were obtained in MS medium supplemented with 1.0 mg/L NAA and 2.0 mg/L BAP. The highest root length increase (4.82 ± 0.22 cm) and root number (2.75 ± 0.17) per shoot bud were observed on agar solidified MS medium with 0.5 mg/L NAA.	Bhowmik and Rahman (2020b)
Cyrtopodium brandonianum	Root tip	Half-strength MS medium supplemented with 0.5 mg/L TDZ showed the best shoot production response (43% of the explants with shoots). Optimal rooting (20% of shoots with an average of 4.3 roots per explants), with no intervening callus, was observed in the half-strength MS medium with 6% sucrose and 1 mg/L NAA.	Flachsland et al. (2011)
Cymbidium aloifolium, *Cymbidium iridioides*	Aerial roots	In *C. aloifolium*, within 20 days of culture, 60% of explants responded best by forming PLBs on MS medium supplemented with 3 µM KN and 3% sucrose. Medium augmented with 3 µM BA and 3% sucrose produced a maximum of 12 shoot buds. In *C. iridioides*, 50% of the explants responded best after 40 days of culture on medium enriched with 3 µM IAA, 3% sucrose, and 0.1% AC. Optimum regeneration was achieved when the medium was supplemented with 3% sucrose, 15% CW, and 100 mg/L casein hydrolysate (CH), producing 20 shoot buds.	Deb and Pongener (2012)

Cyrtopodium paludicolum	Root tip	KC medium fortified with 1.34 μM NAA and 2.27 μM TDZ resulted in better response on PLB formation, subsequent shoot differentiation (55.25 shoots per explant), and better rooting, favoring a high survival rate (90%) after acclimatization. TDZ did not induce PLB formation without NAA. However, the medium supplemented with only NAA (1.34 μM) resulted in 33.50 shoots per explant, indicating a synergistic effect of both NAA and TDZ.	Picolotto et al. (2017)
Cymbidium aloifolium	Root tip	Nearly 100% of the explants responded on Mitra medium enriched with 3 mg/L NAA via shoot bud formation within 13 days of culture. The shoot buds subsequently developed into healthy plantlets within 55 days of culture.	Verma and Pathak (2018)
Rhynchostylis retusa	Root tip	Mitra medium supplemented with 3 mg/L KN, 1 mg/L NAA, and 2% sucrose and agar produced maximum regeneration (31%) in the root segment's proximal region, producing 28 plantlets in 15 weeks.	Sharma (2019b)
Dendrobium nanum	Rhizome	The maximum percentage of callus induction was obtained on MS basal medium with 2.0 μM/L NAA and 1.2 μM/L KN. Highest shoot number (5.00 ± 0.91) and shoot length (15.78 ± 0.37 mm) were witnessed on MS medium supplemented with 5.0 μM/L BAP.	Muthiah et al. (2010)
Eulophia dabia	Rhizome	The best regeneration response was achieved on Mitra medium augmented with 1 mg/L BAP and 0.5 mg/L NAA, producing plantlets with 2–3 leaves and 3–4 roots in 4 weeks. TDZ in the medium at 0.1 mg/L and 1 mg/L also induced multiple shoot buds via PLB formation.	Chauhan et al. (2015)
Cymbidium goeringii	Rhizome	The highest number of shoot buds per explant (21.8 ± 1.8) was recorded on MS medium fortified with 20 μM 2,4-D and 2 μM TDZ. The maximum root induction (100%) with roots (5.3 ± 1.1) per in vitro developed shoot was achieved on a half-strength MS medium incorporating 2 μM NAA.	Park et al. (2018)
Esmeralda clarkei	Protocorms	The maximum regeneration potential was found on the MS medium fortified with either 0.5 or 2.0 mg/L BAP, producing an average of 14 shoots per explant. The highest number of three roots per treatment was noticed in medium supplemented with NAA (0.5 and 1.0 mg/L).	Paudel and Pant (2012)
Orchis catasetum	Protocorms	MS medium with 0.5 mg/L BA and 0.5 mg/L NAA was most favorable for PLB regeneration (20.40 PLBs per plantlet). The maximum root (7.16) and leaf number (10.10) per plantlet, the most extended plant height (114.20 mm) and root length (193.40 mm) were noted on MS medium supplemented with 0.5 mg/L BA and 0.5 mg/L NAA.	Baker et al. (2014)

(continued)

TABLE 10.1 (Continued)

Micropropagation of Orchids Using Different Explants

Name of the species	Explant type	Culture media with PGR combinations producing optimal growth response	References
Dendrobium aqueum	Protocorms	The highest number of 9.4 shoots per explant was recorded on a half-strength MS medium with 3 mg/L NAA. Shoot elongation of 1.52 cm was achieved on the medium supplemented with 7 mg/L NAA. Half-strength MS medium augmented with 5 mg/L IBA produced 8.75 shoots, but the longest root of 1.48 cm was found in medium enriched with 7 mg/L NAA.	Parthibhan et al. (2015)
Vanda pumila	Protocorms	The highest shoot number (9.50 ± 0.29) per culture was developed on a half-strength MS medium incorporating 1.0 mg/L KN and 10% CW. The longest shoots (0.78 ± 0.07 cm) per culture were developed on the medium fortified with 2.0 mg/L BAP and 10% CW. The half-strength MS medium with 0.5 mg/L IAA was the most effective condition for the maximum root production (5 ± 0.00) per culture and root length (0.93 ± 0.07 cm).	Maharjan et al. (2019)
Dendrobium chryseum	Protocorms	The highest shoot multiplication (18.75 ± 0.48 shoots per culture) was recorded on a half-strength MS medium fortified with 2.0 mg/L KN and 10% CW. The longest shoots (2 ± 0.20 cm) and the maximum root number (4.5 ± 0.65) per culture were obtained on a half-strength MS medium with 1.0 mg/L GA_3 and 10% CW. The longest roots (1.28 ± 0.14 cm) were noticed on the medium supplemented with 0.5 mg/L GA_3 and 10% CW. The highest leaf number (10.5 ± 0.29) per culture and the most extended leaves (0.47 ± 0.05 cm) grew on half-strength MS medium enriched with 1.0 mg/L BAP.	Maharjan et al. (2020)
Dendrobium Sonia 'Red Jo'	Protocorm-like bodies (PLBs)	Plantlets cultured on VW media supplemented with 200 mM NaCl and 5 mM $CaSiO_3$ showed 100% survival rate with the highest fresh weights in all the treatments. Moreover, VW medium with 200 mM NaCl and all concentrations of $CaSiO_3$ had plant heights significantly higher than those cultured on VW medium incorporating 200 mM NaCl alone.	Obsuwan et al. (2019)

Paphiopedilum niveum	Protocorm-like bodies (PLBs)	Modified Vacin and Went (MVW) medium supplemented with 0.1 mg/L NAA produced the highest percentage of somatic embryo formation (68.33 ± 11.77%) and the maximum number of somatic embryos per explant (5.19 ± 0.67).	Soonthornkalump et al. (2019)
Eulophia nuda	Tuber	95% of the explants responded best, producing 2 shoots per explant with a length of 3.2 cm, on MS medium augmented with 44.4 μM BA. Maximum tuber formation, shoot multiplication (16.2 ± 1.2), and shoot elongation (7.2 ± 1.2 cm) were achieved on MS medium containing 8.88 μM BA and 4.64 μM KN. The highest root number (7 roots per explant) with a root length of 3.7 cm was recorded in 70% of the rooted plantlets.	Panwar et al. (2012)
Phalaenopsis cv. 'Surabaya'	Nodal flower stalks	The optimum condition for direct somatic embryogenesis was observed in MS medium supplemented with 5 mg/L BAP and 2 mg/L NAA. Half-strength MS with 2 mg/L NAA produced the maximum root number (6.7) per plantlet.	Balilashaki et al. (2015)
Dendrobium jerdonianum	Flower stalks	MS basal medium fortified with 0.5 mg/L 2,4-D, 5 mg/L BAP, and 50 mL coconut milk (CM) was the best combination for shoot multiplication and plantlet formation. MS medium with 0.5 mg/L 2,4-D, 5 mg/L IAA, 500 mg AC, and 50 mL CM gave the best results for in vitro rooting.	Sr Sagaya Marry and Divakar (2016)
Crepidium acuminatum	Floral buds	The best shoot bud regeneration response was observed on Mitra medium augmented with 1 mg/L IAA, 1 mg/L KN, 2% sucrose, and 2 g/L AC. Eight to ten pseudobulbous shoots were developed from a single floral bud after 5 months of culturing.	Vasundhara et al. (2019)
Saccolabium papillosum	Inflorescence segment	The in vitro growth response was observed with undifferentiated floral buds in the upper two-thirds region of the explant. The maximum regeneration (75%) was attained in Mitra medium supplemented with 1 mg/L NAA and 2 g/L AC, producing 30 regenerants per explant.	Kaur and Pathak (2015)

clones at elevated concentrations of BAP. Zhenxun and Hongxian (1997) also reported chromosome number aberrations in banana culture when high concentrations of BAP and adenine were employed. Somaclonal variation may be a valuable source of novel variants with high-yielding and disease-resistant traits. Still, the emergence of variation is a significant concern if true-to-type plants are needed as end products for commercial and conservation of elite genotypes. To achieve the prime objective of micropropagation, it has become obligatory to detect somaclonal variation in the regenerants by assessing the clonal fidelity using different DNA markers.

10.3.2 DNA Markers in Clonal Fidelity Assessment

The crucial goal in the mass propagation of commercial plants like orchids through tissue culture is the generation of elite true-to-type plants. However, the types of explants and their genotypes, plant growth regulator concentrations and combinations, culture duration, and sub-culture cycle number might enhance the genetic variation among micropropagated plants. Therefore, assessing the genetic fidelity of the in vitro clones is critical for producing elite plantlets genetically similar to the mother plants. The genetic integrity of the regenerants can be evaluated by detecting chromosomal alteration through cytological analysis, isozyme-based variation study, and phenotypic identification based on morphological characters (Sabir et al. 1992; Dixit et al. 2003; Mallon et al. 2010). Flow cytometry is also an effective method for cell DNA content analysis to affirm ploidy changes detected by classical cytological examination (Dolezel et al. 2007). The technique has been effectively employed to determine somaclonal variation among in vitro clones (Pinto et al. 2004; Prado et al. 2010; Sopalun et al. 2010) and cryopreserved micropropagated plants (Hirano et al. 2005; Popova et al. 2009, 2010; Ai et al. 2012). But these methods are beset with several limitations, making genetic variation detection less efficient. Reliable DNA markers have been successfully developed in recent times and can be utilized for clonal fidelity assessment of several in vitro propagated plants.

Simple sequence repeat (SSR), amplified fragment length polymorphism (AFLP), restriction fragment length polymorphism (RFLP), and single nucleotide polymorphism (SNP) are important co-dominant markers used in genetic stability testing of regenerants (Singh et al. 2013; Butiuc-Keu et al. 2016; Bandupriya et al. 2017). Random amplified polymorphic DNA (RAPD), inter-simple sequence repeat (ISSR), start codon targeted polymorphism (SCoT), and CAAT box-derived polymorphism (CBDP) are some popular dominant markers employed for clonal fidelity evaluation in plants (Raynalta et al. 2018; Rohela et al. 2019; Dey et al. 2019; Dey et al. 2021). RADP is one of the simplest and most cost-effective DNA markers, efficiently detecting genetic variation in micropropagated plants. But RADP markers have major limitations, with lower reproducibility and reliability (Muralidharan and Wakeland, 1993). ISSRs are more reproducible, polymorphic, and informative than RADP markers (Palombi and Damiano, 2002; Amom et al. 2018). They have been used to confirm the RADP results while determining the genetic stability of several in vitro regenerated plants (Gantait et al. 2010; Rathore et al. 2014; Dey et al. 2019). However, RADP and ISSR are arbitrarily dominant markers, as they mainly target the non-coding regions of the plant genome (Wolfe and Liston, 1998; Amom et al. 2020). Gene targeted markers like SCoT, CBDP, and iPBS (inter-primer binding site) were later introduced to detect DNA polymorphism with great accuracy. SCoT markers are based on the short-conserved regions flanking the ATG start codon in the plant genome (Collard and Mackill, 2009; Tikendra et al. 2021d). The CBDP markers target the regions of the CAAT box of the plant gene promoter that plays a critical role during the transcription process (Singh et al. 2014). The iPBS markers, on the other hand, are based on a conserved sequence located adjacent to the 5′ LTR (long terminal repeat sequence) (Amom and Nongdam, 2017; Kalendar et al. 2018). SCoT, CBDP, and iPBS are effective dominant markers frequently used in genetic diversity studies because they are fast, effective, and do not require prior sequence information of the template DNA. The genetic integrity of the in vitro clones can be tested during different culture stages and after the hardening and transplantation of plants to the field. The procedural steps involved in the clonal fidelity assessment of micropropagated orchids with DNA markers are represented in Figure 10.2.

Khoddamzadeh et al. (2010) studied the somaclonal variation of PLBs and the mother plant of *Phalaenopsis bellina* using RAPD markers. Eight selected RADP primers produced 172 bands, of which 154 were monomorphic and the other 18 were polymorphic. P16 gave the highest number of bands (29 bands), while OPU10 generated the lowest (15 bands). Primers OPU08 and P12 furnished 28 and 23 bands, respectively, with 0% polymorphism, revealing that proliferation for up to 6 months in *P. bellina* did not result in somaclonal variation. Other primers, OPU10, OPU12, OPU16, and P14, showed low polymorphism, ranging from 6% to 12%. RAPD markers were also employed to ascertain the genetic variability of TDZ-induced in vitro propagated *Cymbidium giganteum* (Roy et al. 2012). Eighteen primers, which gave reproducible bands, were selected after screening 40 RAPD primers. Low polymorphism in the regenerants was recorded for 17 primers, while lone primer OPB18

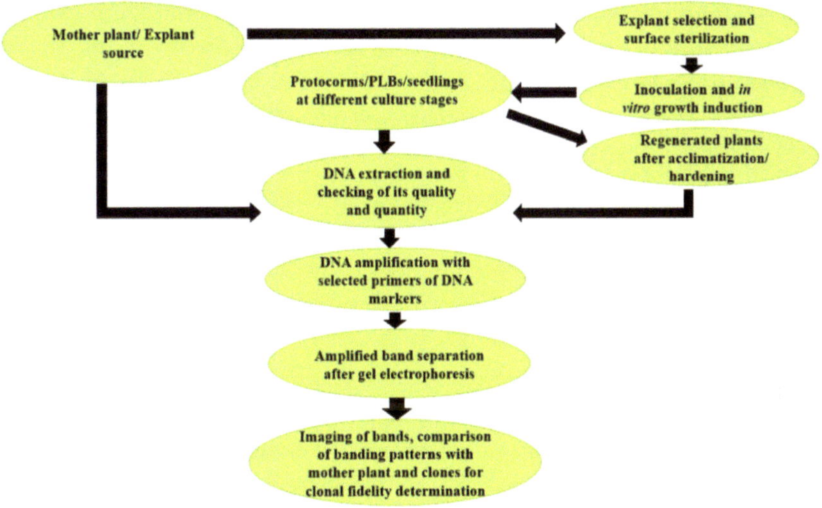

FIGURE 10.2 Steps involved in clonal fidelity assessment of micropropagated orchids using DNA markers.

produced a uniform banding pattern. Maximum genetic variation (16.67%) was detected for OPA11, followed by OPB7 (12.3%) and OPB15 (11.11%). The study revealed low genetic variation (5.81%) among the regenerated orchids, indicating less influence of TDZ on somaclonal variation induction. Mohanty and Das (2013) assessed the genetic fidelity of 45-day-old plants derived from encapsulated microshoots of *Dendrobium densiflorum* using RADP markers. Ten RADP primers produced 39 bands with fragment sizes varying from 0.2 to 1.3 kb. The monomorphic band patterns for all the selected primers indicated the absence of genetic variation between the encapsulated PLB-derived plantlets and the donor mother plants. The investigation revealed that short-term storage of explants possibly did not affect the genetic integrity of regenerated plants. The genetic stability was also tested for regenerated *Paphiopedilum niveum* via direct somatic embryogenesis using RAPD markers (Soonthornkalump et al. 2019). RAPD primers generated 102, 91, and 98 bands for in vitro clones 1, 2, and 3, respectively. There was no genetic variation between the in vitro mother plants (derived from the original protocorm) and the regenerated clones (obtained from primary and secondary somatic embryos) exhibiting identical banding patterns. Ten RADP primers used in evaluating the clonal fidelity of micropropagated *Rhynchostylis retusa* yielded 23 bands with an average of 2.3 bands per primer (Oliya et al. 2021). The number of bands varied from one (OPC-11) to three (OPA-03, OPA-06, OPA-07, OPA-08). Monomorphic bands were generated, and the allele sizes recorded were similar to those of the mother plant, showing maintenance of genetic stability in micropropagated *R. retusa*.

Gantait and Sinniah (2013) evaluated the genetic stability of alginate-encapsulated shoot tips of monopodial orchid hybrid *Aranda* Wan Chark Kuan 'Blue' × *Vanda coerulea* using ISSR markers. Nine ISSR primers produced 51 reproducible monomorphic bands per clone. No polymorphism was detected among oret (control) and the plantlets regenerated from synthetic seeds (ramets) stored for 200 days at 4 to 25 °C. The ISSR marker analysis disclosed maintenance of high genetic uniformity among the in vitro clones. Samarfard et al. (2013) utilized ISSR markers to study the effect of chitosan on the genetic stability of in vitro regenerated PLBs of *Phalaenopsis gigantea*. ISSR markers generated a total of 275 bands with an average of 6.9 bands per primer. The secondary PLBs developed in chitosan-treated liquid medium were genetically uniform and were similar to the mother plant. ISSR markers were also applied by Sherif et al. (2018) to determine the genetic uniformity of direct somatic embryogenesis (DSE) and indirect somatic embryogenesis (IDSE) regenerated plantlets of *Anoectochilus elatus*. Fifteen ISSR primers gave 533 bands in DSE- and 557 bands in IDSE-derived plants. Low polymorphism was noticed, with 30 and 37 polymorphic bands recorded in DSE and IDSE, respectively. The study showed that plants developing from DSE and IDSE possessed 94.22% and 93.05% of genetic homogeneity, respectively. Bose et al. (2017) studied the genetic stability of *Malaxis wallichii* micropropagated through a transverse thin cell layer approach using intron splice junction (ISJ) markers. Fifteen ISJ primers produced 84 bands ranging from 4 to 8, with an average of 5.6 bands per primer. The clonal variability was determined at 4.76% with four polymorphic bands, and the Jaccard's similarity index varied from 0.97 to 1.00. The investigation also revealed no detectable phenotypic variations among in vitro transverse thin cell layer (tTCL)-derived plantlets.

Bhattacharyya et al. (2017) determined the clonal fidelity of micropropagated *Ansellia africana* at various sub-culture stages and after successful acclimatization using SCoT markers. Sixteen SCoT primers generated 56 bands in meta-Topolin-derived plants with 3 polymorphic bands. Sixty-two bands were produced in plants resulting from the N^6-benzyladenine (BA) pathway with four polymorphic bands in the first generation. In the second generation, 72 and 56 bands were generated in meta-Topolin- and BA-derived plants, respectively, with 5 and 4 polymorphic bands. Seventy-five amplified bands were noticed in the hardened plants, of which five bands were polymorphic with 7.14% polymorphism.

The use of a single marker system does not guarantee a precise assessment of the genetic homogeneity of the in vitro clones. It is always appropriate to use multiple markers, as accurate and reliable results are produced by validating the outcome of different marker analyses (Rathore et al. 2014; Dey et al. 2019). Several micropropagated orchids were evaluated for clonal fidelity using combined markers. Bhattacharyya et al. (2016a) successfully examined the clonal fidelity of the micropropagated plants of *Dendrobium nobile* using two marker systems: IRAP (Inter-Retrotransposon Amplified Polymorphism) and SCoT markers. Five IRAP primers produced 26 bands, of which 2 bands were polymorphic, showing 7.6% polymorphism. Eight SCoT primers generated 57 bands, out of which 4 were polymorphic, exhibiting a total variability of 7.01% in the micropropagated plants. Pooled IRAP and SCoT data analysis also revealed a low polymorphism of 7.22%, with a Jaccard's similarity index varying from 0.97 to 1.00. Bhattacharyya et al. (2018a) also assessed the genetic stability of 3-month-old *Ansellia africana* plants developed from encapsulated micro-shoots using SCoT and IRAP markers. Eight SCoT primers yielded 55 bands, with 4 bands showing polymorphism. The 5 IRAP primers produced 26 bands, with 2 bands showing polymorphism. The clonal variability within the regenerated plantlets was determined at 7.40%, with Jaccard's similarity index extending from 0.94 to 1.00. The polymorphic information content (PIC) values recorded for IRAP, SCoT, and IRAP+SCoT primers were 0.77, 0.88, and 0.80, respectively, while resolving power (Rp) ranged from 2.79 to 3.23 for IRAP and from 2.41 to 3.88 for SCoT markers. The Rp and PIC values for IRAP and SCoT primers were significantly high, indicating the suitability of these marker systems for detecting the genetic stability of the micropropagated orchids.

Tikendra et al. (2019a) tested the genetic uniformity of micropropagated *Dendrobium chrysotoxum* using RAPD and ISSR markers. Twelve RAPD primers produced 74 scorable bands with an average of 6.17 bands per primer. Seventy-three bands were monomorphic, ensuing 98.81% monomorphism. Similarly, 11 ISSR primers gave 76 reproducible bands, out of which 73 bands were monomorphic, producing 97.47% monomorphism. The genetic closeness of the mother plant and in vitro clones was high, as they were clustered together in the dendrogram. Tikendra et al. (2019b) similarly determined the genetic homogeneity of micropropagated *D. moschatum* using RAPD and ISSR markers. Each of the10 primers of RADP and ISSR markers yielded 48 and 54 scorable bands, respectively. Forty-five and 53 bands of RAPD and ISSR markers were monomorphic, resulting in 95.2% and 98.0% monomorphism in the regenerants. The ISSR generated a higher average number of bands per primer than RAPD markers, indicating its effectiveness in analyzing genetic polymorphism. The dendrograms revealed a close genetic association of the mother plant with the in vitro progenies. Principal Coordinate Analysis (PCoA) showed the grouping of the regenerants and the mother plant, similarly to the clustering patterns exhibited by dendrograms. The investigation disclosed the effectiveness of the combined marker system of RAPD and ISSR in assessing the genetic integrity of in vitro propagated *D. moschatum*.

Chin et al. (2019) employed 9 ISSR and 11 direct amplification of minisatellite DNA region (DAMD) primers to examine the presence of genetic variation in PLBs of *Dendrobium Sabin* Blue treated with different concentrations of NAA, KN, TDZ, and AC. The PLBs were subcultured for 2 years with different additives to check their influence on the genetic stability of the culture. PLBs under treatment with 1.5 mg/L KN harbored the highest genetic variation, while the protocorms on a medium supplemented with either 4.0 mg/L TDZ or 0.5 g/L AC showed the maximum genetic uniformity. The study showed the importance of DNA marker-assisted clonal fidelity testing of long-term cultured PLBs for producing genetically uniform plants. Khor et al. (2020) performed a somaclonal variation detection study of cryopreserved PLBs of *Aranda* Broga Blue orchid using RADP and SCoT markers. Cryopreserved and non-cryopreserved PLBs were examined for genomic uniformity using 7 RADP and 11 SCoT primers. The band number produced by RAPD primers varied from 6 (OPK-16) to 13 (OPK-14), with band sizes ranging from 200 to 2000 kb. In contrast, the band number generated by SCoT primers ranged from 6 (S32) to 11 (S6, S31, and S33), with band sizes extending from 200 to 1800 kb. The RADP and SCoT primers generated no polymorphism for either cryopreserved or non-cryopreserved PLBs.

Singh and Kumaria (2020) evaluated the genetic variation of micropropagated plantlets of *Coelogyne ovalis* employing SCoT and ISSR markers. Each of the 10 primers of SCoT and ISSR primers produced 44 and 53 reproducible bands, respectively. The micropropagated plantlets showed monomorphic banding profiles almost identical to that of the mother plant. Combined SCoT and ISSR data analysis showed a total variation of 5.15%, with Jaccard's co-efficient varying from 0.95 to 1.00. SCoT and ISSR markers were also applied by Sherif et al. (2020) to test the genetic fidelity of in vitro propagated *Aenhenrya rotundifolia*. Each set of 6 SCoT and ISSR primers generated 145 and 182 distinct and scorable bands, respectively. Three polymorphic bands were produced by SCoT primers, showing 0.51% polymorphism. ISSR primers, on the other hand, gave four polymorphic bands exhibiting only 1.41% polymorphism among the regenerated orchids. The study revealed 99% genetic similarity of the in vitro derived plants and only 1% variance from the mother plant. The summary of investigations on clonal fidelity of micropropagated orchids using different markers is shown in Table 10.2.

10.4 Conclusions

Due to their rapidly declining populations, the highly valued ornamental orchids require effective strategies for conservation. Conventional orchid propagation through seed culture and stem and rhizome cutting cannot produce sufficient planting materials. Micropropagation techniques have been successfully adopted to rapidly propagate several orchids on a large scale. The clonal fidelity assessment of several micropropagated orchids was performed using DNA markers to produce genetically identical superior plants. Genetic polymorphism among the clones can be detected more accurately with two marker systems, as the outcome of one marker analysis can be validated by that of the other. The large-scale propagation of genetically elite plants may help fulfill the global orchid demand by preventing rapid population depletion.

TABLE 10.2

Summary of the Investigation on Clonal Fidelity Assessment of Micropropagated Orchids Using Different DNA Markers

Species	Marker types	Number of primers used	Total bands amplified	Average bands per primer	Number of polymorphic bands	Percentage of polymorphic bands (%)	References
Aenhenrya rotundifolia	ISSR	6	182	22.76	4	1.41	Sherif et al. (2020)
	SCoT	6	145	18.25	1	0.50	
Anoectochilus elatus Lindl.	ISSR	6	420	38.1	10	2.38	Sherif et al. (2017)
Anoectochilus elatus Lindl.	ISSR	15	533 (DSE) 557 (IDSE)	5.76 (DSE) 6.18(IDSE)	30 (DSE) 37 (IDSE)	5.77 (DSE) 6.90 (IDSE)	Sherif et al. (2018)
Ansellia africana	SCoT	16	70	4.37	5	7.14	Bhattacharyya et al. (2017)
Ansellia africana	IRAP	5	26	5.20	2	7.69	Bhattacharyya et al. (2018a)
	SCoT	8	55	6.87	4	7.27	
	IRAP + SCoT	13	81	6.23	6	7.40	
Aranda Wan Kuan × *Vanda coerulea*	ISSR	9	561	5.5	0	0	Gantait and Sinniah (2013)
Bulbophyllum auricomum Lindl.	RADP	12	146	12	101	69	Than et al. (2011)
Bletilla striata (Thunb.) Reichb. f.	ISSR	42	406	9.67	0	0	Wang and Tian (2014)
Coelogyne ovalis Lindl.	SCoT	10	44	4.4	3	6.81	Singh and Kumaria (2020)
	ISSR	10	53	5.3	2	3.77	
	SCoT + ISSR	20	97	4.85	5	5.15	
Dendrobium aphyllum	IRAP	5	26	5.20	2	7.69	Bhattacharrya et al. (2018b)
	ISSR	9	50	5.55	2	4.00	
	IRAP + ISSR	14	76	5.42	4	5.26	

(continued)

TABLE 10.2 (Continued)

Summary of the Investigation on Clonal Fidelity Assessment of Micropropagated Orchids Using Different DNA Markers

Species	Marker types	Number of primers used	Total bands amplified	Average bands per primer	Number of polymorphic bands	Percentage of polymorphic bands (%)	References
Dendrobium Bobby Messina	RADP	10	20 (+LN)	2 (+LN)	2 (+LN)	10 (+LN)	Antony et al. (2012)
			19 (–LN)	1.9 (–LN)	1 (–LN)	5.3 (–LN)	
Dendrobium Bobby Messina	TRAP	8	29 (+LN)	3.62 (+LN)	7 (+LN)	24 (+LN)	Antony et al. (2015)
			31 (–LN)	3.88 (–LN)	7 (–LN)	23 (–LN)	
	SCoT	4	12 (+LN)	3 (+LN)	9 (+LN)	75 (+LN)	
			15 (–LN)	3.75 (–LN)	12 (–LN)	80 (–LN)	
Dendrobium chrysotoxum Lindl	RADP	12	74	6.17	1	1.19	Tikendra et al. (2019a)
	ISSR	11	76	6.91	3	2.53	
	RADP + ISSR	23	111	6.54	4	3.61	
Dendrobium crepidatum	SCoT	8	30	3.75	3	10.00	Bhattacharyya et al. (2016b)
	ISSR	5	18	3.60	_	_	
	SCoT + ISSR	13	48	3.69	3	6.25	
Dendrobium densiflorum Lindl.	RADP	10	39	3.9	0	0	Mohanty and Das (2013)
Dendrobium Earsakul	ISSR	11	173	_	39	22.5	Wannajindaporn et al. (2014)
Dendrobium fimbriatum Lindl. var. Oculatum Hk.f.	RADP	11	53	4.82	1	1.52	Tikendra et al. (2021a)
	ISSR	12	67	5.58	2	2.58	
	SCoT	10	65	6.5	2	3.93	
	RADP + ISSR + SCoT	33	185	5.63	5	2.70	
Dendrobium moschatum Sw.	RADP	10	48	4.8	3	4.8	Tikendra et al. (2019b)
	ISSR	10	54	5.4	1	2	
	RADP + ISSR	20	102	4.6	4	4	

Dendrobium nobile Lindl.	RADP	7	27	3.85	3	11.11	Bhattacharyya et al. (2014)
	SCoT	15	57	3.80	2	3.50	
	RADP + SCoT	22	84	3.81	5	5.95	
Dendrobium nobile Lindl.	SCoT	8	57	7.12	4	7.01	Bhattacharyya et al. (2016a)
	IRAP	5	26	5.20	2	7.69	
	SCoT + IRAP	13	83	6.38	6	7.22	
Dendrobium thyrsiflorum	SCoT	8	31 (DSO)		1 (DSO)	3.22 (DSO)	Bhattacharyya et al. (2015)
	ISSR		50 (ISO)		4 (ISO)	8.00 (ISO)	
		5	22 (DSO)		– (DSO)	–	
	SCoT + ISSR		42 (ISO)		2 (ISO)	4.76 (ISO)	
		13	53 (DSO)	–	–	1.88 (DSO)	
			92 (ISO)		–	6.52 (ISO)	
Malaxis wallichii	ISJ	15	84	5.6	4	4.76	Bose et al. (2017)
Phalaenopsis bellina (Rchb.f.)	RADP	8	172	21.5	18	17.24	Khoddamzadeh et al. (2010)
Phalaenopsis gigantea	ISSR	8	275	6.9	0	0	Samarfard et al. (2013)
Paphiopedilum niveum (Rchb.f)	RADP	10	Clone1-306 Clone2-273 Clone3-294	–	0	0	Soonthornkalump et al. (2019)
Rhynchostylis retusa (L.)	RADP	10	23	2.3	0	0	Oliya et al. (2021)

Note: Cryopreserved = +LN; Non-cryopreserved= −LN; Direct somatic embryogenesis = DSE; Indirect somatic embryogenesis = ISE; Direct shoot organogenesis = DSO; Indirect shoot organogenesis = ISO.

Acknowledgments

Potshangbam Nongdam is thankful to SERB (Science and Engineering Research Board), New Delhi, India, for financial support.

REFERENCES

Agbadje, E.T.A.E., Agbidinoukoun, A., Zandjanakou-Tachin, M., Cacai, G.T.H., Ahanhanzo, C., 2021. Mass production of bananas and plantains (Musa spp.) plantlets through in vitro tissue culture partway: A review. *Eur. J. Biol. Biotech.* 2,229.

Ai, P., Lu, L., Song, J., 2012. Cryopreservation of in vitro grown shoottips of *Rabdosia rubescens* by encapsulation-dehydration and evaluation of their genetic stability. *Plant Cell Tissue Organ Cult.* 108,381–387.

Amom, T., Nongdam, P., 2017. The use of molecular marker methods in plants: A review. *Int. J. Cur. Res. Rev.* 9,1–6.

Amom, T., Tikendra, L., Rahaman, H., Potshangbam, A., Nongdam, P., 2018. Evaluation of genetic relationship between 15 bamboo species of North-East India based on ISSR marker analysis. *Mol. Biol. Res. Commun.* 7,7–15.

Amom, T.,Tikendra, L., Apana, N., Goutam, M., Sonia, P., Koijam, A.S., Potshangbam, A.M., Rahaman, H., Nongdam, P., 2020. Efficiency of RAPD, ISSR, iPBS, SCoT and phytochemical markers in the genetic relationship study of five native and economical important bamboos of North-East India. *Phytochemistry.* 174,112330.

Antony, J.J.J., Poobathy, R., Danial, M., Sinniah, U.R., Subramaniam, S., 2012. Polymorphism analysis of cryopreserved *Dendrobium* Bobby Messina protocorm-like bodies (PLBs) using RAPD markers. *Plant Omics J.* 5(5),427–431.

Antony, J.J.J., Shamshir, R.A., Poobathy, R., Chew, B.L., Subramaniam, S., 2015. Somaclonal variations were not induced by the cryopreservation: Levels of somaclonal variations of in vitro and thawed protocorms of *Dendrobium* Bobby Messina analysed by SCoT and TRAP DNA markers. *S. Afr. J. Bot.* 100,148–157.

Arditti, J., 1992. *Fundamentals of orchid biology*. Wiley, New York.

Arditti, J., Krikorian, A.D., 1996. Orchid micropropagation: The path from laboratory to commercialization and an account of several unappreciated investigators. *Bot. J. Linn. Soc.*122(3),183–241. https://doi.org/10.1111/j.1095-8339.1996.tb02073.x.

Asghar, S., Ahmad, T., Hafiz, I.A., Yaseen, M., 2011. In vitro propagation of orchid (*Dendrobium nobile*) var. Emma white. *Afr. J. Biotechnol.* 10(16),3097–3103. https://doi.org/10.5897/ajb10.401.

Baker, A., Kaviani, B., Nematzadeh, G., Negahdar, N., 2014. Micropropagation of *Orchis catasetum* – A rare and endangered orchid. *Acta Sci. Pol. Hortorum Cultus.* 13(2),197–205.

Balilashaki, K., Ghehsareh, M.G., 2016. Micropropagation of *Phalaenopsis amabilis* var. 'Manila' by leaves obtained from in vitro culturing the nodes of flower stalks. *Not. Sci. Biol.* 8(2). https://doi.org/10.15835/nsb.8.2.9782.

Balilashaki, K., Vahedi, M., Karimi., 2015. In vitro direct regeneration from node and leaf explants of *Phalaenopsis* cv. 'Surabaya'. *Plant Tissue Cult. Biotechnol.* 25(2),193–205. https://doi.org/10.3329/ptcb.v25i2.26254.

Bandupriya, H.D.D., Iroshini, W.W.M.A., Perera, S.A.C.N., Vidhanaarachchi, V.R.M., Fernando, S.C., Santha, E.S., Gunathilake, T.R., 2017. Genetic fidelity testing using SSR marker assay confirms trueness to type of micropropagated coconut (*Cocos nucifera* L) plantlets derived from unfertilized ovaries. *The Open Plant Sci. J.* 10,46–54.

Bhatia, R., Singh, K.P., Jhang, T., Sharma, T.R., 2009. Assessment of clonal fidelity of micropropagated gerbera plants by ISSR markers. *Sci. Hortic.* 119(2),208–211.

Bhattacharyya, P., Kumaria, S., Job, N., Tandon, P., 2015. Phyto-molecular profiling and assessment of antioxidant activity within micropropagated plants of *Dendrobium thyrsiflorum*: A threatened, medicinal orchid. *Plant Cell Tissue Organ Cult.* 122,535–550.

Bhattacharyya, P., Kumaria, S., Tandon, P., 2016a. High-frequency regeneration protocol for *Dendrobium nobile*: A model tissue culture approach for propagation of medicinally important orchid species. *S. Afr. J. Bot.* 104,232–243.

Bhattacharyya, P., Kumaria, S., Job, N., Tandon, P., 2016b. *En-masse* production of elite clones of *Dendrobium crepidatum*: A threatened, medicinal orchid used in Tradition Chinese Medicine (TCM). *J. Appl. Res. Med. Aromat. Plants.* 3,168–176.

Bhattacharyya, P., Kumar, V., Staden, J.V., 2017. Assessment of genetic stability amongst micropropagated *Ansellia africana*, a vulnerable medicinal orchid species of Africa using SCoT markers. *S. Afr. J. Bot.* 108, 294–302. https://doi.org/10.1016/j. sajb.2016.11.007.

Bhattacharyya, P., Kumaria, S., Diengdoh, R., Tandon, P., 2014. Genetic stability and phytochemical analysis of the in vitro regenerated plants of *Dendrobium nobile* Lindl., an endangered medicinal orchid. *Meta Gene.* 2, 489–504. https://doi.org/10.1016/j. mgene.2014.06.003.

Bhattacharyya, P., Kumar, V., Staden, J.V., 2018a. In vitro encapsulation-based short term storage and assessment of genetic homogeneity in regenerated *Ansellia africana* (Leopard orchid) using gene-targeted molecular markers. *Plant Cell Tissue Organ Cult.* 133,299–310. https://doi.org/10.1007/s11240-018-1382-0.

Bhattacharyya, P., Paul, P., Kumaria, S., Tandon, P., 2018b. Transverse thin cell layer (t-TCL) mediated improvised micropropagation protocol for endangered medicinal orchid *Dendrobium aphyllum* Roxb: An integrated phytomolecular approach. *Acta Physiol. Plant.* 40,137.

Bhowmik, T.K., Rahman, M.M., 2020a. In vitro study of medicinally important orchid *Aerides multiflora* Roxb. from nodal and leaf explants. *J. Pharmacogn. Phytochem.* 9(4),179–184.

Bhowmik, T.K., Rahman, M.M., 2020b. Micropropagation of commercially important orchid *Dendrobium palpebrae* Lindl. through in vitro developed pseudobulb culture. *J. Adv. Biotechnol. Exp. Ther.* 3(3),225–232. https://doi.org/10.5455/jabet.2020.d128.

Borah, N.J., Chakraborty, S., Choudhury, S.R., Dutta, B.K., 2015. In vitro propagation of Paphiopedilum spicerianum (Reichb. F.) Pfitz. – A rare and endangered orchid species from Northeast India. *J. Orchid Soc. India.* 29,85–90.

Bose, B., Kumaria, S., Choudhury, H., Tandon, P., 2017. Insights into nuclear DNA content, hydrogen peroxide and antioxidative enzyme activities during transverse thin cell layer organogenesis and *ex-vitro* acclimatization of *Malaxis wallichii*, a threatened medicinal orchid. *Physiol. Mol. Biol. Plants.* 23,955–968. https://doi.org/ 10.1007/s12298-017-0474-3.

Bustam, B.M., Dixon, K.Y., Bunn, E., 2014. In vitro propagation of temperate Australian terrestrial orchids: Revisiting asymbiotic compared with symbiotic germination. *Bot. J. Linn. Soc.* 176(4),556–566. https://doi.org/10.1111/boj.12216.

Butiuc-Keu, A., Farkas, A., Cristea, V., 2016. Genetic stability assessment of in vitro plants by molecular markers. *Biologia.* 61(1),107–114.

Cassels, A.C., Curry, R.F., 2001. Oxidative stress and physiological, epigenetic and genetic variability in plant tissue culture: Implications for micropropagators and genetic engineers. *Plant Cell Tissue Organ Cult.* 64,145–157. doi:10.1023/A:1010692104861

Castillo-Perez, L.J., Martínez-Soto, D., Fortanelli-Martinez, J., Carranza-Alvarez, C., 2021. Asymbiotic seed germination, in vitro seedling development, and symbiotic acclimatization of the Mexican threatened orchid *Stanhopea tigrina*. *Plant Cell Tissue Organ Cult.* 146(2),249–257. https://doi.org/10.1007/s11240-021-02064-9.

Chauhan, S., Pathak, P., Anuprabha, K., Sharma, S., 2015. Teepol regeneration of *Eulophia dabia* through rhizome explants and flowering: A study in vitro. *J. Orchid Soc. India.* 29,61–65.

Chen, B., Trueman, S.J., Li, J., Li, Q., Fan, H., Zhang, J., 2014. Micropropagation of the endangered medicinal orchid, *Dendrobium officinale*. *Life Sci.* 11(9),526–530.

Chen, J.-T., Chang, W.-C., 2000. Plant regeneration via embryo and shoot bud formation from flower-stalk explants of *Oncidium* Sweet Sugar. *Plant Cell Tissue Organ Cult.* 62(2),95–100.

Chen, J.-T., Chang, W.-C., 2001. Effects of auxins and cytokinins on direct somatic embryogenesis on leaf explants of *Oncidium* 'Gower Ramsey'. *Plant Growth Regul.* 34(2),229–232. https://doi.org/10.1023/A:1013304101647.

Chin, C.K., Lee, Z.H., Mubbarakh, S.A., Antony, J.J.J., Chew, B.L., Subramaniam, S., 2019. Effects of plant growth regulators and activated charcoal on somaclonal variations of protocorm-like bodies (PLBs) of *Dendrobium sabin* blue orchid. *Biocatal. Agric. Biotechnol.* 22,101426.

Chongtham, N., Nongdam, P., Tewari, R., 2006. Biotechnological and molecular approaches for improvement of orchids. *Plant Cell Biotechnol. Mol. Biol.* 7(1),1–10.

Chookoh, N., Chiu, Y.-T., Chang, C., Hu, W.-H., Dai, T.-E., 2019. Micropropagation of *Tolumnia* orchids through induction of protocorm-like bodies from leaf segments. *Hort. Sci.* 54(7),1230–1236. https://doi.org/10.21273/HORTSCI13747-18.

Chugh, S., Guha, S., Rao, I.U., 2009. Micropropagation of orchids: A review on the potential of different explants. *Sci. Hortic.* 122(4),507–520.

Collard, B.C.Y., Mackill, D.J., 2009. Start codon targeted (SCoT) polymorphism: A simple, novel DNA marker technique for generating gene targeted markers in plants. *Plant Mol. Biol. Rep.* 27,86–93.

D'Amato, F., 1985. Cytogenetics of plant cell and tissue cultures and their regenerants. *CRC Crit. Rev. Plant Sci.* 3,73–112.

Das, S., Choudhury, M.D., Mazumder, P.B., 2013. In vitro propagation of *Arundina graminifolia* D. Don. Hochr – A bamboo orchid. *Asian J. Pharm. Clin. Res.* 6(5),156–158.

David, D., Jawan, R., Marbawi, H., Gansau, J.A., 2015. Organic additives improves the in vitro growth of native orchid *Vanda helvola* Blume. *Not. Sci. Biol.* 7(2),192–197. https://doi.org/10.15835/nsb.7.2.9546.

De, K.K., Sil, S., 2015. Protocorm-like bodies and plant regeneration from foliar explants of *Coelogyne flaccida*, a horticulturally and medicinally important endangered orchid of Eastern Himalaya. *Lankesteriana.* 15(2),151–158. https://doi.org/10.15517/lank.v15i2.20747.

Deb, C.R., Arenmongla, T., 2014. Development of cost-effective in vitro regeneration protocol of *Malaxis acuminata* D. Don a therapeutically important orchid using pseudobulbs as explant source. *J. Plant Stud.* 3(2),13–22. https://doi.org/10.5539/jps.v3n2p13.

Deb, C.R., Jakha, H.Y., 2020. Factors affecting asymbiotic immature seed culture and in vitro propagation of *Paphiopedilum insigne* (Wall. Ex. Lindl.) Pfitzer – A horticultural important vulnerable orchid. *Plant Cell Biotechnol. Mol. Biol.* 21(15&16),129–141.

Deb, C.R., Pongener, A., 2012. Studies on the in vitro regenerative competence of aerial roots of two horticultural important *Cymbidium* species. *J. Plant Biochem. Biot.* 21(2),235–241. https://doi.org/10.1007/s13562-011-0099-5.

Decruse, S.W., Gangaprasad, A., 2018. Restoration of *Smithsonia maculata* (Dalz.) Saldanha, an endemic and vulnerable orchid of Western Ghats through in vitro propagation. *J. Orchid Soc.* 32,25–32.

Devi, H.S., Devi, S.I., Singh, T.D., 2013. High frequency plant regeneration system of *Aerides odorata* Lour. through foliar and shoot tip culture. *Not. Bot. Horti Agrobot. Cluj-Napoca.* 41(1),169–176. https://doi.org/10.15835/nbha4119007.

Dey, A., Nongdam, P., Nandy, S., Mukherjee, S., Tikendra, L., Hazra, A.K., Pandey, D.K., 2021. Polyamine elicited aristolochic acid production in in vitro clonally fidel *Aristolochia indica* L.: An ISSR and RAPD markers and HPTLC based study. *S. Afr. J. Bot.* 140, 326–335.

Dey, A., Nongdam, P., Nandy, S., Tikendra, L., Mukherjee, A., Banerjee, S., Mukherjee, S., Pandey, D.K., 2019. Enhanced bacoside content in polyamine treated in vitro raised *Bacopa monnieri* (L.) *Wettst. S. Afr. J. Bot.* 123, 259–269.

Dixit, S., Mandal, B.B., Sangeeta, A., Srivastava, P.S., 2003. Genetic stability assessment of plants regenerated from cryopreserved embryogenic tissues of *Dioscorea bulbifera* l. using RAPD, biochemical and morphological analysis. *Cryo Letters.* 24,77–84.

Do, V.N.T., Hsu, S.-T., Lee, Y.-I., 2019. Clonal propagation in vitro of *Paphiopedilum* hybrids from adult plants. *HortScience.* 54(9),1565–1569. https://doi.org/10.21273/HORTSCI13791-18.

Dohling, S., Kumaria, S., Tandon, P., 2012. Multiple shoot induction from axillary bud cultures of the medicinal orchid, Dendrobium longicornu. *AoB Plants.* https://doi.org/10.1093/aobpla/pls032.

Dolezel, J., Greilhuber, J., Suda, J., 2007. Flow cytometry with plants: An overview. In: Dolezel, J., Greilhuber, J., Suda, J., (eds), *Flow cytometry with plant cells: Analysis of genes, chromosomes and genomes.* Wiley-VCH Verlag Gmbh& Co, Weinheim. 41–66.

Duncan, R. R., 1997. Tissue culture-induced variation and crop improvement. *Adv. Agron.* 58,201–240.

Flachsland, E., Terada, G., Fernandez, J.M., Medina, R., Cacace, A., Rey, H., Mroginski, L., 2011. Plant regeneration from root-tip culture of *Cyrtopodium brandonianum* Barb. Rodr. (Orchidaceae). *Propag. Ornam. Plants.* 11(4),184–188.

Gamborg, O.A., Miller, R.A., Ojima, K., 1968. Nutrient requirements of suspension culture of soybean root cells. *Exp. Cell Res.* 50(1),151–158.

Gantait, S., Sinniah, U.R., 2012. Rapid micropropagation of monopodial orchid hybrid (*Aranda* Wan Chark Kuan 'Blue' × *Vanda Coerulea* Grifft. Ex. Lindl.) through direct induction of protocorm-like bodies from leaf segments. *Plant Growth Regul.* 68(2),129–140. https://doi.org/10.1007/s10725-012-9698-y.

Gantait, S., Sinniah, U.R., 2013. Storability, post-storage conversion and genetic stability assessment of alginate-encapsulated shoot tips of monopodial orchid hybrid *Aranda* Wan Chark Kuan 'Blue' × *Vanda coerulea* Grifft. ex. Lindl. *Plant Biotechnol. Rep.* 7,257–266. https://doi.org/10.1007/s11816-012-0257-9.

Gantait, S., Mandai, N., Bhattacharyya, S., Das, P.K., 2010. Determination of genetic integrity in long-term micropropagated plantlets of *Allium ampeloprasum* L. using ISSR markers. *Biotechnology.* 9,218–223.

Gao, X., Yang, D., Cao, D., Ao, M., Sui, X., Wang, Q., Kimatu, J., Wang, L., 2010. In vitro micropropagation of Freesia hybrid and the assessment of genetic and epigenetic stability in regenerated plantlets. *J. Plant Growth Regul.* 29,257–267.

Gutiérrez, R.M.P., 2010. Orchids: A review of uses in traditional medicine, its phytochemistry and pharmacology. *J. Med. Plants Res.* 4(8),592–638.

Hajong, S., Kumaria, S., Tandon, P., 2013. Effect of plant growth regulators on regeneration potential of axenic nodal segments of *Dendrobium chrysanthum* Wall. Ex Lindl. *J. Agric. Sci. Tech.* 15,1425–1435.

Hardjo, P.H., Savitri, W.D., 2017. Somatic embryo from basal leaf segments of *Vanda tricolor* Lindl. Var. Pallida. *KnE Life Sci.* 3(5),173. https://doi.org/10.18502/kls.v3i5.990.

Hirano, T., Godo, T., Mii, M., Ishikawa, K., 2005. Cryopreservation of immature seeds of *Bletilla striata* by vitrification. *Plant Cell Rep.* 23,534–539.

Hsiao, Y.-Y., Pan, Z.-J., Hsu, C.-C., Yang, Y.-P., Hsu, Y.-C., Chuang, Y.-C., Shih, H.-H., Chen, W.-H., Tsai, W.-C., Chen, H.-H., 2011. Research on orchid biology and biotechnology. *Plant Cell Physiol.* 52(9),1467–1486. https://doi.org/10.1093/pcp/pcr100

Jain, S.M., 2001. Tissue culture derived variation in crop improvement. *Euphytica.* 118,153–166.

Jainol, J.E., Gansau, J.A., 2017. Embryogenic callus induction from leaf tip explants and protocorm-like body formation and shoot proliferation of *Dimorphorchis lowii*: Borneon endemic orchid. *Agrivita.* 39(1),1–10. https://doi.org/10.17503/agrivita.v39i1.895.

Kalendar, R., Amenov, A., Daniyarov, A., 2018. Use of retrotransposon-derived genetic markers to analyse genomic variability in plants. *Funct. Plant Biol.* 40(1),15–29. doi:10.1071/fp18098.

Kang, H., Kang, K.W., Kim, D.H., Sivanesan, I., 2020. In vitro propagation of *Gastrochilus matsuran* (Makino) Schltr., an endangered epiphytic orchid. *Plants.* 9(4),1–10. https://doi.org/10.3390/plants9040524.

Karp, A., 1995. Somaclonal variation as a tool for crop improvement. *Euphytica.* 85,295–302.

Kaur, S., 2017. In vitro regeneration of shoots from nodal explants of *Dendrobium chrysotoxum* Lindl. *J. Hortic. Res.* 25(1),27–34. https://doi.org/10.1515/johr-2017-0003.

Kaur, S., Bhutani, K.K., 2009. In vitro propagation of *Vanda testacea* (Lindl.) Reichb. f. - A rare orchid of high medicinal value. *Plant Tissue Cult. Biotechnol.* 19(1),1–7. https://doi.org/10.3329/ptcb.v19i1.4077.

Kaur, S., Bhutani, K.K., 2010. Micropropagation of *Malaxis acuminata* d. Don: A rare orchid of high therapeutic value. *J. Med. Aromat. Plants.* 1(2),29–33.

Kaur, S., Pathak, P., 2015. Reversion of reproductive phase to vegetative phase in the inflorescence segments of *Saccolabium papillosum* Lindl.- A study in vitro. *J. Orchid Soc. India.* 29,75–79.

Kerbauy, G.B., 1984. Regeneration of protocorm-like bodies through in vitro culture of root-tips of *Catasetum* (orchidaceae). *Z. Pflanzenphysiol.* 113,287–291.

Khoddamzadeh, A.A., Sinniah, U.R., Kadir, M.A., Kadzimin, S.B., Mahmood, M., Sreeramanan, S., 2010. Detection of somaclonal variation by random amplified polymorphic DNA analysis during micropropagation of *Phalaenopsis bellina* (Rchb.f.) Christenson. *Afri. J. Biotechnol.* 40(9),6632–6639.

Khoddamzadeh, A.A., Sinniah, U.R., Kadir, M.A., Kadzimin, S.B., Mahmood, M., Sreeramanan, S., 2011. In vitro induction and proliferation of protocorm-like bodies (PLBs) from leaf segments of *Phalaenopsis bellina* (Rchb.f.) Christenson. *Plant Growth Regul.* 65(2),381–387. https://doi.org/10.1007/s10725-011-9611-0.

Khor, S.P., Yeow, L.C., Poobathy, R., Zakaria, R., Chew, B.L., Subramaniam, S., 2020. Droplet-vitrification of *Aranda* Broga Blue orchid: Role of ascorbic acid on the antioxidant system and genetic fidelity assessments via RADP and SCoT markers. *Biotechnol. Rep.* e00448.

Kim, J.-H., Son, S.W., Kim, S.Y., Jeong, M.J., 2021. Asymbiotic seed germination and in vitro seedling development of *Pelatantheria scolopendrifolia*, a rare epiphytic orchid native to Korea. *Rhizosphere.* 19,100371. https://doi.org/10.1016/j.rhisph.2021.100371.

Knudson, L., 1946. A new nutrient solution for germination of orchid seed. *Am. Orchid Soc. Bull.* 15,214–217.

Kraus, J.E., Kerbuay, G.B., 1992. Formation of protocorm-like bodies from root apices of *Catasetum pileatum* (Orchidaceae) cultivated in vitro. II. Some non-hormonal requirements involved in the regeneration. *Bolet. Bot. USP.* 13,31.

Kunakhonnuruk, B., Inthima, P., Kongbangkerd, A., 2018. In vitro propagation of *Epipactis flava* Seidenf., an endangered rheophytic orchid: A first study on factors affecting asymbiotic seed germination, seedling development and greenhouse acclimatization. *Plant Cell Tissue Organ Cult.* 135(3),419–432. https://doi.org/10.1007/s11240-018-1475-9.

Larkin, P.J., Scowcroft, W.R., 1981. Somaclonal variation-a novel source of variability from cell cultures or plant improvement. *Theor. Appl. Genet.* 60,197–214.

Ma, N.L., Khoo, S.C., Lee, J.X., Soon, C.F., Shukor, N.A.A.B., 2020. Efficient micropropagation of *Dendrobium aurantiacum* from shoot explant. *Plant Sci. Today.* 7(3),476–482.

Maharjan, S., Pradhan, S., Thapa, B.B., Pant, B., 2019. In vitro propagation of endangered orchid, *Vanda pumila* Hook.f. through protocorms culture. *Am. J. Plant Sci.* 10(07),1220–1232. https://doi.org/10.4236/ajps.2019.107087.

Maharjan, S., Thakuri, L.S., Thapa, B.B., Pradhan, S., Pant, K.K., Joshi, G.P., Pant, B., 2020. In vitro propagation of the endangered orchid *Dendrobium chryseum* Rolfe from protocorms culture. *Nepal J. Sci. Technol.* 19(1),39–47. https://doi.org/10.3126/njst.v19i1.29737.

Mallon, R., Rodriguez-Oubina, J., Gonzalez, M., 2010. In vitro propagation of the endangered plant *Centaurea ultreiae*: Assessment of genetic stability by cytological studies, flow-cytometry and RAPD analysis. *Plant Cell Tissue Organ Cult.* 101,31–39.

Mamun, A., Islam, M., Ahmed, M., Ghosh, G., 2018. In vitro mass propagation from shoot tip of *Dendrobium* Red Bull - An endangered epiphytic orchid. *Plant Tissue Cult. Biotech.* 28(2),161–169. https://doi.org/10.3329/ptcb.v28i2.39675

Martin, K.P., Madassery, J. 2006. Rapid in vitro propagation of Dendrobium hybrids through direct shoot formation from foliar explants, and protocorm-like bodies. *Sci. Hortic.* 108(1),95–99.

Mitra, G.C., Prasad, R.N., Roychowdhury, A., 1976. Inorganic salts and differentiation of protocorms in seed callus of orchid and correlative changes in its free amino acid content. *Indian J. Exp. Biol.* 14,350–351.

Mohanty, P., Das, J., 2013. Synthetic seed technology for short term conservation of medicinal orchid *Dendrobium densiflorum* Lindl. Ex Wall and assessment of genetic fidelity of regenerants. *Plant Growth Regul.* 70,297–303.

Mondal, T., Aditya, S., Banerjee, N., 2013. In vitro axillary shoot regeneration and direct protocorm-like body induction from axenic shoot tips of *Doritis pulcherrima* Lindl. *Plant Tissue Cult. Biotech.* 23(2),251–261.

Morel, G., 1960. Producing virus-free *Cymbidiums*. *Am. Orchid Soc. Bull.* 29,495–497.

Muralidharan, K., Wakeland, E.K., 1993. Concentration of primer and template qualitatively affects products in random-amplified polymorphic DNA PCR. *Bio Techniques.* 14,362–364.

Murashige, T., Skoog, F., 1962. A revised medium for rapid growth and bio assays with tobacco tissue culture. *Physiol. Plant.* 15(3),473–497.

Muthiah, M., Mahesh, R., Raju, G., Benniamin, A., Muthuchelian, K., 2010. In vitro propagation of *Dendrobium nanum* through rhizome bud culture. *Int. J. Biol. Technol.* 1(2),50–54.

Naing, A.H., Chung, J.D., Lim, K.B., 2011. Plant regeneration through indirect somatic embryogenesis in *Coelogyne cristata* orchid. *Am. J. Plant Sci.* 2(2),262–267. https://doi.org/10.4236/ajps.2011.22028.

Nongdam, P., Chongtham, N., 2011. In vitro rapid propagation of *Cymbidium aloifolium* (L.) Sw.: A medicinally important orchid via seed culture. *J. Biol. Sci.* 11(3),254–260.

Nongdam, P., Chongtham, N., 2012. In vitro seed germination and mass propagation of *Cymbidium dayanum* Reichb.: An important ornamental orchid of North-East India. *Trends Hortic. Res.* 2(2),28–37.

Nongdam, P., Chongtham, N., Tewari, R., 2006. In vitro multiplication of *Cymbidium* orchids via embryo culture. *Plant Cell Biotechnol. Mol. Biol.* 7(3),145–150.

Nongdam, P., Tikendra, L., 2014. Establishment of an efficient in vitro regeneration protocol for rapid and mass propagation of *Dendrobium chrysotoxum* Lindl. using seed culture. *Sci. World J.* https://doi.org/10.1155/2014/740150.

Nuraini, I., Shaib, M.J., 1992. Micropropagation of orchids using scape nodes as the explant material. *Acta Hortic.* 292, 169–172.

Obsuwan, K., Seraypheap, K., Thepsithar, C., 2019. Effects of calcium silicate and proline-induced salt tolerance on the in vitro propagation of *Dendrobium* Sonia 'Red Jo'. *Acta Hortic.* 1262,87–91.

Oliya, B.K., Chand, K., Thakuri, L.S., Baniya, M.K., Sah, A.K., Pant, B., 2021. Assessment of genetic stability of micropropagated plants of *Rhynchostylis retusa* (L.) using RAPD markers. *Sci. Hortic.* 281,110008. https://doi.org/10.1016/j.

Palombi, M., Damiano, C., 2002. Comparison between RAPD and SSR molecular markers in detecting genetic variation in kiwi fruit (*Actinidia deliciosa* A. Chev). *Plant Cell Rep.* 20,1061–1066.

Pant, B., Thapa, D., 2012. In vitro mass propagation of an epiphytic orchid, *Dendrobium primulinum* Lindl. through shoot tip culture. *Afr. J. Biotechnol.* 11(42),9970–9974. https://doi.org/10.5897/ajb11.3106.

Panwar, D., Ram, K., Harish, Shekhawat, N.S., 2012. In vitro propagation of *Eulophia nuda* Lindl., an endangered orchid. *Sci Hortic.*139,46–52. https://doi.org/10.1016/j.scienta.2012.01.011.

Park, H.Y., Kang, K.W., Kim, D.H., Sivanesan, I., 2018. In vitro propagation of *Cymbidium goeringii* Reichenbach Fil. through direct adventitious shoot regeneration. *Physiol. Mol. Biol. Plants.* 24(2),307–313. https://doi.org/10.1007/s12298-017-0503-2.

Parthibhan, S., Rao, M.V., Kumar, T.S., 2015. In vitro regeneration from protocorms in *Dendrobium aqueum* Lindley - An imperiled orchid. *Genet. Eng. Biotechnol.* 13(2),227–233. https://doi.org/10.1016/j.jgeb.2015.07.001.J.

Pathak, P., Verma, S., Prakash, A., Mahant, K.C., 2017. Regenerative competence of an ornamentally important epiphytic orchid, *Rhynchostylis gigantea* (Lindl.) Ridl. through leaf segments: A study in vitro. *J. Orchid Soc. India.* 31,97–101.

Paudel, M.R., Pant, B., 2012. In vitro micropropagation of rare orchid (*Esmeralda clarkei* Rchb. f.) from shoot tip section. *Int. J. Biol. Pharm. Allied Sci.* 1(11),1587–1597.

Paul, S., Kumaria, S., Tandon, P., 2012. An effective nutrient medium for asymbiotic seed germination and large-scale in vitro regeneration of *Dendrobium hookerianum*, a threatened orchid of Northeast India. *AoB Plants*. 12(1), 1–7. https://doi.org/10.1093/aobpla/plr032.

Picolotto, D.R.N., Paiva, V.B.D., Barros, F.D., Padilha, D.R.C., Cruz, A.C.F.D., Otoni, W.C., 2017. Micropropagation of *Cyrtopodium paludicolum* (Orchidaceae) from root tip explants. *Crop Breed. Appl. Biotechnol.* 17,191–197. http://dx.doi.org/10.1590/1984-70332017v17n3a30.

Pinto, G., Louveiro, T., Santos, C., 2004. Analysis of the genetic stability of *Eucaliptus globulus* Labill. somatic embryos by flow cytometry. *Theor. Appl. Genet.* 109,580–587.

Poobathy, R., Zakaria, R., Murugaiyah, V., Subramaniam, S., 2019. Surface sterilization and micropropagation of *Ludisia discolor*. *Biocatal. Agric. Biotechnol.* 22,101380. https://doi.org/10.1016/j.bcab.2019.101380.

Popova, E.V., Kima, H., Paek, H., 2010. Cryopreservation of coriander (*Coriandrum sativum* L.) somatic embryos using sucrose preculture and air desiccation. *Sci. Hortic.* 124,522–528.

Popova, E.V., Lee, E., Wu, C., Hahn, E., Paek, K., 2009. A simple method for cryopreservation of *Ginkgo biloba* callus. *Plant Cell Tissue Organ Cult.* 97,337–343.

Pradhan, S., Paudel, Y.P., Pant, B., 2013. Efficient regeneration of plants from shoot tip explants of *Dendrobium densiflorum* Lindl., a medicinal orchid. *Afr. J. Biotechnol.* 12(12),1378–1383. https://doi.org/10.4314/ajb.v12i12.

Prado, M.J., Rodriguez, E., Rey, L., Gozalez, M.V., Santos, C., Rey, M., 2010. Detection of somaclonal variants in somatic embryogenesis regenerated plants of *Vitis vinifera* by flow cytometry and microsatellite markers. *Plant Cell Tissue Organ Cult.* 102,49–59.

Preeta, L.K., Shylaraj, K.S., Rohini, P.C., 2017. An improved method for rapid propagation of *Phalaenopsis* hybrids via culture of longitudinally bisected shoot tips. *J. Trop. Agric.* 55(1),45–51.

Priyakumari, I., Sabrina George,T., Rajmohan, K., 2013. Influence of plant growth regulators on in vitro clonal propagation of *Dendrobium* Sonia 'Earsakul'. *J. Bio. Innov.* 2(2),51–58.

Rani, V., Raina, S.N., 2000. Genetic fidelity of organized meristem-derived micropropagated plants: A critical reappraisal. *In Vitro Cell Dev. Biol. Plant* 36,319–330.

Rathore, N.S., Rai, M.K., Phulwaria, M., Rathore, N., Shekhawat, N., 2014. Genetic stability in micropropagated *Cleome gynandra* revealed by SCoT analysis. *Acta Physiol. Planta.* 36(2),555–559. https://doi.org/10.1007/s11738-013-1429-0.

Raynalta, E., Elina, J., Sudarsono, S.D., 2018. Clonal fidelity of micro propagated *Phalaenopsis* plantlets based on assessment using eighteen Ph-Pto SNAP marker loci. *Agrivita J. Agri. Sci.* 40(3),390–402.

Riva, S.S., Islam, A., Hoque, M.E., 2017. In vitro regeneration and rapid multiplication of *Dendrobium bensoniae*, an indigenous ornamental orchid. *The Agriculturists.* 14(2),24–31.

Rodrigues, L.A., Paiva, V.B.D., Boaretto, A.G., Oliveira, J.F.D., Torrezan, M.D.A., Lima, S.F.D., Otoni, W.C., 2015. In vitro propagation of *Cyrtopodium saintlegerianum* Rchb. f. (Orchidaceae), a native orchid of the Brazilian Savannah. *Crop Breed. Appl. Biotechnol.* 15(1),10–17. https://doi.org/10.1590/1984-70332015v15n1a2.

Rohela, G.K., Jogam, P., Bylla, P., Reuben, C., 2019. Indirect regeneration and assessment of genetic fidelity of acclimated plantlets by SCoT, ISSR and RAPD markers in *Rauwolfia tetraphylla* L.: An endangered medicinal plant. *Biomed. Res. Int.* 3698742.

Rotor, G., 1949. A method of vegetative propagation of *Phalaenopsis* species and hybrids. *Am. Orchid Soc. Bull.* 18,738–739.

Roy, A.R., Sanjeev, S., Pattanayak, A., Deka, B.C., 2012. TDZ induced micropropagation in *Cymbidium giganteum* Wall. Ex Lindl. and assessment of genetic variation in the regenerated plants. *Plant Growth Regul.* 68,435–445.

Sabir, A., Newbury, H.J., Todd, G., Catty, J., Ford-Lloyd, B.V., 1992. Determination of genetic stability using isozymes and RFLPs in beet plants generated in vitro. *Theor. Appl. Genet.* 84,113–117.

Salvi, N.D., George, L., Eapen, S., 2001. Plant regeneration from leaf base callus of turmeric and random amplified polymorphism DNA analysis of regenerated plants. *Plant Cell Tissue Organ Cult.* 66,113–119.

Samarfard, S., Kadir, M.A., Kadzimin, S.B., Ravanfar, S., 2013. Genetic stability of in vitro multiplied *Phalaenopsis gigantea* protocorm-like bodies as affected by chitosan. *Not. Bot. Horti. Agrobot.* 41(1),177–183. https://doi.org/10.15835/nbha4118315.

Samarfard, S., Kadir, M.A., Kadzimin, S.B., Saud, H.M., Ravanfar, S.A., Danaee, M., 2014. In vitro propagation and detection of somaclonal variation in *Phalaenopsis gigantea* as affected by chitosan and thidiazuron combinations. *HortScience.* 49(1),82–88. https://doi.org/10.21273/hortsci.49.1.82.

Seon, K.M., Kim, D.H., Kang, K.W., Sivanesan, I., 2018. Highly competent in vitro propagation of *Thrixspermum japonicum* (Miq.) Rchb.f., a rare epiphytic orchid. *In Vitro Cell. Dev. Biol.—Plant.* 54(3),302–308. https://doi.org/10.1007/s11627-018-9890-5.

Sharma, V., 2012. Regenerative competence in root explants of *Cattleya* Hybrid, an endangered genera: A study in vitro. *Int. J. Sci. Res.* 2(12),2011–2013.

Sharma, V., 2019a. Regenerative competence in root explants of *Rhynchostylis gigantea*, an endangered genera: An in vitro study. *Int. J. Conserv. Sci.* 10(1),133–138.

Sharma, V., 2019b. In vitro conversion of root meristem into protocorm-like bodies in *Rhynchostylis retusa*-an endangered Fox Tail orchid. *Eco. Env. Cons.* 25(4),1812–1816.

Shen, H.-J., Chen, J.-T., Chung, H.-H., Chanf, W.-C., 2018. Plant regeneration via direct somatic embryogenesis from leaf explants of *Tolumnia* Louise Elmore 'Elsa'. *Bot. Stud.* 59(4). https://doi.org/10.1186/s40529-018-0220-3.

Sherif, N.A., Benjamin, J.F., Kumar, T.S., Rao, M.V., 2018. Somatic embryogenesis, acclimatization and genetic homogeneity assessment of regenerated plantlets of *Anoectochilus elatus* Lindl., an endangered terrestrial jewel orchid. *Plant Cell Tissue Organ Cult.* 132(2),303–316. https://doi.org/10.1007/s11240-017-1330-4.

Sherif, N.A., Kumar, T.S., Rao, M.V., 2017. In vitro propagation and genetic stability assessment of an endangered terrestrial jewel orchid *Anoectochilus elatus* Lindl. *Indian J. Ex. Biol.* 55(12),853–863.

Sherif, N.A., Kumar, T.S., Rao, M.V., 2020. DNA barcoding and genetic fidelity assessment of micropropagated *Aenhenrya rotundifolia* (Blatt.) C.S. Kumar and F.N. Rasm.: A critically endangered jewel orchid. *Physiol. Mol. Biol. Plants.* 26,2391–2405. https://doi.org/10.1007/s12298-020-00917-9.

Singh, N., Kumaria, S., 2020. A combinational phytomolecular-mediated assessment in micropropagated plantlets of *Coelogyne ovalis* Lindl.: A horticultural and medicinal orchid. *Proc. Natl. Acad. Sci., India, Sect. B Biol. Sci.* 90(2),455–466.

Singh, S.R., Dalal, S., Singh, R., Dhawan, A.K., Kalia, R.K., 2013. Ascertaining clonal fidelity of micropropagated plants of *Dendrocalamus hamiltonii* Nees et Arn. ex Munro using molecular markers. *In Vitro Cell Dev. Biol.* 49,572–583.

Singh, A.K., Rana, M.K., Singh, S., Kumar, S., Kumar, R., Singh, R., 2014. CAAT box-derived polymorphism (CBDP): A novel promoter-targeted molecular marker for plants. *J. Plant Biochem. Biotechnol.* 23, 175–183. https://doi.org/10.1007/s13562-013-0199-5

Sinha, P., Jahan, M.A.A., 1970. Clonal propagation of *Phalaenopsis amabilis* (L.) BL. Cv. 'Golden Horizon' through in vitro culture of leaf segments. *Bangladesh J. Sci. Ind. Res.*46(2),163–168. https://doi.org/10.3329/bjsir.v46i2.8182.

Soonthornkalump, S., Nakkanong, K., Meesawat, U., 2019. In vitro cloning via direct somatic embryogenesis and genetic stability assessment of *Paphiopedilum niveum* (Rchb.f.) Stein: The endangered Venus's Slipper Orchid. *In Vitro Cell. Dev. Biol.—Plant.* 55(3),265–276. https://doi.org/10.1007/s11627-019-09981-7.

Sopalun, K., Thammasiri, K., Ishikawa, K., 2010. Micropropagation of the Thai orchid *Grammatophyllum speciosum* Blume. *Plant Cell Tissue Organ Cult.* 101(2),143–150. https://doi.org/10.1007/s11240-010-9671-2.

Sr Sagaya Mary, B., Divakar, K.M., 2016. In vitro propagation of *Dendrobium jerdonianum* Wight through flower stalk, leaf, nodes/ internodes as explants. *Am. J. Life Sci.* 4(3),76–81.

Srivastava, D., Gayatri, M.C., Sarangi, S.K., 2015. In vitro seed germination and plant regeneration of an epiphytic orchid *Aerides ringens* (Lindl.). *Fischer. Indian J. Biotechnol.* 14(4),574–580.

Stover, R.H., 1987. Somaclonal variation in Grand Naine and Saba banana in nursery and field. *ACIAR Proc. Series, Australian Center for Intl. Agric. Res.* 21,136–139.

Tan, B.C., Chin, C.F., Alderson, P., 2011. Optimisation of plantlet regeneration from leaf and nodal derived callus of *Vanilla planifolia* Andrews. *Plant Cell Tissue Organ Cult.* 105(3),457–463. https://doi.org/10.1007/s11240-010-9866-6.

Teixeira da Silva, J.A., Winarto, B., Dobránszki, J., Cardoso, J.C., Zeng, S.J., 2016. Tissue disinfection for preparation of Dendrobium in vitro culture. *Folia Hortic.* 28,57–75.

Than, M.M.M., Majumder, A., Pal, A., Jha, S., 2011. Genomic variations among in vitro regenerated *Bulbophyllum auricomum* Lindl. plants. *Nucleus.* 54(1),9–17. https://doi.org./ 10.1007/s13237-011-0025-1.

Tikendra, L., Amom, T., Nongdam, P., 2018. Effect of phytohormones on rapid in vitro propagation of *Dendrobium thyrsiflorum* Rchb.f.: An endangered medicinal orchid. *Pharmacogn. Mag.* 14,495–500.

Tikendra, L., Koijam, A.S., Nongdam, P., 2019a. Molecular markers based genetic fidelity assessment of micropropagated *Dendrobium chrysotoxum* Lindl. *Meta Gene.* 20,100562. https://doi.org/10. 1016/j.mgene.2019.100562.

Tikendra, L., Amom, T., Nongdam, P., 2019b. Molecular genetic homogeneity assessment of micropropagated *Dendrobium moschatum* Sw.-A rare medicinal orchid, using RAPD and ISSR markers. *Plant Gene.* 19,100196. https://doi.org/10.1016/j.plgene.2019.100196.

Tikendra, L., Potshangbam, A.M., Amom, A., Dey, A., Devi, T.R., Sahoo, M.R., Nongdam, P., 2021a. RAPD, ISSR, and SCoT markers based genetic stability assessment of micropropagated Dendrobium fimbriatum Lindl. var.oculum Hk.f. – an important endangered orchid. Physiol. Mol. Biol. Plts. 27,341–357.

Tikendra, L., Choudhary, R., Sanayaima, R., Dey, A., Potshangbam, A.M., Nongdam, P., 2021b. Micropropagation of bamboos and clonal fidelity assessment using molecular markers. In: Ahmad, Z., Ding, Y., Shahzad, A. (eds), *Biotechnological advances in bamboo*. Springer, Singapore. https://doi.org/10.1007/978-981-16-1310-4_7.

Tikendra, L., Apana, N., Potshangbam, A.M., Amom, T., Choudhary, R., Sanayaima, R., Dey, A., Nongdam, P., 2021c. *Dendrobium* sp.: In vitro propagation of genetically stable plants and ethnomedicinal uses. In: Merillon, J.M., Kodja, H. (eds), *Orchids phytochemistry, biology and horticulture*. Reference series in phytochemistry. Springer, Cham. https://doi.org/10.1007/978-3030-11257-8_30-1

Tikendra, L., Potshangbam, A.M., Amom, A., Dey, A., Nongdam, P., 2021d. Understanding the genetic diversity and population structure of *Dendrobium chrysotoxum* Lindl.- An endangered medicinal orchid and implication for its conservation. *S. Afr. J. Bot.* 138,364–376. https://doi.org/10.1016/j.sajb.2021.01.002

Trujillo, I., Garcia, E., 1996. Strategies for obtaining somaclonal variants resistant to yellow Sigatoka (*Mycosphaerella musicola*). *Infomusa*. 5(2),12–13.

Utami, E.S.W., Hariyanto, S., 2019. In vitro seed germination and seedling development of a rare Indonesian native orchid *Phalaenopsis amboinensis* J.J.Sm. *Scientifica*. 2019,1–6. https://doi.org/10.1155/2019/8105138.Sci.

Vacin, E.F., Went, F.W., 1949. Some pH changes in nutrient solutions. *Bot. Gaz.* 110,605–613.

Vanyushin, B.F., 1984. Replicative DNA methylation in animals and higher plants. *Curr. Top. Microbiol. Immunol.* 108,99–114.

Vasundhara, Promila, P., Prakash, A., 2019. In vitro shoot induction and regeneration potential of floral buds in *Crepidium acuminatum* (D . Don) Szlach., A medicinal ayurvedic plant from Northwestern Himalayas. *J. Orchid Soc.* 33,43–48.

Vázquez, A.M., 2001. Insight into somaclonal variation. *Plant Biosyst.* 135,57–62.

Verma, S., Pathak, P., 2018. Regenerative competence in root explants of medicinally important orchid *Cymbidium aloifolium*: An in vitro study. *Vegetos*. 31(1),100–103. doi:10.5958/2229-4473.2018.00014.9.

Wang, C.X., Tian, M., 2014. Callus-mediated and direct protocorm-like body formation of *Bletilla striata* and assessment of clonal fidelity using ISSR markers. *Acta Physiol. Plant*. 36,2321–2330. https://doi.org/10.1007/s11738-014-1594-9.

Wannajindaporn, A., Poolsawat, O., Chaowiset, W., Tantasawat, P.A., 2014. Evaluation of genetic variability in in vitro sodium azide-induced *Dendrobium* 'Earsakul' mutants. *Genet. Mol. Res.* 13(3),5333–5342.

Warghat, A.R., Bajpai, P.K., Srivastava, R.B., Chaurasia, O.P., Chauhan, R.S., Sood, H., 2014. In vitro protocorm development and mass multiplication of an endangered orchid, *Dactylorhiza hatagirea*. *Turk. J. Bot.* 38(4),737–746. https://doi.org/10.3906/bot-1308-48.

Wimber, D.E., 1965. Additional observations on clonal multiplication of Cymbidiums through culture of shoot meristems. *Cymbidium Soc. News*. 20,7–10.

Winarto, B., Samijan, 2018. Axillary shoots derived from shoot tips in in vitro mass propagation of *Anoectochilus formosanus* Hayata. *Agraarteadus*. 29(2),121–130. https://doi.org/10.15159/jas.18.11.

Wolfe, A.D., Liston, A., 1998. Contributions of PCR-based methods to plant systematics and evolutionary biology. In Soltis, D.E., Soltis, P.S., Doyle, J.J., (eds), *Molecular systematics of plants II, DNA sequencing*. Kluwer Academic Publishers, New York, 43–86.

Wu, K., Zeng, S., da Silva, J.A.T., Chen, Z., Zhang, J., Yang, Y., Duan, J., 2012. Efficient regeneration of *Renanthera* Tom Thumb 'Qilin' from leaf explants. *Sci Hortic*.135,194–201. https://doi.org/10.1016/j.scienta.2011.11.028.

Wu, K., Zeng, S., Lin, D., da Silva, J.A.T., Bu, Z., Zhang, J., Duan, J., 2014. In vitro propagation and reintroduction of the endangered *Renanthera imschootiana* Rolfe. *PLoS ONE*. 9(10). https://doi.org/10.1371/journal.pone.0110033.

Yam, T.W., Arditti, J., 2017. *Micropropagation of orchids*. John Wiley & Sons, UK.

Zhenxun, W., Hongxian, L., 1997. Chromosome aberration in banana micropropagation. *Acta Genet. Sin.* 24,550–560.

11

Tissue Culture Studies in Lamiaceae: A Review

A.V. Deepa
Central University of Kerala
Kerala, India

Dennis T. Thomas
Central University of Kerala
Kerala, India

CONTENTS

DOI: 10.1201/9781003239932-11

11.1 Introduction

The Lamiaceae (formerly Labiatae) family is also known as the mint family or the dead nettle family. The name Labiatae originated from its bilipped corolla, which is a characteristic feature of the family. Lamiaceae is one of the largest dicot families, which comprises about 236 genera and 7200 species. The largest genera of Lamiaceae include *Salvia* with 900 species, followed by *Scutellaria* (360), *Coleus* (325), *Plectranthus* (300), *Hyptis* (280), *Teucrium* (250), *Thymus* (220) and *Nepeta* (200) (Harley et al. 2004). They are highly evolved with their epipetalous stamens and gamopetalous flowers. Most of the members of this family are highly aromatic and are widely used for their medicinal and culinary properties (Saraç and Uğur 2007). While lavender, mint, basil, rosemary, etc. are some important herbs that produce highly valued essential oils, oregano, thyme, savory, sage, etc. are important culinary herbs of the family Lamiaceae. Though most of the members of Lamiaceae are herbs and shrubs, they also include some trees and vines. This family is well known for the production of a wide array of alkaloids, essential oils, saponins, tannins and organic acids (Giuliani and Bini 2008).

Though some of these plants are cultivated, many of them are still collected from their wild habitat. This has caused over exploitation of the plants in the wild for their essential oils and other by-products. This has caused an urgent need for their conservation and cultivation. Micropropagation is an alternative means of propagation that can be employed for the mass multiplication of plants in a relatively shorter time. Recent techniques of propagation have been developed to facilitate large-scale production of true-to-type plants and for the improvement of the species using genetic engineering techniques in the next century. This review is a consolidation of tissue culture-mediated propagation and conservation of selected economically and medicinally important Lamiaceae members to date.

11.2 *Agastache foeniculum* (Pursh) Kuntze

Agastache foeniculum (Pursh) Kuntze (common name: anise hyssop) is a perennial herb, which is highly medicinal and aromatic (Ayres and Widrlechner 1994). The major attraction of the genus *Agastache* is that its aerial parts contain a variety of essential oils (Charles et al. 1991). *A. foeniculum* contains several other compounds such as monoterpenes, phenyl propanoids and phenolic compounds, including rutin, galangin, apigenin, rosmarinic acid and chlorogenic acid (Nourozi et al. 2014). Methyl chavicol and linalool are some of the essential oils isolated from *Agastache*, which are used in the pharmaceutical, flavoring, sanitary and perfume industries (Mazza and Kiehn 1992; Wilson et al. 1992). Traditional medical practitioners use *A. foeniculum* for the treatment of cough and lung diseases (Nourozi et al. 2014).

Moharami et al. (2014) developed a micropropagation protocol for *A. foeniculum*. Shoot induction was successfully obtained in cotyledons, hypocotyls, shoot tips and nodal segments when cultured on MS medium supplemented with a combination of 6-benzylaminopurine (BAP) and 1-naphthaleneacetic acid (NAA). Shoot regeneration frequency was highest in nodal explants, followed by cotyledonary, shoot tip and hypocotyl explants. Nodal explants gave maximum shoot induction (53.7 shoots) on Murashige and Skoog (MS) medium supplemented with 8.8 μM BAP and 1 μM indole-3-acetic acid (IAA). On half-strength MS medium containing 1.1 μM indole-3-butyric acid (IBA), 92% of shoots produced roots. The plantlets were then moved to *ex vitro* conditions and were eventually established in the field with 100% survival.

11.3 *Ajuga bracteosa*

Ajuga bracteosa (common name: Bungle) is a perennial, erect, hairy herb distributed in higher altitudes of temperate and subtropical countries such as India, China, Afghanistan, Pakistan, Bhutan, etc. (Kirtikar and Basu 1918). The herb is used in the treatment of rheumatism, palsy, gout, amenorrhea, hypertension, sore throat and jaundice, and as a blood purifier (Hamayun et al. 2006; Islam et al. 2006; Chopra et al. 1986). Leaves of *A. bracteosa* are used locally as a remedy for pimples, boils, burns, headache, measles and stomach acidity (Sharma et al. 2004). There are several reports on anthelmintic, anti-malarial, anti-cancerous and anti-inflammatory activity of *A. bracteosa* (Singh et al. 2006; Njoroge and Bussmann 2006; Kuria et al. 2002).

An investigation on indirect organogenesis from leaf, petiole and intermodal explants of *A. bracteosa* was conducted by Jan et al. (2014). Explant type and the type and concentration of plant growth regulator had a significant effect on callus formation. Leaf explants gave maximum callus induction on MS medium supplemented with 22.2 μM BAP within 19 days of culture. While petiole explants gave optimum callusing on MS medium with 11.41 μM and 17.12 μM IAA after 51 days, internodal explants responded well on MS medium with 8.78 μM BAP and 26.84 μM NAA after 35 days of culture. Maximum multiples of shoots were formed after 28 days of culture of the callus in MS medium containing 22.2 μM BAP.

An effective micropropagation method for *A. bracteosa* has been developed by Kaul et al. (2013). Among the three explants (leaf, root and petiole) cultured on MS medium supplemented with various cytokinin (kinetin (KN), BAP) and auxin (IAA) combinations, leaf gave optimum results followed by petiole and root. Leaf explants gave an average of 41 shoots of 8.4 cm height when cultured on MS medium augmented with 22.2 μM BAP and 11.42 μM IAA. Optimum rooting (100%) was obtained at 2.46 μM IBA and eventually hardened with 82% survival rate.

11.4 *Calamintha nepeta*

Calamintha nepeta is an important aromatic and floricultural plant of Mediterranean origin. It is an erect, pubescent and highly branched perennial shrub distributed in the rocky areas of Mediterranean countries (Vlachou et al. 2016). Its beautiful inflorescence of white lilac flowers blooms between June and October (Blamey and Grey-Wilson 2000). Essential oils isolated from the shoots of *C. nepeta* are used in the perfumery and pharmaceutical industries (Marongiu et al. 2010; Riela et al, 2008; Hammer et al. 2005). The plant also possesses antimicrobial and snake-repellent properties and is also used in Italian cuisine (Kitic et al. 2005; Flamini et al. 1999; Panizzi et al. 1993).

An efficient protocol for direct regeneration of *C. nepeta* from shoot tip explants was standardized by Vlachou et al. (2016). MS medium containing 4.44 μM BAP was suitable for multiple shoot induction from shoot tip explants. Further, successive subcultures in 4.44 μM BAP, 0.88 μM BAP and 4.56 μM zeatin (ZN) increased shoot induction. The micro shoots thus obtained rooted well (90–100%) on culture in half-strength MS medium for 6 weeks or on full strength MS medium with IBA for 1 week followed by 5 weeks' culture on half-strength basal MS medium. Successful acclimatization (79%) was obtained on a 1:1 peat–perlite mixture.

11.5 *Coleus* spp.

11.5.1 *Coleus blumei*

Coleus blumei (Miana) is a common ornamental herb with colorful leaves. Besides its ornamental potential, *C. blumei* is also used in several remedies, including those for cough, asthma, bronchitis, diarrhea, boils and bruises. Coleus species are well known for their phytochemical constituents, such as rosmarinic acid, β-sitosterol, stigmasterol alkaloids, flavonoids, tannins, terpenoids and saponins (Alfermann and Petersen 1988; Petersen and Simmonds 2003). Studies have shown that Miana leaves possess antibacterial, antiviral, antioxidant, anti-inflammatory, housefly-repellent and anti-mutagenic properties (Nagpal et al. 2008; Bismelah et al. 2019; Surahmaida and Umarudin 2019).

Rani et al. (2006) established an efficient method for micropropagation of *Coleus blumei* Benth. using shoot tip and nodal explants. Explants were inoculated on MS medium augmented with different concentrations and combinations of IAA, IBA, NAA or BAP. Both explants gave optimum shoot induction on MS medium supplemented with 8.88 μM BAP and 5.37 μM NAA. Shoot tip explants gave a comparatively higher percentage of shoot induction and shoot number. MS medium with 9.84 μM IBA induced optimum rooting in regenerated shoots.

A different approach for indirect regeneration of *C. blumei* using leaf explants was reported by Jing et al. (2008). Optimum callus induction was obtained from leaf explants inoculated on half-strength MS medium supplemented with 8.88 μM BAP and 5.37 μM NAA. *In vitro* developed shoots rooted best on half-strength MS medium supplemented with 0.49 μM IBA.

11.5.2 *Coleus forskohlii* Briq.

Coleus forskohlii Briq. is a highly aromatic medicinal herb indigenous to India. It is distributed on the barren hills of subtropical countries with its origin from the Indian subcontinent (Valdes et al. 1987; Willemse 1985). *C. forskohlii* is used in traditional medicine as well as Ayurveda for the treatment of various ailments such as asthma, bronchitis, heart diseases, convulsions, abdominal colic, intestinal disorders, constipation, insomnia, burning sensation, angina and epilepsy (Ammon and Müller 1985). The brownish-red fasciculate tuberous roots of *C. forskohlii* are the only source of a diterpenoid alkaloid, forskolin, which is a potential drug for spasmolysis, congestive cardiomyopathy, hypertension, constipation, obesity, cancer, glaucoma and painful urination (Vivek et al. 2015). Being the only source for forskolin, *C. forskohlii* is overexploited, leading to depletion of its population (Vishwakarma et al. 1988).

George et al. (2001) developed a protocol for indirect regeneration of *C. forskohlii* using leaf explants. Leaf explants with abaxial surface touching the medium gave better results when placed in MS medium supplemented with 0–10.74 μM NAA and 4.44 μM BAP. Calli 50–60 days old produced shoots when transferred to MS medium with

4.44 μM BAP. The maximum number of long roots was obtained in half-strength basal MS medium. On average, 80% of plants survived when transferred to the greenhouse. Another simple protocol for indirect regeneration of *C. forskohlii* was also described by Reddy et al. (2001). According to their protocol, optimum callus induction was obtained in MS medium supplemented with 2.4 μM KN, which is different from the findings of George et al. (2001). Shoot induction was obtained from callus in MS medium containing 4.6 μM KN and 0.5 μM NAA. Further subcultures increased shoot number, and the highest number was obtained at the sixth subculture (158 shoots). *In vitro* shoots rooted well in half-strength MS medium without any hormones.

Indirect regeneration from leaf-derived callus was also studied by Sreedevi et al. (2013). Callus induction was observed when explants were cultured on Gamborg medium (B_5) containing 9.05 μM 2,4-D. Regeneration of shoots from callus was obtained in MS medium supplemented with 8.88 μM BAP and 5.37 μM NAA. More than 2000 shoots/callus clump were obtained by the sixth subculture. Vibhuthi and Kumar (2019) also reported the effect of 8.88 μM BAP on callus induction followed by multiple shoot induction from shoot tip explants of *C. forskohlii.*

An efficient protocol for quick regeneration of *C. forskohlii* using stem tip explants was developed by Bhattacharyya and Bhattacharya (2001). Optimum shoot induction (90%) with the highest shoot number (12.5) was obtained in MS medium supplemented with 0.57 μM IAA and 0.46 μM KN. All the shoots were rooted in the same medium and successfully transferred to soil after hardening in a 1:1 mix of soilrite and loamy soil for 25 days. In yet another method, developed by Krishna et al. (2010), direct shoot regeneration was obtained from leaf explants. When leaf explants excised into proximal, middle and distal segments were cultured on MS medium containing 22.2 μM BAP, direct shoot regeneration was obtained from all the leaf segments, with the maximum number (45.0) of shoots in the distal segment. A combination of 0.44 μM BAP and 0.57 μM IAA was helpful in the elongation of the regenerated shoots. Elongated shoots produced roots when cultured in half-strength MS medium with 1.5% sucrose and were successfully transferred to the soil after acclimatization.

Another study by Sahai and Shahzad (2013) reported direct regeneration of *C. forskohlii* from nodal explants when cultured in MS medium supplemented with 5.0 μM BAP. Further subculture in 5.0 μM BAP and 0.1 μM NAA resulted in shoot multiplication. Half-strength MS medium with and without auxins induced profuse rooting in micro shoots, with the highest number (11.6) of roots in 1.0 μM NAA. Rooted shoots were transferred to a potting mixture containing garden manure, garden soil and sand in a 1:2:1 ratio with 70% survival.

An excellent method for direct regeneration from nodal explants of *C. forskohlii* was standardized by Janarthanam and Sumathi (2020). Nodal explants, when cultured on MS medium augmented with 4.44 μM BAP, produced the optimum response with the highest number (24.3 shoots) of shoots within 30 days of culture. These micro shoots produced maximum rooting (7.8 roots/shoot) when cultured in half-strength MS medium containing 2.46 μM IBA.

11.6 *Hyssopus officinalis* L.

Hyssopus officinalis is a perennial shrub with high medicinal value. It is often used in traditional medicine as a tonic, expectorant, cough reliever and antiseptic compound. All these properties are due to the presence of bioactive compounds such as α- and β-pinene, diosmin, flavonoids, pinocamphone, sesquiterpenes and tannins (Fathiazad and Hamedeyazdan 2011).

The first ever report on micropropagation of *H. officinalis* was by Hosseini et al. (2016). When the nodal explants were inoculated in MS medium fortified with different concentrations of hormones, both BAP and thidiazuron (TDZ) in combination with 1.0 µM IAA induced shoot bud formation. The highest percentage of shoot formation with the highest number of shoots was obtained in MS medium supplemented with 4.4 µM TDZ and 1.0 µM IAA. Optimum rooting was observed in MS medium supplemented with 9.84 µM IBA. In a similar study, Bulavin et al. (2021) found that the nodal explants regenerated in MS medium containing 2.22 µM BAP. Also, 0.49 µM IBA was sufficient for optimum root induction.

Maslova et al. (2021) studied the effect of different hormones on callus induction and *in vitro* propagation of *H. officinalis* using seedling explants. They found that MS medium with 4.92 µM IBA alone and a combination of 2.68 µM NAA, 0.46 µM KN and 4.44 µM BAP produced optimum callus induction. MS medium containing 11.42 µM IAA and 0.93 µM KN was found to be effective in *in vitro* cultivation of the seedlings. Similar results were reported by Shoja and Shishavani (2021), where the *in vitro* derived nodal explants produced optimum shoot regeneration in MS medium supplemented with TDZ.

11.7 *Lavandula* spp.

11.7.1 *Lavandula angustifolia* Mill.

Lavandula angustifolia Mill. is a perennial shrub and widely used essential oil-yielding plant. Lavander oil is mainly used in the food, perfumery and cosmetic industries. It is also an important ornamental plant with high medicinal value. The high demand for lavender oil made it essential to cultivate it in large quantities. A protocol was optimized in *Lavandula angustifolia* for indirect organogenesis from *in vitro* leaf explants (Devasigamani et al. 2020). Maximum (92%) callus induction and callus growth were obtained when *in vitro* leaf segments were cultured in MS medium fortified with 8.88 µM BAP and 5.36 µM NAA. Optimum shoot regeneration (93%) from callus occurred when the calli were subcultured in MS with 4.44 µM BAP, 4.64 µM Kn and 2.68 µM NAA. Microshoots rooted efficiently after 30 days' culture in MS basal medium containing a combination of 8.88 µM BAP, 4.92 µM IBA and 2.68 µM NAA.

In another report on callus induction and morphogenesis in *L. angustifolia*, Yegorova et al. (2020) cultured *in vitro* leaf explants on MS medium supplemented with various concentrations and combinations of BAP, KN, TDZ and gibberellic acid (GA_3). A combination of 5.37 µM NAA and 2.22 µM BAP was most efficient in callogenesis. The highest frequency (39.5–43.2%) of morphogenesis was obtained on MS medium supplemented with 4.44 µM BAP and 4.64 µM KN. In addition, they also studied the content of some endogenous hormones in morphogenic and non-morphogenic calli.

Kumar et al. (2015) studied the influence of growth regulators on callus induction and indirect regeneration from nodal explants of *L. angustifolia*. Callus cultures were established by inoculating nodal segments on MS fortified with 0.57 µM IAA, 0.08 µM BAP and 0.9 µM 2,4-D. Maximum numbers of shoots regenerated from organogenic calli in MS medium containing 2.85 µM IAA and 8.88 µM BAP. The tallest shoots were obtained in basal MS medium, but these were clustered and not properly differentiated.

Nodal segment culture was effectively used for *in vitro* propagation of *L. angustifolia* by Hamza et al. (2011). Shoot proliferation was obtained through axillary branching from nodal segments cultured on MS medium with BA and TDZ. Maximum shoot induction frequency (30.55 shoots/explant) was obtained on MS medium fortified with 0.9 μM TDZ. However, a comparatively lower result (16.5 shoots/explant) was obtained with 3.55 μM BAP. Regenerated shoots rooted (21.25 roots) in half-strength MS medium containing 5.37 μM NAA and successfully acclimatized in soil with 90% survival.

Li et al. (2019) also standardized a micropropagation protocol for direct regeneration of *L. angustifolia* from a 2 cm-long stem with buds as explant. With a disinfection time of 15 minutes in 0.1% $HgCl_2$, the 2 cm-long stem explants proved to be highly efficient in multiple shoot induction (354 shoot buds/explant) when cultured on MS medium supplemented with 4.44 μM BAP and 1.47 μM IBA. For root induction, White medium supplemented with 2.15 μM NAA was found to be optimum (66.7%; 9.7 roots/shoot). Rooted plants were successfully transplanted to a peat–perlite mixture (1:1) with 66.7% survival.

11.7.2 *Lavandula dentata* L.

Lavandula dentata L. is an evergreen plant with melliferous and ornamental uses. Essential oil of *L. dentata* L. is widely used in perfumery and aromatherapy as well as in the food industry (Kim and Lee 2002; Ghelardini et al. 1999). It is also used as a therapeutic agent for its antibacterial, antiviral, spasmolytic and sedative activities (Gamez et al. 1990).

Echeverrigaray et al. (2005) established an *in vitro* propagation protocol for *Lavandula dentata* L. using axillary buds from adult field-grown plants. While evaluating the effect of plant growth regulators (PGRs) and culture media on axillary bud explants, the maximum rate of multiplication was obtained from explants cultured in MS medium augmented with a combination of 2.2 μM BAP and 2.5 μM IBA. Optimum root induction was recorded in MS medium supplemented with 2.5 μM NAA. After acclimatization, the rooted plants were successfully transferred to soil. They also noticed that long-term cultures (more than 1 year) show a low frequency of non-heritable morphological changes. A similar method for micropropagation of *L. dentata* using nodal explants was standardized by Jordan et al. (1998). Multiple shoot induction with maximum number of shoots was obtained when nodal explants cultured in MS medium supplemented with 5.0 μM BAP or 20 μM KN were transferred to MS medium containing 5.0 μM BAP and 15% coconut water. According to their observation, further subculturing drastically reduced the frequency of shoots. Optimum root induction was obtained in half-strength MS medium devoid of growth regulators. Successfully acclimatized plants were transplanted to soil.

Attia et al. (2017) established a micropropagation protocol for medium-term preservation of *L. dentata* using a slow growth technique. Direct shoot induction with 90% response was obtained when nodal explants were cultured in MS medium supplemented with 6.66 μM BAP and 2.46 μM IBA. Shoot tips and axillary buds of *in vitro* plants were used as explants for slow growth culture initiation. The highest survival rate (90%) was recorded on MS medium augmented with 15 g/L sucrose and 10 g/L sorbitol. They also evaluated and confirmed the genetic stability of preserved plants using random amplified polymorphic DNA (RAPD) analysis.

11.8 *Melissa officinalis* L.

Melissa officinalis L. (lemon balm) is a perennial aromatic plant used in traditional medicine to treat insect bites, heart failure, dyspepsia, insomnia, irritability, hysteria, melancholy and depression (Moradkhani et al. 2010). These properties of *M. officinalis* are due to the presence of phenolic compounds, including flavonoids, tannins and rosmarinic acid (Barros et al. 2013).

An efficient protocol for multiple shoot induction from cotyledonary nodes of *M. officinalis* was standardized by Tavares et al. (1996). Cotyledonary nodes collected from 10-day-old seedlings of *M. officinalis*, when cultured on MS medium fortified with various concentrations of BAP, underwent shoot differentiation. Subcultures further increased shoot induction. The maximum number of shoots (24 shoots/explant) was achieved after two inoculations in 8.88 μM BAP. The highest shoot length was obtained in 0.88 μM BAP, and a further increase in hormone concentrations had a negative effect on shoot height. Rooting was achieved by transferring 30-day-old shoots to MS medium augmented with 0–19.68 μM IBA or NAA alone. *In vitro* plants were successfully transplanted to soil.

Tantos et al. (1999) studied the effect of triacontanol on micropropagation of *M. officinalis*. Triacontanol, when added to shoot multiplication and root induction medium, induced increased shoot multiplication and rooting. While 11.39 μM triacontanol was found to be optimal for shoot multiplication, 4.55 μM was the best for root induction. It also enhanced the shoot growth, chlorophyll content, and number and length of roots, as well as the fresh weight, but the dry weight remained unaltered.

An efficient shoot regeneration protocol in *Melissa officinalis* L. using shoot tip explants was developed by Meftahizade et al. (2010). Four different land races were investigated for establishing a stable regeneration system. Among the different hormones tested, a combination of BAP and NAA gave the highest rate of shoot induction in all the races. Callus induction was best in a combination of NAA with IAA and KN. Comparatively higher rooting was obtained with 5.37 μM NAA than with other auxins tested. An approximately 20-fold increase in fresh weight (5.48 g) and dry weight (0.407 g) of callus, as well as the highest number of cells, was obtained with 4.52 μM 2,4-D and 2.22 μM BAP.

By using *in vitro* stem explants, Moradkhani (2012) developed a protocol for *in vitro* regeneration of *M. officinalis*. MS medium was found to be better in shoot induction compared with woody plant medium and B5 medium. The highest frequency of regeneration (74.5%), with 12.3 shoots per explant, was observed in MS medium supplemented with 2% glucose, 29.7 μM BA and 5.8 μM NAA. Optimum rooting was obtained in quarter-strength MS medium fortified with 15.4 μM IBA. Almost 100% of rooted plants were acclimatized and transferred to soil.

A method for micropropagation of *M. officinalis* using cotyledonary leaf explants was reported by Aasim et al. (2018). The efficacy of cotyledonary leaves in adventitious regeneration was studied by culturing explants excised from 8–10-day-old *in vitro* seedlings in MS medium supplemented with various concentrations and combinations of TDZ and IBA. Necrosis of explants was controlled by subculturing them in the same medium containing 1.0 mg/l PVP (polyvinylpyrrolidone) after 2 to 3 weeks. The highest frequency (83.33%) of callus induction was obtained with 0.45 µM TDZ, but TDZ alone was found to be least effective in shoot regeneration from callus. A combination of 1.81 µM TDZ and 0.49 µM IBA recorded maximum shoot regeneration frequency (61.11%) with 9.40 shoots per explant. Within 4 weeks of culture, the *in vitro* shoots rooted well in MS medium containing 0.0–4.92 µM IBA. Though numerous rooted plants were produced, only 20% survived during acclimatization in peat due to fungal infection.

Ulgen et al. (2020) tried to enhance the regeneration of *M. officinalis* through a different approach of applying magnetic fields. They designed two separate experiments to find the most efficient explant and to find the effect of magnetic field on *in vitro* regeneration. Among seven different explants cultured on MS minimal organic medium supplemented with various growth regulators, only those with meristematic cells (i.e., axillary buds, shoot tip and cotyledonary buds) regenerated. In the second experiment, axillary buds, shoot tips and cotyledonary buds were cultured on media containing BAP in combination with IAA or NAA. These cultures were exposed to two different magnetic fields (50 mT and 100 mT) for 1 hour, resulting in enhanced regeneration compared with the control. The best results were obtained with axillary bud explant in 6.66 µM BAP exposed to100 mT magnetic field for 1 hour.

Petrova et al. (2021) studied the effectiveness of plant growth regulators on micropropagation of *M. officinalis. In vitro* derived stem segments from 1-month-old seedlings were cultured on MS medium fortified with various concentrations and combinations of PGRs. A combination of 4.44 µM BAP and 0.49 µM IBA, as well as 6.66 µM BAP and 2.68 µM NAA, was found to be best for shoot induction. Optimum rooting (100%) was obtained when the *in vitro* shoots were cultured in a half-strength basal MS medium with 2% sucrose. Successful acclimatization (98%) was recorded in a 2:1:1:1 mixture of soil, perlite, peat and sand. They also assessed the total phenol content as well as metabolic profiles of *M. officinalis*. The highest total phenol content (13.92 mg/g extract) was obtained in shoots grown on MS medium containing 4.44 µM BAP and 0.49 µM IBA. The results indicate the ability of PGRs to enhance the accumulation of metabolites in *in vitro* plants.

11.9 *Mentha* spp.

11.9.1 *Mentha arvensis*

Mentha arvensis (common mint) is a small herbaceous plant with white flowers. It is usually used as a home remedy for its beneficial effect on digestion and its antiseptic properties. Fresh leaves are used in flavoring cooked foods, salads and drinks. Herbal tea is often made from dried leaves of *M. arvensis*. Leaves possess about 0.2% essential oil, which is used as a flavoring agent in beverages and sweets (Islam and Bari 2012).

In order to meet the growing demand, an efficient protocol for micropropagation of *M. arvensis* using axillary buds has been developed by Maity et al. (2011). The study showed that the axillary bud explants produced multiple shoots when cultured on MS medium fortified with BAP and AdS (adenine sulfate). Optimum shoot proliferation (8.81 shoots/explant) was observed with 4.44 µM BAP. The highest root induction was achieved when *in vitro* shoots were transferred to MS medium with 4.92 µM IBA. Successfully acclimatized plants were similar as per their cytological and biochemical studies.

Islam and Bari (2012) standardized an efficient protocol for synthetic seed production and its regeneration using shoot tip explants of *M. arvensis. In vitro* derived shoot tips and nodal segments were encapsulated using sodium alginate as gelling agent along with $CaCl_2$ solution. Among various concentrations of PGRs tested, synthetic seeds produced the highest (80%; 9.87 shoots per synseed) shoot induction in MS medium augmented with 8.88 µM BAP and 1.07 µM NAA. The highest shoots (6.27 cm) were obtained in MS medium containing 4.44 µM BAP along with 2.68 µM NAA.

A reliable protocol for *in vitro* mass propagation of *M. arvensis* was established by Maity (2013) through callus culture. Nodal explants cultured on MS medium supplemented with 2.22 µM BAP and 1.07 µM NAA produced green compact callus within 4 weeks of culture. The highest number (268) of globular somatic embryoids was recorded with 0.88 µM BAP. Transferring the calli to PGR-free MS medium induced high-frequency shoot and root formation (96.43 shoots/culture) with a shoot length of 5.61 cm. The regenerated plantlets were acclimatized in the greenhouse by transferring them to the soil.

11.9.2 *Mentha piperita* L.

Jullien et al. (1998) developed a micropropagation protocol for *M. piperita* through protoplast culture. Protoplasts of three varieties of *M. piperita*, namely Mitcham Digne38, Mitcham Ribecourt19 and Todd's, initiated cell division in the presence of 2,4-D. Optimum microcallus formation was recorded in MS medium supplemented with 1 µM 2,4-D in combination with 2.5 µM NAA and 4.0 µM BA. A solid medium was more efficient than a liquid medium in forming microcalli. Shoot regeneration was observed in microcalli when cultured in a low auxin (0.5 µM NAA) and high cytokinin concentration. Maximum bud formation (n = 434) was observed with 2.3 µM TDZ and 8.8 µM BA in 61% calli. Genotypic differences slightly influenced the regeneration capacity and shoot regeneration pathway.

An efficient method for *in vitro* regeneration from *M. piperita* using nodal explants was standardized by Laslo et al. (2011). Multiple shoots (23.9 shoots/explant) were successfully induced from nodal explants on MS medium fortified with 2.28 µM ZN and 2.85 µM IAA. While optimum rooting (10.2 roots/shoot) was recorded in MS medium supplemented with 4.44 µM BAP and 5.7 µM IAA, the highest root length was observed with 2.28 µM ZN and 2.85 µM IAA.

The impact of sanitation and *in vitro* clonal micropropagation on the essential oil content of *M. piperita* was studied by Shkopynska et al. (2019). *In vitro* sanitation increased the yield of air-dried leaves of peppermint breeding samples by 2.9–7.1% and rhizomes by 2.2–3.8%. Sanitation increased the yield of air-dried leaves and rhizome of the Chornolysta variety of peppermint by 51.4% and 28.5%, respectively. Further, they reported a notable increase in essential oil yield of breeding samples (4.0–9.9 kg/ha) as well as the Chornolysta variety (28.6 kg/ha).

Łyczko et al. (2020) studied the influence of PGRs on the composition of volatile organic compounds in *M. piperita*. Highest overall effects in shoot (83%) and root induction (97% in apical meristems and 100% for in meristems), as well as increased content of odor-active compounds, were obtained from explants cultured in MS medium supplemented with 2.46 μM IBA. The content of aromatic compounds such as menthofurolactone and cis-carvone oxide increased considerably.

A rapid and reliable method for *in vitro* micropropagation of *M. piperita* was developed by Ayaz and Memon (2021). Optimum shoot induction, as well as root induction, was reported in nodal explants cultured in MS medium fortified with 100 μL/L NAA and 600μl/L IBA followed by four successive subcultures of 60 days each in the same media. Rooted plantlets were successfully acclimatized and transplanted to soil.

11.10 *Ocimum* spp.

11.10.1 *Ocimum basilicum* L.

Ocimum basilicum or sweet basil is an economically important medicinal plant distributed in the tropical and subtropical regions of Africa, Asia, and Central and South America. The plant is aromatic, and the essential oil obtained from the plant is insect repellent, nematocidal and possesses antioxidant, antibacterial and antifungal activities (Lee et al. 2005; Juliani and Simon 2002).

An efficient micropropagation protocol of *O. basilicum* using nodal segments was standardized by Saha et al. (2010). The nodal segments were cultured on MS medium supplemented with various concentrations of BAP and KN (0–10.0 μM). Explants on the medium containing 2.22 μM BAP produced the maximum number of shoots (6.2 shoots) with the highest average length (3.7 cm). Half-strength MS medium supplemented with 5.37 μM NAA was found to be optimum for the rooting of shoots. The plants were then transferred to 1:1 soil and vermiculite mixture and then to the field with a 90% success rate.

Siddique and Anis (2008) tested MS medium supplemented with BAP, TDZ, KN and 2-isopentenyl adenine (2ip) for the micropropagation of *O. basilicum* using nodal segments. Half-strength MS medium supplemented with the combination of 2.5 μM BA and 0.5 μM IAA produced the highest rate of shoot multiplication. MS medium with 1.0 μM IBA was found better for rooting compared with NAA- or IAA-supplemented medium. The propagated plants were then transferred to the field with a 90% success rate. Sharma et al. (2014) used 2,4-D as the sole growth promoter for callus induction from nodal explants. MS medium supplemented with 9.05 μM 2,4-D was found optimum (94.44% callus induction) for this purpose. BAP, KN and IAA, alone or in combinations at different concentrations, were tested for direct shoot induction from nodal segments, and MS medium with 4.44 μM BAP produced the highest number of shoots (7.74 shoots). Repeated subculture of proliferated shoots for about 4–5 months yielded 25-30 shoots from a single explant. For rooting, half-strength MS medium with 4.92 μM IBA produced the highest result (92.50%). The plants were successfully established in garden soil, farmyard manure and sand (1:1:1) with an 85% survival rate. An improved method for the *in vitro* propagation from nodal explant was optimized by Shahzad (2012). Different concentrations and combinations of BAP, 2-isopentanyl adenine and L-glutamine were tested. The highest shoot induction frequency (100%) and mean number of shoots (13.4 shoots) were obtained on media supplemented with 10.0 μM BA and 30 mg/L L-glutamine. The formed shoots then rooted successfully on MS medium supplemented with 5.0 μM IBA and were then transferred to the field.

GA_3 at 1.15 μM added to the MS medium along with 4.44 μM BAP markedly enhanced the frequency of bud break from nodal explants when cultured, according to Sahoo et al. (1997). Maintaining the formed shoots on the proliferation medium (1.11 μM BAP) for a longer duration produced *in vitro* inflorescence and flowers in this case. The shoots formed were then rooted in half-strength MS medium supplemented with 4.92 μM IBA and successfully transferred to the field. Manan et al. (2016) standardized an efficient micropropagation and *in vitro* flowering protocol using shoot tips collected from aseptic seedlings of *O. basilicum*. MS medium supplemented with various concentrations of BAP and GA_3, alone or in combination with NAA, was used for shoot induction, and 4.44 μM BAP produced the maximum number of shoots. All the plants cultured on 2.89 μM GA_3 flowered *in vitro*. Compared with the mother plant, the *in vitro* plants matured early. There was no seed formation, low essential oil content, few fully filled peltate glandular trichomes and higher methyl chavicol content in *in vitro* flowering plants. The micro propagated plants that flowered *ex vitro* showed similar characteristics to the mother plant and flowered at an intermediate time.

An efficient protocol was developed for the shoot regeneration from epicotyl, hypocotyl and shoot tip explants isolated from *in vitro* grown seedlings of *O. basilicum* (Ekmekci and Aasim 2014). The explants were cultured on MS medium supplemented with 3.63–10.88 μM TDZ alone or in combination with 0.49 μM IBA along with 1.0 mg/L PVP and 3.0 g/L activated charcoal. One hundred percent callus induction was observed on all the cultures. The maximum number of shoots from shoot tip (3.58) and epicotyl (3.22) was obtained on MS medium with 10.88 μM TDZ and 0.49 μM IBA. For hypocotyl, MS medium with 9.08 μM TDZ induced the maximum number of shoots (5.17). Relatively short roots were observed on all explants, but the plants were successfully acclimatized and planted in the garden. A procedure for shoot induction from the cotyledon explants was standardized by Dode et al. (2003). The explants were cultured on MS medium supplemented with different concentrations (0–22.2 μM) of BAP alone or in combination with NAA (1.07 μM). The maximum number of shoots was obtained on MS medium with 22.2 μM BAP and 1.07 μM NAA. The presence of NAA inhibited rooting in formed plantlets, and the plantlets rooted immediately when transferred to a hormone-free MS medium.

Siddique and Anis (2007) standardized a protocol for rapid propagation of *O. basilicum* from shoot tip. Shoot tips were treated in liquid MS medium supplemented with various concentrations (5.0–100.0 μM) TDZ for different time durations (4.0–16.0 days). The explants treated in 50 μM TDZ for 8 days and then moved to hormone-free medium showed the maximum shoot induction rate (78%) and mean number of shoots (11.6 ± 1.16) and highest shoot length (4.8 ± 0.43 cm) per explant. MS medium containing 1.0 μM IBA gave the best result in rooting. The plants successfully acclimatized with a 95% survival rate.

11.10.2 *Ocimum kilimandscharicum* Guerke

Ocimum kilimandscharicum Guerke is a perennial herb with high economic and medicinal value. It is widely used in folk medicine as a remedy for many ailments, including cough, cold, diarrhea, abdominal pain and measles (Obeng-Ofori et al. 1998). Essential oils of *O. kilimandscharicum* possess antibacterial, antioxidant and insect-repellent activity (Kweka et al. 2008; Hakkim et al. 2014). Camphor is one of the most important bioactive components obtained from the seed oil of *O. kilimandscharicum* (Jembere et al. 1994).

As an alternative to conventional propagation through seeds, an efficient *in vitro* regeneration system and subsequent rooting method were developed for the medicinal plant *O. kilimandscharicum* (Saha et al. 2010). Nodal explants, when cultured on MS medium supplemented with various concentrations of cytokinins, induced axillary shoot bud proliferation, with the maximum (6.09 shoots; 3.83 cm height) in 4.44 μM BAP. Shoot multiplication was maintained by further subculture in shoot induction medium. *In vitro* shoots rooted well in half-strength MS medium containing 2% sucrose and 7.38 μM IBA. Well-rooted plantlets were transferred to soil after acclimatization with 81.13% survival. The genetic fidelity of micro propagated plants was confirmed through RAPD analysis.

11.10.3 *Ocimum sanctum*

Ocimum sanctum, commonly known as holy basil, is considered a sacred plant by different cultures and used for religious purposes apart from its remarkable medicinal and insecticidal properties. Several studies have been conducted to reveal its therapeutic potential. Traditionally, *O. sanctum* is used to treat bronchitis, various skin, gastric and hepatic disorders. The plant shows antioxidant, anti-pyretic, anti-inflammatory, analgesic, anti-asthmatic (Pathania et al. 2020), antifungal (Awuah and Ellis 2002), antibacterial and antiviral (Zahran et al. 2020) properties.

The *in vitro* propagation of *O. sanctum* has been accomplished using young inflorescence, nodal segments and leaves as explants. Shukla et al. (2021) standardized a procedure for the micropropagation of a selected germplasm line with high antioxidant potential called Vrinda, using nodal explants. MS medium supplemented with 1.1 μM BAP, 0.3 μM GA_3 and 0.6% activated charcoal was found to be optimum for the multiplication and growth of shoots. MS supplemented with 0.44 μM BAP and 1.43 μM IAA was found to be optimum for both shoot multiplication and successful rooting in this study. For the acclimatization and hardening, soil: farmyard manure (75:25) mixture and 100% sand were found to be the best supporting medium with an 87% survival rate of the transferred plantlets.

Mishra (2015) optimized an efficient plant regeneration protocol via organogenesis from callus derived from leaves of *O. sanctum.* MS medium supplemented with picloram at a concentration of 3.0 mg/L produced the highest amount of organogenic callus. The highest percentage of shoot organogenesis (82%) from callus, with a mean of 23.8 ± 0.23 shoots, was obtained on MS medium with 4.44 μM BAP and 2.85 μM IAA. MS medium containing 7.38 μM IBA was found to be optimum for rooting of *in vitro* grown shoots. The plants were then acclimatized and transferred to the field with a 90% survival rate.

Micropropagation of *O. sanctum* from young inflorescence was optimized by Singh and Sehgal (1999). MS medium supplemented with 4.44 μM BAP and 0.28 μM IAA produced high-frequency shoot induction without callus formation. The elongated shoots were rooted on MS basal medium and transferred to the field.

11.11 *Origanum vulgare* L.

Origanum vulgare (oregano) is a perennial woody bush with attractive pinkish-purple or white flowers. High demand for the plant is due to its essential oil content as well as culinary application. *O. vulgare* possesses antiseptic, antimicrobial, antispasmodic, carminative and sudorific activity (Matłok et al. 2020). Kumari and Saradhi (1992) standardized a protocol for *in vitro* regeneration of *O. vulgare* from callus cultures. Cotyledons, hypocotyl and root segments collected from 15-day-old seedlings were used as explants. Among the three explants cultured on Gamborg's B 5 medium augmented with 10^{-7} M 2,4-D, cotyledonary explants produced the maximum compact and nodulated callus. The cotyledonary calli, when transferred to MS medium containing 10^{-6} M BAP and 10^{-6} M NAA, produced the maximum number of shoots. Both IBA (10^{-6} M) and NAA (10^{-6} M) in half-strength B_5 liquid medium were equally effective in root induction from 2 cm-long micro shoots. Rooted shoots were acclimatized under controlled conditions and transferred to soil.

A method for callus-mediated indirect regeneration was developed for *O. vulgare* (Leelavathi and Kuppan 2013). Apical bud (0.5–1.0 cm) excised from *in vitro* cultured *O. vulgare* was cultured on MS medium supplemented with 8.88 μM BAP in combination with various concentrations of NAA and 2,4-D. Whitish-green compact callus formation was recorded in MS medium with 8.88 μM BAP and 2.26 μM 2,4-D. Further subcultures in the same medium resulted in shoot induction (20–30 shoots) after 42 days of culture, followed by rooting in 70 days. *In vitro* rooted plants were then hardened and transferred to soil with 90% survival.

Habibi et al. (2016) established an efficient protocol for *Agrobacterium rhizogenes*-mediated gene transfer and plant regeneration via hairy roots. Co-cultivation of *in vitro* leaf explants with *A. rhizogenes* strains K599 and ATCC15834 on MS medium (modified) resulted in high-frequency transformation (91.3%). The highest frequency of callus induction (81.18%) was observed in hairy root segments cultured in MS medium with 1.13 μM 2,4-D. Maximum shoot regeneration (85.18 %) occurred in calli subcultured in MS medium supplemented with 1.11 μM BAP. Root induction was achieved after 15 days of culture in MS medium containing 12.3 μM IBA.

Premi et al. (2021) studied the effect of chitosan (CHT), BAP and KN on *in vitro* seed germination and shoot development in *O. vulgare*. MS medium fortified with CHT and KN greatly enhanced the rate of seed germination and shoot development. On the contrary, BAP had a negative impact on it. Shoot elongation and leaf production were highly triggered by CHT compared with other cytokinins. MS medium with 0.75 mg/L CHT induced the highest leaf number (17.71 leaves/shoot) and shoot length (4.38 cm). Although KN (9.29 μM and 18.58 μM) enhanced shoot production, cluster analysis proved that 0.50 and 0.75 mg/L CHT were much better than BAP and KN for enhancing leaf development and shoot growth in *O. vulgare*.

11.12 *Pogostemon cablin* Benth.

Pogostemon cablin Benth. (patchouli) is an aromatic herb with commercial importance. Patchouli plants have been used in Indian and Chinese traditional medicine for the treatment of various ailments. Major bioactive compounds obtained from *P. cablin* include patchouli alcohol, pogostone, pogostol, eugenol, and α- and β-patchoulene (Zhang et al. 1998). It is very effective in the treatment of headaches, colds, fever, vomiting, nausea, abdominal pain, diarrhea, insect bite and snake bites. Patchouli oil is also used in aromatherapy to get rid of stress and depression, calm the nerves and control appetite (Kalra et al. 2006; Swamy and Anuradha 2011).

A micropropagation protocol using nodal explants was standardized for *P. cablin* by Mayerni (2020). The study aimed to identify the best possible concentrations of BAP and KN that produce a synergistic effect on multiple shoot induction. MS medium supplemented with 2.32 μM KN and BAP, 2.32 μM KN and 4.44 μM BAP, and 4.64 μM KN and BAP induced *in vitro* shoot organogenesis in nodal explants, indicating their interactive role in shoot induction. Micropropagation from nodal explants of *P. cablin* was also reported by Jin et al. (2014). Optimum shoot induction (100%) with the maximum number of shoots (129.7–138.1) was obtained in MS medium fortified with BAP (0.44–0.88 μM). The highest root induction (49.3 roots/shoot) in regenerated shoots was achieved in half-strength MS medium containing 0.98 μM IBA. Over 90% of rooted plants were acclimatized successfully and transferred to soil. Maulia et al. (2021) reported optimum shoot induction from nodal explants of *P. cablin* when cultured in MS medium supplemented with 4.44 μM BAP and 10.74 μM NAA. This combination of BAP and NAA was found to be more efficient in terms of time for growth, height, and number of shoots and leaves. While 8.88 μM BAP induced optimum rooting, a combination of 8.88 μM BAP and 5.37 μM NAA was best for increased root length.

Widoretno (2016) standardized a protocol for the induction of tetraploid plants of *P. cablin.* Polyploidy was induced by culturing leaf segments in MS medium supplemented with different concentrations of colchicine for 3 weeks. Colchicine at 60 mg/L was most effective in inducing polyploidy. Explants were then transferred to MS medium containing 0.54 μM NAA and 1.33 μM BAP for shoot regeneration. Regenerated shoots were transferred to rooting medium and then acclimatized in the greenhouse. Tetraploid plants were identified by chromosome number counting. Significant morphological differences in leaf size, stem diameter, stomatal size and plant size were observed between diploid and tetraploid plants, indicating that tetraploids could possibly be a superior variety. In a similar study by Yan et al. (2016), octoploid plants of *P. cablin* were obtained with 0.05% colchicine treatment on explant for 72 hours, followed by culture in shoot and root induction medium. Morphological studies showed that octoploids had larger leaves and stomata compared with tetraploids and diploids. Also, the patchouli alcohol content was much higher in octoploid plants, adding to its medicinal potential.

Lalthafamkimi et al. (2020) reported a rapid regeneration protocol for *P. cablin* using meta-topolin (mT). Leaf, petiole and nodal explants cultured in MS medium supplemented with 1.0 mg/L mT for 4 weeks produced the highest number of shoots (45 shoots/explant) with an optimum rate of proliferation. Maximum root induction was achieved when the micro shoots were transferred to MS medium containing 7.38 μM IBA. Direct rooting was obtained at varying degrees when the explants were treated with different concentrations of NAA, IAA and IBA. Clonal fidelity of the regenerants was confirmed with Start Codon Targeted Polymorphism (SCoT) and Inter-Retrotransposon Amplified Polymorphism (IRAP) analysis. The study stresses the role of mT and other aromatic cytokinins in high-frequency micropropagation.

Swamy et al. (2014) studied the effect of PGRs and some natural supplements on the regeneration of *P. cablin*. Among various PGRs tested, a combination of 2.22 μM BAP and 2.32 μM KN in MS medium was best for multiple shoot induction (63.20 shoots/explant; 5.27 cm). In the case of natural supplements, coconut water (10%) showed the best response. Also, 10% of tomato, carrot and papaya extract as well as 20% banana extract in MS medium notably increased the number, length and fresh weight of shoots. There was a considerable increase in the total protein, total carbohydrate and chlorophyll content in the plants treated with natural supplements. The highest frequency (93%) of rooting was observed on half-strength MS medium supplemented with 100 mg/L activated charcoal (AC). Rooted plantlets were successfully transferred to soil after hardening. Genetic fidelity of the *in vitro* regenerated plants was confirmed with RAPD analysis.

A method for the mass propagation of three genotypes of *P. cablin* (POG002, POG014 and POG021) has been developed by Santos et al. (2010). Direct regeneration from leaf explants (1.0 cm^2) was achieved in MS medium supplemented with low concentrations of KN and IAA. MS medium fortified with 4.64 μM KN and 2.85 μM IAA induced the highest shoot regeneration in POG014 (175 shoots) and POG021 (154 shoots) genotypes. Optimum regeneration (41 shoots) in genotype POG002 was achieved with 4.64 μM KN. Vermiculite supplemented with MS medium salts was best for acclimatization of the plantlets. They also noted that essential oil and the patchouli alcohol content were much higher in micropropagated plants of all three genotypes. Paul et al. (2010) also reported a rapid *in vitro* regeneration method for *P. cablin* from leaf explants. Direct organogenesis was observed from leaf explants cultured on MS medium supplemented with NAA and BAP for 4 weeks. Origin, leaf position and age of the donor plant greatly influenced shoot induction. The highest shoot induction (94.6 shoots/explant) was given by almost 96.2% of leaf explants excised from the second node of 3-month-old *in vivo* plants and cultured in MS medium supplemented with 2.5 μM BAP and 0.5 μM NAA. Shoot elongation was further increased (1.8-fold) by the addition of 1.0 μM GA_3 to 95% of cultures. Rapid rooting was observed in half-strength basal MS medium followed by acclimatization and transfer to soil with 96–100% survival. RAPD and gas chromatography–mass spectrometry (GC-MS) analysis confirmed the genetic stability and essential oil content stability, respectively, of the micropropagated plants.

11.13 *Prunella vulgaris* L.

Prunella vulgaris L. (all heal) is a perennial herb that grows at high altitudes of India. It is called "all heal" for its high medicinal potential. Phytochemicals such as anionic polysaccharide, betulinic acid, flavonoids, triterpenoids, oleanolic acid, tannins and urosolic acid (Ryu et al. 2000; Xu et al. 1999) impart properties such as antiallergic, anti-inflammatory, antioxidant, anticancer and antiestrogenic activity to the plant (Chiu et al. 2004; Shin et al. 2001).

An *in vitro* regeneration protocol for *P. vulgaris* has been standardized by Turker et al. (2009). Seedling-derived shoot tips gave the highest frequency of multiple shoot induction when cultured in MS medium supplemented with a combination of

13.32 μM BAP and 0.57 μM IAA. Other explants such as internodes, leaf, petiole and root explants did not respond in this media combination. Micro shoots quickly rooted in MS medium containing various concentrations of auxins, with the best being 17.12 μM IAA or 14.76 μM IBA. Rooted shoots were then acclimatized in vermiculite for 2 weeks and successfully transplanted to soil.

Kour et al. (2014) reported a simple and reproducible protocol for high-frequency regeneration of *P. vulgaris*. Callogenesis was induced from various explants when cultured in MS medium fortified with combinations of TDZ and 2,4-D. Maximum callus induction was obtained from internodal explants in a medium containing 4.54 μM TDZ and 0.23 μM 2,4-D. Callus thus obtained proliferated well to form green nodular callus on media containing 4.54 μM TDZ and 0.28 μM IAA. Shoot induction was observed when this callus was subcultured into MS medium augmented with 6.66 μM BAP and 0.057 μM IAA. They also reported direct shoot induction from nodal explants when cultured in MS medium supplemented with 2.22 μM BAP and 0.28 μM IAA. Optimum rooting was recorded with 2.46 μM IBA. Rooted plants were hardened and transferred to the field.

11.14 *Rosmarinus officinalis* L.

Rosmarinus officinalis L. (rosemary) is a perennial aromatic herb with needle-like leaves and pinkish-white flowers. It is distributed in the Mediterranean countries but cultured all around the world. It originated in the Mediterranean region but is now grown worldwide. This plant is an important source of various phytochemicals, including essential oils. Rosemary oil contains carsanol, carnosic acid, rosmarinic acid and volatile oils (Okamura et al. 1994), giving the plant antiviral, antimicrobial, antioxidant, antispasmodic, analgesic, diuretic, hepatoprotective, anticonvulsant and anticarcinogenic properties (Saltan and Ozaydin 2013; Bozin et al. 2008).

An efficient protocol for micropropagation of *R. officinalis* L. was reported by Darwesh and Alayafi (2018) using *in vitro* seedling explants. Shoots (1–1.5 cm) excised from *in vitro* germinated seedlings were cultured in MS medium fortified with various PGRs (BAP and KN) and coconut water. Results indicated that 13.32 μM BAP was most efficient in multiple shoot induction. A combination of BAP (13.32 μM) and coconut water (5 ml/L) proved to be best for increased number of leaves as well as shoot length. The highest concentrations of phenolics (10.45 mg/g) and chlorophyll b (0.67 mg/g) were also obtained with 13.32 μM BAP. On the other hand, chlorophyll a was highest (0.64 mg/g) in the presence of 22.2 μM BAP and 5.0 ml/L coconut water. Anthocyanin content was at its peak in a combination of BAP (13.32 μM) and coconut water (5.0 ml/L).

Misra and Chaturvedi (1984) also reported *in vitro* shoot regeneration using shoot tip and nodal explants of *R. officinalis* L. Nodal explants responded better than shoot tips in MS medium supplemented with BAP. The best results (14 shoots/explant) were observed in 0.88 μM BAP after 30 days of culture. Subculturing the shoots in 0.25 mg/L indolepropionic acid (IPA) induced optimum rooting (80%). Plantlets thus obtained were successfully transferred to soil after hardening.

Mascarello et al. (2017) employed tissue culture techniques for the production of secondary metabolites from *R. officinalis*. Shoot multiplication was achieved by culturing *in vitro* derived seedlings in basal MS medium. Although basal MS medium induced rooting (50%) from *in vitro* shoots, the highest percentage of rooting was obtained in the presence of 2.85 μM IAA (75% rooting) and 5.7 μM IAA(78.6% rooting). Almost 64% of plants were acclimatized and transferred to the greenhouse and then to the field. Green friable calli were initiated from leaf explants of *R. officinalis* after 2 months of culture in NAA-containing medium. These calli were used to initiate cell suspension culture in a liquid medium for extraction of metabolites.

Yet another excellent protocol for *in vitro* mass propagation of *R. officinalis* through somatic embryogenesis was standardized by Aman and Afrasiab (2014). Although young leaf explants produced somatic embryos in modified Woody plant medium (WPM) supplemented with various concentrations of 2,4-D in combination with BAP, the best results were obtained in medium supplemented with 2.25 μM 2,4-D and 0.45 μM BAP. There was almost 100% somatic embryogenesis from these calli. These primary embryos, when subcultured on modified basal WPM, formed clusters with secondary somatic embryos and embryogenic calli. On further subculturing (4-week interval) these clusters, the secondary somatic embryos developed into plantlets (average 10% response) in each subculture. Addition of abscisic acid (ABA) to the same medium induced the formation of secondary somatic embryos and embryogenic calli.

Boix et al. (2013) studied the chemico-morphological characters and anatomy of *in vitro* derived calli of *R. officinalis*. Callus culture was produced by culturing leaf explants of *R. officinalis* in MS medium containing 2,4-D and TDZ. Light, as well as electron, microscopic studies revealed the presence of peltate and capitate glandular trichomes in callus. Gas chromatographic analysis has proved that 2,4-D has a positive influence on *R. officinalis* callus to produce volatile compounds. Notable quantities of pinene and camphor were produced in a medium containing 2.26 μM 2,4D.

11.15 *Salvia* spp.

11.15.1 *Salvia officinalis* L.

Salvia officinalis L. (common sage) is an important perennial woody shrub with high medicinal value, native to the Mediterranean countries. Traditionally, plant extracts of *S. officinalis* have been used against cataracts, bronchial asthma, ischemic heart disease, inflammatory conditions, hepatotoxicity, atherosclerosis, insufficient sperm mobility and cancer (Baricevic et al. 2000). Studies have shown the potential of its essential oils in improving memory, and thus, it is a potential candidate for the treatment of Alzheimer's disease (Perry et al. 1999). As dried leaves of *S. officinalis* are a raw material of the perfumery industry, it is widely cultivated around the world (Santos-Gomes et al. 2002).

Somatic embryogenesis from leaf explants of *S. fruticose* was reported by Kintzios et al. (1999). Leaf explants cultured on MS medium supplemented with a combination of 1.8–18 μM 2,4-D and KN or 10.5–21 μM NAA and BAP (10.5–21 μM) produced embryogenic callus with several globular somatic embryos. Only younger explants were capable of induction of somatic embryos at low (50 $\mu mol/m^2/s$) light intensities. Further development of somatic embryos occurred in the same medium. Once the heart- and torpedo-shaped embryos were formed, they were subcultured on basal MS medium for further development and maturation. Accumulation of rosmarinic acid was highest (25.9 g/L) in calli cultured with 4.5 μM 2,4-D and 4.5 μM Kin.

Jafari et al. (2017) standardized a protocol for callus-mediated regeneration of *S. officinalis* L. Callus culture was initiated from internode and leaf explants in MS medium containing 2.22 μM BAP and 10.74 μM NAA. Internode explants responded better than leaf explants in the callus induction medium. The highest percentage of shoot induction (70%) was reported with 0.5 mg/mL TDZ. High-frequency rooting was observed in micro shoots transferred to half-strength MS medium with 4.92 μM IBA added. These *in vitro* rooted plantlets were successfully acclimatized. Callus-mediated regeneration of *S. officinalis* was also reported by Tawfik and Mohamed (2007). Callus induction was observed on the basal cut end of shoot tip explants cultured on MS medium fortified with TDZ (4.5, 13.5 or 22.5 μM) for 1 week in the dark, followed by 4 weeks in the light. The highest percentage of callus induction with large callus size was observed in 4.5 μM TDZ. MS medium containing 4.5 μM TDZ and 0.45 mM ascorbic acid was optimum for maintaining the callus culture. The highest shoot regeneration was recorded in a medium containing BAP (4.4 or 8.8 μM). A further increase in shoot number was obtained with the addition of 0.45 mM ascorbic acid. Optimum root induction was obtained with 4.9 μM IBA. Almost 75% of plantlets were hardened and transferred to soil. Stages of adventitious bud formation and shoot development were also confirmed through histological studies.

An excellent method for micropropagation of *S. officinalis* was developed by Petrova et al. (2015). *In vitro* seedling-derived nodal segments showed maximum shoot proliferation when cultured in MS medium supplemented with 2.22 μM BAP and 0.57 μM IAA. Successful root induction was obtained in half-strength MS medium fortified with 20 mg/L yeast extract, 10 mg/L ascorbic acid and 4.92 μM IBA. *In vitro* rooted plantlets were then acclimatized in a 1:1:1:2 mixture of peat, perlite, sand and soil. They also estimated the total flavonoid content and antioxidant potential of *in vitro* derived leaves. Thin layer chromatography (TLC) revealed the presence of six flavonoid aglycones. Also, the methanolic extract of *in vitro* plant showed significant radical-scavenging activity, with half maximal inhibitory concentration (IC_{50}) = 22.18 μg/mL.

A novel protocol for liquid shoot culture of *S. officinalis* in MS liquid medium supplemented with 0.57 μM IAA and 1.99 μM BAP was reported by Grzegorczyk and Wysokińska (2008). An average of three shoots were formed per shoot tip explant in 3 weeks of culture. Shoots thus obtained produced 8.2 mg diterpenoids and 31.2 mg rosmarinic acid. Addition of triacontanol to the medium further increased shoot proliferation as well as diterpenoid production (30–50% increase).

11.15.2 *Salvia miltiorrhiza* Bunge

Salvia miltiorrhiza Bunge (Chinese sage or Danshen) is a ubiquitous component of traditional Chinese medicine (Wang and Wu 2010). Most of its medicinal constituents are present in the roots. Tanshinones are the major bioactive constituents isolated from the roots of *S. miltiorrhiza* (Wang and Wu 2010). It is mainly used in the treatment of cardiovascular diseases, such as heart pain (Zhang and Wang 2006; Wang et al. 2013). As it is such a valuable plant, tissue culture techniques have been employed for the propagation as well as biosynthesis of active compounds from *S. miltiorrhiza*.

Tsai et al. (2016) reported micropropagation of *S. miltiorrhiza* via leaf explant culture. Leaf explants exhibited high morphogenetic plasticity, i.e., direct and indirect shoot formation, as well as direct and indirect rhizogenesis. TDZ (0.45 or 2.27 µM) was most efficient in direct shoot organogenesis. Optimum root induction was recorded in PGR-free MS medium. Indirect organogenesis was mostly dependent on the ratio of BAP to NAA used in the medium. When a higher ratio promoted shoot induction, a lower ratio induced rooting from callus. After 45 days of acclimatization, *in vitro* regenerated plants showed a 100% survival rate.

Another protocol for the production of polyploid plants of *S. miltiorrhiza* was reported by Gao et al. (1996). *In vitro* seedling-derived shoot tips were cultured in MS medium supplemented with a combination of 4.44 µM BAP and 2.85 µM IAA. Bud clumps thus induced were transferred to MS medium containing 10 ppm colchicine for optimum polyploidy induction. Surviving buds were then transferred to media containing 4.44 µM BAP and 2.85 µM IAA for further development of shoots. B5 solid medium fortified with 0.98 µM IBA was most efficient for root induction from *in vitro* plants. Tetraploids were selected, transferred to the field and screened for 15 agronomic traits. The study also reported that the tanshinone content was much higher in tetraploid than in diploid plants.

Wang et al. (2013) reported a different approach to the regeneration of *S. miltiorrhiza* via indirect somatic embryogenesis from hairy roots. *Agrobacterium rhizogenes* R1601 was efficient in inducing hairy roots from leaf explants of *S. miltiorrhiza.* Embryogenic callus induction was observed from hairy roots cultured on MS medium supplemented with 4.52 µM 2,4-D and 2.22 µM BAP. Embryogenic calli were then subcultured in MS medium with 0.27 µM NAA and 2.22 µM BAP, resulting in the formation of somatic embryos, which developed into micro plants. Polymerase chain reaction (PCR) analysis confirmed the origin of plants from hairy roots. They had morphological variations such as wrinkled leaves and thick and longer taproots. A notable increase in biomass of root (2.7-fold) and shoot (1.5-fold), as well as tanshinone content (79.5%), was observed in the hairy root-derived plants compared with wild type. It was also observed that the increase in tanshinone content is due to the upregulation of two genes that code for key enzymes (3-hydroxy-3-methylglutaryl CoA reductase and 1-deoxy-D-xylulose 5-phosphate reductoisomerase) in the tanshinone biosynthesis pathway.

11.16 *Thymus vulgaris* L.

Thymus vulgaris L. (German thyme) is an invaluable aromatic perennial herb distributed in the Mediterranean countries, North Africa as well as southern Europe. A member of the thyme genus, *T. vulgaris* is used in folk medicine, tonics, and herbal teas and is antitussive, carminative, antifungal, anti-inflammatory and antimicrobial. Most of its activities are due to the essential oils present in the plant (Fachini-Queiroz et al. 2012; Giordani et al. 2004). Increased demand for the plant can be fulfilled with the help of micropropagation of *T. vulgaris*.

In order to meet the growing demand, an efficient protocol for *in vitro* propagation of *T. vulgaris* was standardized by Karalija and Parić (2011). *In vitro* shoot induction was obtained from *in vitro* seedling explants. Multiple shoot induction was best in MS medium supplemented with 8.88 μM BAP and 0.49 μM IBA (14.3 shoots) or 17.76 μM BAP and 0.49 μM IBA (13.3 shoots). Varying concentrations of hormones caused quantitative changes in chlorophyll a and b and carotenoid contents of *in vitro* plants. In addition, the presence of 2.22 μM BAP and 0.49 μM IBA in the medium elicited the production of phenolic compounds in the plantlets. Similarly, 8.88 to 17.76 μM BAP enhanced the flavonoid content but reduced monomeric anthocyanins.

An efficient method for *in vitro* propagation of *T. vulgaris* was developed by Radomir and Stan (2020). Seeds were inoculated in basal MS medium and incubated in dark or light conditions for 4 weeks. While 81% of seeds germinated with light incubation, only 50% germinated with dark incubation. Seed-derived *in vitro* seedlings, when cultured in basal MS medium, produced an average of 5.3 shoots/explant with a mean shoot length of 6.5 cm. Effective root induction from *in vitro* shoots was obtained in half-strength MS medium supplemented with 9.84 μM IBA. Well-rooted plants were successfully acclimatized with a 96% success rate.

El Ansari et al. (2019) studied the effect of various macronutrients as well as PGRs in the organogenesis of *T. vulgaris* and also reported an excellent protocol for its micropropagation. Multiple shoot induction was obtained from nodal explants cultured on MS medium supplemented with Shah and Dalal macronutrients, 0.7% bacterial agar and 0.46 μM KN. Further experiments were carried out with the *in vitro* derived shoot tips. Testing of various hormones and macronutrients revealed that addition of 0.46 μM or 0.93 μM BAP, 0.46 μM KN, 0.46 μM ZN, 0.46 μM 2iP, 0.46 μM adenine (AD) and 0.46 μM 1,3-diphenylurea (DPU) to MS medium enhanced the growth and multiplication of *in vitro* shoots. Also, MS medium supplemented with combinations of 0.46 μM KN and 0.57 μM IAA or NAA, 0.46 μM AD and 0.57 μM IBA, as well as 0.46 μM DPU and 0.57 μM IBA, was capable of inducing optimum rooting from *in vitro* shoots. Successfully acclimatized plants were transferred to the field.

El-Banna (2017a) developed an *in vitro* regeneration protocol for *T. vulgaris* from shoot tip and nodal explants. Nodal segments proved to be better than shoot tips for regeneration in MS medium supplemented with 8.88 μM BAP. For shoot elongation, a combination of 8.88 μM BAP and 1.44 μM GA_3 was best. *In vitro* shoots rooted well in MS medium fortified with 8.05 μM NAA. *In vitro* rooted plants successfully hardened in soil: peat moss (1: 1) mixture with 100% survival. An indirect micropropagation protocol for *T. vulgaris* using shoot tip explants was also developed by El-Banna (2017b). Optimum callus induction (100%) was obtained from shoot tip explants when cultured in MS medium fortified with a combination of 9.05 μM 2,4-D and 4.64 μM KN. The highest shoot induction (100%; 23.42 shoots/explant) was obtained on MS medium containing 11.1 μM BAP. MS medium augmented with 8.05 μM NAA was preferable (100%; 19.83 roots/shoot) for root induction from *in vitro* shoots. Rooted plants were successfully (98% survival) transplanted into pots containing a mixture of soil: peat moss (1:1).

In another report on micropropagation of *T. vulgaris*, Kulpa et al. (2018) cultured shoot explants derived from *in vitro* grown seedlings and evaluated the essential oil content of *in vitro* plants. Shoot explants were cultured on MS medium fortified with BAP, KN or 2iP. Maximum shoot multiplication was obtained with 5 mg/dm^3 2iP. Root induction was best in MS medium containing IBA. Essential oils were extracted using hydrodistillation in Clevenger and Deryng apparatus. GC-MS analysis disclosed the presence of 54 components, including oxygenated monoterpenes and monoterpene hydrocarbons.

A recent study intended to optimize the culture media for clonal propagation of *T. vulgaris* was reported by Tevfik and Yegorova (2020). The results conclude that MS medium fortified with 2.88 μM GA_3 and 4.64 μM KN induced optimum regeneration, with an average of 2.2 shoots per explant having an average length of 1.9 cm, while 4.64 μM KN was most effective in shoot multiplication. Rooting of *in vitro* shoots was best with 4.92 μM IBA or IAA. Well-rooted plantlets were acclimatized in a peat and perlite (1:1) mixture and transferred to soil with a 89.5% survival rate.

Ozudogru et al. (2011) proposed a protocol for *in vitro* propagation and production of genetically stable plants of *T. vulgaris* from *in vitro* derived shoot tips. Optimum regeneration (97%; 8.6 shoots/explant) was achieved when the explants were cultured in semi-solid MS medium containing 4.64 μM KN and 0.86 μM GA_3. Micro shoots rooted well (92.5%; 19 roots/shoot) in semi-solid MS medium fortified with 0.02 μM 2,4-D. RAPD analysis revealed the genetic stability of micropropagated plants. Rooted plants were transferred to soil after acclimatization.

REFERENCES

Aasim, M., Kahveci, B., Korkmaz, E., Doğanay, F., Bakırcı, Ş., Sevinc, C., Akin, F., Kirtiş, A., 2018. TDZ-IBA induced adventitious shoot regeneration of water balm (*Melissa officinalis* L.). *Journal of Global Innovation on Agriculture and Social Sciences*, 6, 35–39.

Aman, N., Afrasiab, H., 2014. Primary and secondary somatic embryogenesis from leaf explants of rosemary (Rosmarinus officinalis L. - Lamiaceae), *Pakistan Journal of Botany*, 46(3), 903–909.

Ammon, H.P., Müller, A.B., 1985. Forskolin: From an ayurvedic remedy to a modern agent. *Planta Medica*, 51(6), 473–477.

Attia, A.O., Dessoky, E.D.S., Al-Sodany, Y.M., Ismail, I.A., 2017. Ex situ preservation for some endemic and rare medicinal plants in Taif, KSA. *Biotechnology & Biotechnological Equipment*, 31(5), 912–920.

Awuah, R.T., Ellis, W.O., 2002. Effects of some groundnut packaging methods and protection with Ocimum and Syzygium powders on kernel infection by fungi. *Mycopathologia*, 154(1), 29–36.

Ayaz, E., Memon, A., 2021. Development of the aromatic medicinal plants, *Mentha × piperita* L. and *Mentha pulegium* L. through *in vitro* Callus Induction and Micropropagation. *Turkish Journal of Agriculture - Food Science and Technology*, 9(1), 159–165.

Ayres, G. S., Widrlechner, M. P., 1994. The genus *Agastache* as bee forage: A historical perspective. *American Bee Journal*, 134(5), 341.

Baricevic, D., Bartol, T., 2000. The biological/pharmacological activity of the Salvia genus. In S.E. Kintzios (Ed.), *Sage: The Genus Salvia*. Harwood Academic Publishers, Amsterdam, pp. 143–184.

Barros, L., Dueñas, M., Dias, M.I., Sousa, M.J., Santos-Buelga, C., Ferreira, I.C., 2013. Phenolic profiles of cultivated, *in vitro* cultured and commercial samples of *Melissa officinalis* L. infusions. *Food Chem.*, 136, 1–8.

Bhattacharyya, R., Bhattacharya, S., 2001. *In vitro* multiplication of *Coleus forskohlii* Briq.: An approach towards shortening the protocol. *In Vitro Cell Dev Biol - Plant*, 37, 572–575.

Bismelah, N.A., Ahmad, R., Kassim, Z.M., Ismail, N.H., 2019. *Coleus blumei* extract as a potential antibacterial oral rinse. In *IOP Conference Series: Earth and Environmental Science* (Vol. 269, No. 1, p. 012015). IOP Publishing.

Blamey, M., Grey-Wilson, C., 2000. *Wild Flowers of the Mediterranean: A Complete Guide to the Flowers of Mediterranean Coasts and Islands, Native and Introduced over 2000 Illustrated*, HarperCollins, USA, pp. 396–397.

Boix, Y.F., Arruda, R.O., Defaveri, A.C.A., Sato, A., Lage, C.L.S., Victorio, C.P., 2013. Callus in *Rosmarinus officinalis* L.(Lamiaceae): A morphoanatomical, histochemical and volatile analysis. *Plant Biosystems-An International Journal Dealing with all Aspects of Plant Biology*, 147(3), 751–757.

Bozin, B., Mimica-Dukic, N., Bogavac, M., Suvajdzic, L., Simin, N., Samojlik, I., Couladis, M., 2008. Chemical composition, antioxidant and antibacterial properties of *Achillea collina* Becker ex Heimerl s. I. and *A. pannonica* Scheele essential oils. *Molecules*, 13(9), 2058–2068.

Bulavin, I.V., Ivanova, N.N., Mitrofanova, I.V., 2021. In vitro regeneration of Hyssopus officinalis L. and plant genetic similarity. *Dokl Biol Sci*, 499, 109–112.

Charles, D.J., Simon, J.E., Widrlechner, M.P., 1991. Characterization of the essential oil of Agastache species. *Journal of Agricultural and Food Chemistry*, 39(11), 1946–1949.

Chiu, L.C.M., Zhu, W., Ooi, V.E.C., 2004. A polysaccharide fraction from medicinal herb *Prunella vulgaris* downregulates the expression of herpes simplex virus antigen in Vero cells. *Journal of Ethnopharmacology*, 93(1), 63–68.

Chopra, R.N., Nayar, S.L., Chopra, I.C., 1986. *Glossary of Indian Medicinal Plants (Including the Supplement)*. Council of Scientific and Industrial Research, New Delhi, p. 330.

Darwesh, H.Y., Alayafi, A.A., 2018. In vitro propagation response of *Rosmarinus officinalis* L. to biotic and abiotic elicitors on phenolic content and photosynthetic pigments. *Journal of Agricultural Science*, 10(2), 301.

Devasigamani, L., Devarajan, R., Loganathan, R., Rafath, H., Padman, M., GovindaRaju, M.V., Giridhar, L., Chetan, H.C., Kuppan, N., 2020. *Lavandula angustifolia* L. plants regeneration from *in vitro* leaf explants-derived callus as conservation strategy. *Biotecnología Vegetal*, 20(2), 75-82.

Dode, L.B., Bobrowski, V.L., Bolacel Braga, E.J., Seixas, F.K., Schuch, M.W., 2003. *In vitro* propagation of *Ocimum basilicum* L. (Lamiaceae). *Acta Scientiarum - Biological Sciences*, 25(2), 435–437.

Echeverrigaray, S., Basso, R., Andrade, L.B., 2005. Micropropagation of *Lavandula dentata* from axillary buds of field-grown adult plants. *Biologia Plantarum*, 49(3), 439–442.

Ekmekci, H., Aasim, M., 2014. *In vitro* plant regeneration of Turkish sweet basil (*Ocimum basilicum* L.). *Journal of Animal and Plant Sciences*, 24(6), 1758–1765.

El Ansari, Z.N., El Mihyaoui, A., Boussaoudi, I., Benkaddour, R., Hamdoun, O., Tahiri, H., Badoc, A., El Oualkadi, A., Lamarti, A., 2019. Effect of macronutrients, cytokinins and auxins, on *in vitro* organogenesis of *Thymus vulgaris* L. *American Journal of Plant Sciences*, 10(09), 1482–1502.

El-Banna, H., 2017a. Micropropagation of thyme plant (*Thymus vulgaris*). *Journal of Plant Production*, 8(11), 1221–1227.

El-Banna, H., 2017b. Indirect micropropagation of *Thymus vulgaris* plant. *Journal of Plant Production*, 8(11), 1241–1246.

Fachini-Queiroz, F.C., Kummer, R., Estevao-Silva, C.F., Carvalho, M.D.D.B., Cunha, J.M., Grespan, R., Bersani-Amado, C.A., Cuman, R.K.N., 2012. Effects of thymol and carvacrol, constituents of *Thymus vulgaris* L. essential oil, on the inflammatory response. *Evidence-Based Complementary and Alternative Medicine*, 2012, 657026. doi: 10.1155/2012/657026.

Fathiazad, F., Hamedeyazdan, S., 2011. A review on *Hyssopus officinalis* L.: Composition and biological activities. *African Journal of Pharmacy and Pharmacology*, 5(17), 1959–1966.

Flamini, G., Cioni, P.L., Puleio, R., Morelli, I., Panizzi, L., 1999. Antimicrobial activity of the essential oil of *Calamintha nepeta* and its constituent pulegone against bacteria and fungi. *Phytother Res*, 13(4), 349–351.

Gamez, M.J., Jimenez, J., Navarro, C., Zarzuelo, A., 1990. Study of the essential oil of *Lavandula dentata* L. *Pharmazie*, 45(1), 69–70.

Gao, S.L., Zhu, D.N., Cai, Z.H., Xu, D.R., 1996. Autotetraploid plants from colchicine-treated bud culture of *Salvia miltiorrhiza* Bge. *Plant Cell, Tissue and Organ Culture*, 47(1), 73–77.

George, M.M., Subramanian, R.B., Prajapati, H., 2001. Regeneration and selection of root resistant *Coleus forskohlii*: A threatened medicinal plant. *Plant Resour*, 4, 65–74.

Ghelardini, C., Galeotti, N., Salvatore, G., Mazzanti, G., 1999. Local anaesthetic activity of the essential oil of *Lavandula angustifolia*. *Planta Medica*, 65(08), 700–703.

Giordani, R., Regli, P., Kaloustian, J., Mikail, C., Abou, L., Portugal, H., 2004. Antifungal effect of various essential oils against *Candida albicans*. Potentiation of antifungal action of amphotericin B by essential oil from *Thymus vulgaris*. *Phytotherapy Research*, 18(12), 990–995.

Giuliani, C., Bini, L.M., 2008. Insight into the structure and chemistry of glandular trichomes of Labiatae, with emphasis on subfamily Lamioideae. *Plant Systematics and Evolution*, 276, 199–208.

Grzegorczyk, I., Wysokińska, H., 2008. Liquid shoot culture of *Salvia officinalis* L. for micropropagation and production of antioxidant compounds; effect of triacontanol. *Acta Societatis Botanicorum Poloniae*, 77(2), 99–104.

Habibi, P., Grossi de Sa, F.M., Lopes da Silva, A.L., Makhzoum, A., da Luz Costa, J., Borghetti, I.A., Soccol, C.R., 2016. Efficient genetic transformation and regeneration system from hairy root of *Origanum vulgare*. *Physiology and Molecular Biology of Plants*, 22, 271–277.

Hakkim, F.L., Arivazhagan, G., Boopathy, R., 2014. Antioxidant property of selected Ocimum species and their secondary metabolite content. *Journal of Medicinal Plants Research*, 2(9), 250–257.

Hamayun, M., Afzal, S., Khan, M.A., 2006. Ethnopharmacology, indigenous collection and preservation techniques of some frequently used medicinal plants of Utror and Gabral, district Swat, Pakistan. *African Journal of Traditional, Complementary and Alternative Medicines*, 3(2), 57–73.

Hammer, K., Laghetti, G., Pistrick, K., 2005. *Calamintha nepeta* (L.) Savi and *Micromeria thymifolia* (Scop.) Fritsch cultivated in Italy. *Genet. Resour. Crop Evol.* 52(2), 215–219.

Hamza, A.M., El-Kafie, A., Omaima, M., Kasem, M.M., 2011. Direct micropropagation of English lavender (*Lavandula angustifolia* Munstead) Plant. *Journal of Plant Production*, 2(1), 81–96.

Harley, R.M., Atkins, S., Budantsev, A.L., Cantino, P.D., Conn, B.J., Grayer, R., Harley, M.M., De Kok, R.D., Krestovskaja, T.D., Morales, R., Paton, A.J., 2004. Labiatae. In *Flowering Plants. Dicotyledons*. Springer, Berlin, Heidelberg, pp. 167–275.

Hosseini, B., Alizadeh, M., Hassani, A., Jafari, M., Rahimi, A., 2016. High-frequency *in vitro* direct shoot regeneration from nodal explants of hyssop plant (*Hyssopus officinalis* L.). *Journal of Medicinal plants and By-products*, 5(2), 187–193.

Islam, M.S., Bari, M.A., 2012. *In vitro* regeneration protocol for artificial seed production in an important medicinal plant *Mentha arvensis* L. *Journal of Bio-Science*, 20(0), 99–108.

Islam, M., Ahmad, H., Rashid, A., Razzaq, A., Akhtar, N., Khan, I., 2006. Weeds and medicinal plants of Shawar valley, district Swat. *Pak. J. Weed Sci. Res.*, 12(1–2), 83–86.

Jafari, S., Daneshvar, M.H., Salmi, M.S., Jalal-Abadi, A.L., 2017. Indirect organogenesis and plant regeneration in common sage (*Salvia officinalis* L.): An important medicinal plant of Iran. *Modern Applied Science*, 11(5), 22.

Jan, M., Singh, S., Kaloo, Z.A., Maqbool, F., 2014. Callus induction and multiple shoot regeneration in *Ajuga bracteosa* Wall. ex Benth.—an important medicinal plant growing in Kashmir Himalaya. *J. Sci. Innov. Res.*, 3, 319–324.

Janarthanam, B., Sumathi, E., 2020. *In vitro* plant regeneration from nodal explants of Coleus. *PTC & B*, 30, 143–148.

Jembere, B., Hassanali, A., Obeng-Ofori, D., 1994. Scientific rational of traditional post-harvest insect control practices: Lessons from studies on Ocimum species. In *Proceedings 3rd International Conference on Tropical Entomology*. 30 October–4 November, Nairobi, Kenya.

Jin, H., Deng, Z.C., He, H., 2014. Effect of explant types and plant growth regulators on direct regeneration in medicinal plant *Pogostemon cablin*. *Plant OMICS*, 7(5), 322–327.

Jing, W., Ye, C., Rongmei, L., Baozhong, H., 2008. Study on the tissue culture of *Coleus blumei*. *Journal of Northeast Agricultural University*, 15, 14–17.

Jordan, A.M., Calvo, M.C., Segura, J., 1998. Micropropagation of adult *Lavandula dentata* plants. *The Journal of Horticultural Science and Biotechnology*, 73(1), 93–96.

Juliani, H.R., Simon, J.E., 2002. Antioxidant activity of basil. In: J. Janic and A. Whipkey (eds), *Trends in New Crops and New Uses*, ASHS Press, Alexandria, VA. pp. 575–579.

Jullien, F., Diemer, F., Colson, M., Faure, O., 1998. An optimising protocol for protoplast regeneration of three peppermint cultivars (*Mentha* x *piperita*). *Plant Cell, Tissue and Organ Culture*, 54(3), 153–159.

Kalra, A., Prakasa Rao, E.V.S., Khanuja, S.P.S., 2006. Cultivation and processing technologies of patchouli (*Pogostemon cablin*). *J. Med. Arom. Plants Sci*, 28, 414–419.

Karalija, E., Parić, A., 2011. The effect of BA and IBA on the secondary metabolite production by shoot culture of *Thymus vulgaris* L. *Biologica Nyssana*, 2(1), 29–35.

Kaul, S., Das, S., Srivastava, P.S., 2013. Micropropagation of *Ajuga bracteosa*, a medicinal herb. *Physiology and Molecular Biology of Plants*, 19(2), 289–296.

Kim, N.S., Lee, D.S., 2002. Comparison of different extraction methods for the analysis of fragrances from Lavandula species by gas chromatography–mass spectrometry. *Journal of Chromatography A*, 982(1), 31–47.

Kintzios, S., Nikolaou, A., Skoula, M., 1999. Somatic embryogenesis and *in vitro* rosmarinic acid accumulation in *Salvia officinalis* and *S.* fruticosa leaf callus cultures. *Plant Cell Reports*, 462–466.

Kirtikar, K.R., Basu, B.D., 1918. *Indian Medicinal Plants*. 2nd Edition. Lalit Mohan Basu, Allahabad, India, pp. 1158–1159.

Kitic, D., Stojanovic, G., Palic, R., Randjelovic, V., 2005. Chemical composition and microbial activity of the essential oil of *Calamintha nepeta* (L.) Savi ssp. nepeta var. subisodonda (Borb.) Hayek from Serbia. *J. Essent. Oil Res.*, 17(6), 701–703.

Kour, B., Azhar, M., Kaul, S., Dhar, M.K., 2014. *In vitro* regeneration and mass multiplication of *Prunella vulgaris* L. *National Academy Science Letters*, 37(1), 81–86.

Krishna, G., Reddy, P.S., Nair, N.A., Ramteke, P.W., Bhattacharya, P.S., 2010. *In vitro* direct shoot regeneration from proximal, middle and distal segment of *Coleus forskohlii* leaf explants. *Physiology and Molecular Biology of Plants*, 16(2), 195–200.

Kulpa, D., Wesolowska, A., Jadczak, P., 2018. Micropropagation and composition of essentials oils in garden thyme (*Thymus vulgaris* L.). *Notulae Botanicae Horti Agrobotanici Cluj-Napoca*, 46(2), 525–532.

Kumar, A., Kaushal, S., Sharma, S., 2015. Studies on influence of growth regulators in micropropagation of *Lavandula angustifolia. International Journal of Applied Biology and Pharmaceutical Technology*, 6(2), 73–77.

Kumari, N., Saradhi, P.P., 1992. Regeneration of plants from callus cultures of *Origanum vulgare* L. *Plant Cell Reports*, 11(9), 476–479.

Kuria, K.A., Chepkwony, H., Govaerts, C., Roets, E., Busson, R., De Witte, P., Zupko, I., Hoornaert, G., Quirynen, L., Maes, L., Janssens, L., Laekeman, G., 2002. The antiplasmodial activity of isolates from *Ajuga remota. Journal of Natural Products*, 65(5), 789–793.

Kweka, E.J., Mosha, F., Lowassa, A., Mahande, A.M., Kitau, J., Matowo, J., Mahande, M.J., Massenga, C.P., Tenu, F., Feston, E., Lyatuu, E.E., 2008. Ethnobotanical study of some of mosquito repellent plants in north-eastern Tanzania. *Malaria Journal*, 7(1), 1–9.

Lalthafamkimi, L., Bhattacharyya, P., Bhau, B.S., Wann, S.B., Banik, D., 2020. Direct organogenesis mediated improvised mass propagation of Pogostemon cablin: A natural reserve of pharmaceutical biomolecules. *South African Journal of Botany*, 140, 375–384.

Laslo, V., Zapartan, M., Vicas, S., Agud, E., 2011. Use of nodal explants in '*in vitro*' micropropagation of *Mentha piperita* L. *Analele Universității din Oradea, Fascicula Protecția Mediului, XVI* (December 2015), 247–251.

Lee, S.J., Umano, K., Shibamota, T., Lee, K.G., 2005. Identification of volatile components in basil (*Ocimum basilicum* L.) and thyme leaves (*Thymus vulgaris* L.) and their antioxidant properties. *Food Chem*, 91, 131–137.

Leelavathi, D., Kuppan, N., 2013. Callus induction and regeneration of multiple shoots from in-vitro apical bud explant of *Origanum vulgare*, an important medicinal plant. *Int. J. Res. Pharm. Chem*, 3, 898–903.

Li, N.Y., Tang, H.R., Ge, C., Mo, F., Xiao, Y.H., Luo, Y., 2019. Tissue culture of *Lavandula angustifolia* L. In *AIP Conference Proceedings* (Vol. 2079, No. 1, p. 020014). AIP Publishing LLC.

Łyczko, J., Piotrowski, K., Kolasa, K., Galek, R., Szumny, A., 2020. *Mentha piperita* L. micropropagation and the potential influence of plant growth regulators on volatile organic compound composition. *Molecules*, 25(11), 2652. https://doi.org/10.3390/molecules25112652

Maity, S.K., 2013. *In vitro* clonal propagation of *Mentha arvensis* through callus culture. *International Journal of Advanced Research in Engineering and Applied Sciences*, 2(8), 1–9.

Maity, S.K., Kundu, A.K., Tiwary, B.K., 2011. Rapid and large scale micropropagation of true to type clone of *Mentha arvensis* Linn (Lamiaceae) - valuable medicinal plant. *Indian J. Applied & Pure Bio.*, 26(2), 193–198.

Manan, A.A., Taha, R.M., Mubarak, E.E., Elias, H., 2016. *In vitro* flowering, glandular trichomes ultrastructure, and essential oil accumulation in micropropagated

Ocimum basilicum L. *In Vitro Cellular and Developmental Biology - Plant*, 52(3), 303–314.

Marongiu, B., Piras, A., Porcedda, S., Falconieri, D., Maxia, A., Gonçalves, M.J., Cavaleiro, C., Salgueiro, L., 2010. Chemical composition and biological assays of essential oils of *Calamintha nepeta* (L.) Savi subsp. nepeta (Lamiaceae). *Nat. Prod. Res.*, 24(18), 1734–1742.

Mascarello, C., Sacco, E., Pamato, M., Di Silvestro, D., Bassolino, L., Cervelli, C., Ruffoni, B., 2017. *Rosmarinus officinalis* L.: Micropropagation and callus induction for cell biomass development. *Acta Horticulturae*, 1155, 631–636.

Maslova, E., Gulya, N., Perelugina, T., Semykina, V., Kalashnikova, E., 2021. Introduction of *Hyssopus officinalis* L. into *in vitro* culture to optimize the conditions for obtaining callus tissues and microclonal propagation as a promising method of innovative agrobiotechnologies. In *BIO Web of Conferences* (Vol. 30, p. 05006). EDP Sciences.

Matłok, N., Stępień, A.E., Gorzelany, J., Wojnarowska-Nowak, R., Balawejder, M., 2020. Effects of organic and mineral fertilization on yield and selected quality parameters for dried herbs of two varieties of oregano (*Origanum vulgare* L.). *Applied Sciences*, 10(16), 5503.

Maulia, E., Zuyasna, Basyah, B., 2021. Growth of patchouli shoots (*Pogostemon cablin* Benth) with several concentrations of growth regulator substances *in vitro*. *Journal of Agriculture and Veterinary Science*, 14(1), 38–46.

Mayerni, R., 2020. The direct organogenesis in local clones of patchouli plant (*Pogostemon cablin* Benth) *in vitro*. *JERAMI Indonesian Journal of Crop Science*, 3(1), 16–19.

Mazza, G., Kiehn, F.A., 1992. Essential oil of *Agastache foeniculum*, a potential source of methyl chavicol. *Journal of Essential Oil Research*, 4(3), 295–299.

Meftahizade, H., Moradkhani, H., Naseri, B., Lofti, M., Naseri, A., 2010. Improved *in vitro* culture and micropropagation of different *Melissa officinalis* L. genotypes. *Journal of Medicinal Plants Research*, 4(3), 240–244.

Mishra, T., 2015. Protocol establishment for multiplication and regeneration of 'Holy Basil'(*Ocimum sanctum* Linn). An important medicinal plant with high religious value in India. *J. Med. Plants. Stud.*, 3(4), 16–19.

Misra, P., Chaturvedi, H.C., 1984. Micropropagation of Rosmarinus officinalis L.*, *Plant Cell, Tissue and Organ Culture*, 168 (195), 163–168.

Moharami, L., Hosseini, B., Ravandi, E.G., Jafari, M., 2014. Effects of plant growth regulators and explant types on *in vitro* direct plant regeneration of *Agastache foeniculum*, an important medicinal plant. *In Vitro Cellular & Developmental Biology-Plant*, 50(6), 707–711.

Moradkhani, H., 2012. Investigation of adventitious shoot regeneration from *in vitro* stem explants of *Melissa officinalis* L. *Journal of Medicinal Plants Research*, 6(16), 3217–3221.

Moradkhani, H., Sargsyan, E., Bibak, H., Naseri, B., Sadat-Hosseini, M., Fayazi-Barjin A., Meftahizade, H., 2010. *Melissa officinalis* L., a valuable medicine plant: A review. *J. Med. Plants Res.*, 4, 2753–2759.

Nagpal, A., Singh, B., Sharma, S., Rani, G., Virk, G.S., 2008. *Coleus* spp.: Micropropagation and *in vitro* production of secondary metabolites. *Medicinal and Aromatic Plant Science and Biotechnology*, 2(1), 1–7.

Njoroge, G.N., Bussmann, R.W., 2006. Diversity and utilization of antimalarial ethnophytotherapeutic remedies among the Kikuyus (Central Kenya). *Journal of Ethnobiology and Ethnomedicine*, 2(1), 1–7.

Nourozi, E., Hosseini, B., Hassani, A., 2014. A reliable and efficient protocol for induction of hairy roots in *Agastache foeniculum*. *Biologia*, 69(7), 870–879.

Obeng-Ofori, D., Reichmuth, C.H., Bekele, A.J., Hassanali, A., 1998. Toxicity and protectant potential of camphor, a major component of essential oil of *Ocimum kilimandscharicum*, against four stored product beetles. *International Journal of Pest Management*, 44(4), 203–209.

Okamura, N., Haraguchi, H., Hashimoto, K., Yagi, A., 1994. Flavonoids in *Rosmarinus officinalis* leaves. *Phytochemistry*, 37(5), 1463–1466.

Ozudogru, E.A., Kaya, E., Kirdok, E., Issever-Ozturk, S., 2011. *In vitro* propagation from young and mature explants of thyme (*Thymus vulgaris* and *T. longicaulis*) resulting in genetically stable shoots. *In Vitro Cellular and Developmental Biology - Plant*, 47(2), 309–320.

Panizzi, L., Flamini, G., Cioni, P.L., Morelli, I., 1993. Composition and antimicrobial properties of essential oils of four Mediterranean Lamiaceae. *J. Ethnopharmacol.*, 39(3), 167–170.

Pathania, M., Bhardwaj, P., Pathania, N., Rathaur, V.K., 2020. A review on exploring evidence-based approach to harnessing the immune system in times of corona virus pandemic: Best of modern and traditional Indian system of medicine. *Journal of Family Medicine and Primary Care*, 9(8), 3826.

Paul, A., Thapa, G., Basu, A., Mazumdar, P., Chandra, M., Sahoo, L., 2010. Rapid plant regeneration, analysis of genetic fidelity and essential aromatic oil content of micropropagated plants of Patchouli, *Pogostemon cablin* (Blanco) Benth. – An industrially important aromatic plant. *Industrial Crops & Products,* 32(3), 366–374.

Perry, E.K., Pickering, A.T., Wang, W.W., Houghton, P.J., Perry, N.S., 1999. Medicinal plants and Alzheimer's disease: From ethnobotany to phytotherapy. *Journal of Pharmacy and Pharmacology*, 51(5), 527–534.

Petersen, M., Simmonds, M.S., 2003. Rosmarinic acid. *Phytochemistry,* 62(2), 121–125.

Petrova, M., Nikolova, M., Dimitrova, L., Zayova, E., 2015. Micropropagation and evaluation of flavonoid content and antioxidant activity of *Salvia officinalis* L. *Genetics and Plant Physiology*, 5(1), 48–60.

Petrova, M., Nikolova, M., Dimitrova, M., Dimitrova, L., 2021. Assessment of the effect of plant growth regulators on *in vitro* micropropagation and metabolic profiles of *Melissa officinalis* L. (lemon balm). *Journal of* Microbiology, Biotechnology *and* Food Sciences, 11, 1–5.

Premi, N., Acemi, A., Özen, F., 2021. Cytokinin-like effects of chitosan on *in vitro* culture of *Origanum vulgare* L. *Italus Hortus*, 28(1), 100–108.

Radomir, A.-M., Stan, R., 2020. Micropropagation of *Thymus vulgaris* L., an important medicinal and aromatic plant. *Journal of Horticulture, Forestry and Biotechnology*, 24(1), 6–12.

Rani, G., Talwar, D., Nagpal, A., Virk, G.S., 2006. Micropropagation of *Coleus blumei* from nodal segments and shoot tips. *Biologia Plantarum*, 50, 496–500.

Reddy, P.S., Rodrigues, R., Rajasekharan, R., 2001. Shoot organogenesis and mass propagation of *Coleus forskohlii* from leaf derived callus. *Plant Cell, Tissue and Organ Culture*, 66(3), 183–188.

Riela, S., Bruno, M., Formisano, C., Rigano, D., Rosselli, S., Saladino, M.L., Senatore, F., 2008. Effects of solvent-free microwave extraction on the chemical composition of essential oil of *Calamintha nepeta* (L.) Savi compared with the conventional production method. *J. Sep. Sci.*, 31(6–7), 1110–1117.

Ryu, S.Y., Oak, M.H., Yoon, S.K., Cho, D.I., Yoo, G.S., Kim, T.S., Kim, K.M., 2000. Anti-allergic and anti-inflammatory triterpenes from the herb of *Prunella vulgaris*. *Planta Medica*, 66(04), 358–360.

Saha, S., Ghosh, P.D., Sengupta, C., 2010. An efficient method for micropropagation of *Ocimum basilicum* L. *Indian Journal of Plant Physiology*, 15(2), 168–172.

Sahai, A., Shahzad, A., 2013. High frequency *in vitro* regeneration system for conservation of *Coleus forskohlii*: A threatened medicinal herb. *Acta. Physiol. Plant.*, 35, 473–481.

Sahoo, Y., Pattnaik, S.K., Chand, P.K., 1997. *In vitro* clonal propagation of an aromatic medicinal herb *Ocimum basilicum* L. (sweet basil) by axillary shoot proliferation. In vitro *Cellular and Developmental Biology - Plant,* 33(4), 293–296.

Saltan, F.Z., Ozaydin, O., 2013. Ethnobotany of Eskisehir and its environs. *Pakistan Journal of Botany*, 45(1), 207–214.

Santos, A.V., Arrigoni-blank, M.F., Blank, A.F., Tavares, F.F., Pereira, R., Fernandes, M., Cesar, H., Jesus, R., Alves, P.B., 2010. Mass multiplication of *Pogostemon cablin* (Blanco) Benth genotypes and increase of essential oil and patchoulol yield. *Industrial Crops & Products*, 32(3), 445–449.

Santos-Gomes, P.C., Seabra, R.M., Andrade, P.B., Fernandes-Ferreira, M., 2002. Phenolic antioxidant compounds produced by *in vitro* shoots of sage (*Salvia officinalis* L.). *Plant Science*, 162(6), 981–987.

Saraç, N., Uğur, A., 2007. Antimicrobial activities and usage in folkloric medicine of some Lamiaceae species growing in Mugla, Turkey. *EurAsia J. Bio. Sci.*, 1, 28–34.

Shahzad, A., 2012. An efficient system for *in vitro* multiplication of *Ocimum basilicum* through node culture. *African Journal of Biotechnology*, 11(22), 6055–6059.

Sharma, N.K., Choudhary, R.C., Kumar, M., 2014. Effect of phytohormones on *in vitro* regeneration of *Ocimum basilicum* L. *Medicinal Plants,* 6(3), 163–168.

Sharma, P., Mohan, L., Srivastava, C.N., 2004. Larval susceptibility of *Ajuga remota* against anopheline and culicine mosquitos. *Southeast Asian Journal of Tropical Medicine and Public Health*, 35, 608–610.

Shin, T.Y., Kim, Y.K., Kim, H.M., 2001. Inhibition of immediate-type allergic reactions by *Prunella vulgaris* in a murine model. *Immunopharmacology and Immunotoxicology*, 23(3), 423–435.

Shkopynska, T.Ye., Kolomiets, Yu.V., Hrygoriuk, I.P., Kucenko, N.I., 2019. Economic valuable traits of promising breeding samples and 'Chornolysta' variety of *Mentha piperita* L. after *in vitro* sanitation and micropropagation. *Plant Varieties Studying and Protection*, 15(4), 424–433.

Shoja, H., Shishavani, H., 2021. Effects of different hormonal treatments on growth parameters and secondary metabolite production in organ culture of *Hyssopus officinalis* L. *BioTechnologia*, 102(1), 33–41.

Shukla, M.R., Kibler, A., Turi, C.E., Erland, L.A.E., Alan Sullivan, J., Murch, S.J., Saxena, P.K., 2021. Selection and micropropagation of an elite melatonin rich tulsi (*Ocimum sanctum* L.) germplasm line. *Agronomy*, 11(2), 207. https://doi.org/10.3390/agronomy11020207

Siddique, I., Anis, M., 2007. Rapid micropropagation of *Ocimum basilicum* using shoot tip explants pre-cultured in thidiazuron supplemented liquid medium. *Biologia Plantarum*, 51(4), 787–790.

Siddique, I., Anis, M., 2008. An improved plant regeneration system and ex vitro acclimatization of *Ocimum basilicum* L. *Acta Physiologiae Plantarum*, 30(4), 493–499.

Singh, N., Mahmood, U., Kaul, V.K., Jirovetz, L., 2006. A new phthalic acid ester from *Ajuga bracteosa*. *Natural Product Research*, 20(06), 593–597.

Singh, N.K., Sehgal, C.B., 1999. Micropropagation of 'holy basil' (Ocimum sanctum Linn.) from young inflorescences of mature plants, *Plant Growth Regulation*, 29, 161–166.

Sreedevi, E., Anuradha, M., Pullaiah, T., 2013. Plant regeneration from leaf-derived callus in *Plectranthus barbatus* Andr. [Syn.: *Coleus forskohlii* (Wild.) Briq.]. *African J. Biotechnol.*, 12, 2441–2448.

Surahmaida, S., Umarudin, U., 2019. Toxicity of miana leaf (*Coleus blumei*) extract against houseflies (*Musca domestica*). *Biosaintifika: Journal of Biology & Biology Education*, 11(2), 249–255.

Swamy, M.K., Anuradha, M., 2011. Analysis of genetic variability in patchouli cultivars (*Pogostemon cablin* Benth.) by using RAPD markers. *Research in Biotechnology*, 2(6), 64–71.

Swamy, M.K., Mohanty, S.K., Anuradha, M., 2014. The effect of plant growth regulators and natural supplements on *in vitro* propagation of *Pogostemon cablin* Benth. *Journal of Crop Science and Biotechnology*, 17(2), 71–78.

Tantos, A., Meszaros, A., Kissimon, J., Horvath, G., Farkas, T., 1999. The effect of triacontanol on micropropagation of balm, *Melissa officinalis* L. *Plant Cell Reports*, 19(1), 88–91.

Tavares, A.C., Pimenta, M.C., Goncalves, M.T., 1996. Micropropagation of *Melissa officinalis* L. through proliferation of axillary shoots. *Plant Cell Reports*, 15(6), 441–444.

Tawfik, A.A., Mohamed, M.F., 2007. Regeneration of salvia (*Salvia officinalis* L.) via induction of meristematic callus. *In Vitro Cellular and Developmental Biology - Plant*, 43 (1), 21–27.

Tevfik, A.S., Yegorova, N.A., 2020. Clonal micropropagation of *Thymus vulgaris* L. *E3S Web of Conferences*, 224, 1–9.

Tsai, K.L., Chen, E.G., Chen, J.T., 2016. Thidiazuron-induced efficient propagation of *Salvia miltiorrhiza* through *in vitro* organogenesis and medicinal constituents of regenerated plants. *Acta Physiologiae Plantarum*, 38(1), 1–9.

Turker, A.U., Yucesan, B., Gurel, E., 2009. Effect of phytohormones on micropropagation of self-heal (*Prunella vulgaris* L.). *Journal of Herbs, Spices and Medicinal Plants*, 15(4), 293–302.

Ulgen, C., Yildirim, A., Turker, A., 2020. Enhancement of plant regeneration in lemon balm (*Melissa officinalis* L.) with different magnetic field applications. *Int. J. Second. Metab.*, 7, 99–108.

Valdes, L.J., Mislankar, S.G., Paul, A.G., 1987. *Coleus barbatus* (*C. forskohlii*) (Lamiaceae) and the potential new drug forskolin (Coleonol). *Economic Botany*, 41(4), 474.

Vibhuti, R.K., Kumar, D., 2019. Effect of 6-BAP on callus culture and shoot multiplication of *Coleus forskohlii* (syn Plectranthus forskohlli wild) briq. *Res. J. Life Sci. Bioinform. Chem. Sci.*, 5, 574–581.

Vishwakarma, R.A., Tyagi, B.R., Ahmed, B., Husain, A., 1988. Variation in forskolin content in the roots of *Coleus forskohlii*. *Planta Medica*, 54(05), 471–472.

Vivek, K.B., Yong, H.P., Pooja, A., 2015. Studies on phytochemical analysis, antioxidant and lipid peroxidation inhibitory effects of a medicinal plant, *Coleus forskohlii*. *Frontiers in Life Science*, 8(2), 139–147.

Vlachou, G., Papafotiou, M., Bertsouklis, K.F., 2016. *In vitro* propagation of *Calamintha nepeta*. *Acta Hortic.*, 1113, 189–194.

Wang, J.W., Wu, J.Y., 2010. Tanshinone biosynthesis in *Salvia miltiorrhiza* and production in plant tissue cultures. *Applied Microbiology and Biotechnology*, 88(2), 437–449.

Wang, Q.J., Zheng, L.P., Yuan, H.Y., Wang, J.W., 2013. Propagation of *Salvia miltiorrhiza* from hairy root explants via somatic embryogenesis and tanshinone content in obtained plants. *Industrial Crops and Products*, 50, 648–653.

Widoretno, W., 2016. *In vitro* induction and characterization of tetraploid patchouli (*Pogostemon cablin* Benth.) plant. *Plant Cell, Tissue and Organ Culture (PCTOC)*, 125(2), 261–267.

Willemse, R.H., 1985. Notes on East African Plectranthus species (Labiatae). *Kew Bulletin*, 1, 93–96.

Wilson, L.A., Senechal, N.P., Widrlechner, M.P., 1992. Headspace analysis of the volatile oils of Agastache. *Journal of Agricultural and Food Chemistry*, 40(8), 1362–1366.

Xu, H.X., Lee, S.H., Lee, S.F., White, R.L., Blay, J., 1999. Isolation and characterization of an anti-HSV polysaccharide from *Prunella vulgaris*. *Antiviral Research*, 44(1), 43–54.

Yan, H.J., Xiong, Y., Zhang, H.Y., He, M.L., 2016. *In vitro* induction and morphological characteristics of octoploid plants in *Pogostemon cablin*. *Breeding Science*, 66(2), 169–174.

Yegorova, N., Kruglova, N., Galin, I., Stavtzeva, I., 2020. Induction of morphogenesis in the callus culture of *Lavandula angustifolia* Mill. In *BIO Web of Conferences* (Vol. 24, p. 00098). EDP Sciences.

Zahran, E.M., Abdelmohsen, U.R., Khalil, H.E., Desoukey, S.Y., Fouad, M.A., Kamel, M.S., 2020. Diversity, phytochemical and medicinal potential of the genus Ocimum L. (Lamiaceae). *Phytochemistry Reviews*, 19, 907–953.

Zhang, D., Xiao, L.Y., Cheng, Y.W., Li, H.L., Feng, Z.M., Lin, P.Y., Wu, W.Y., Huang, K.R., 1998. Pharmacological action of Baoji Pill. *Tradit. Chin. Drug. Res. Clin. Pharmacol.*, 9, 212–214.

Zhang, H.S., Wang, S.Q., 2006. Salvianolic acid B from *Salvia miltiorrhiza* inhibits tumor necrosis factor-α (TNF-α)-induced MMP-2 upregulation in human aortic smooth muscle cells via suppression of NAD (P) H oxidase-derived reactive oxygen species. *Journal of Molecular and Cellular Cardiology*, 41(1), 138–148.

12

Cinnamomum tamala: A Review of its Traditional Uses, Phytochemistry and Pharmacological Properties, and Micropropagation

Priyanka Chaudhary
DPG Degree College
Gurugram, India

Shivika Sharma
Sardar Swaran Singh National Institute of Bio-Energy
Kapurthala Punjab, India

Vikas Sharma
Lovely Professional University
Phagwara-Jalandhar, India

CONTENTS

DOI: 10.1201/9781003239932-12

12.1 Introduction

Cinnamomum is a genus of shrubs and evergreen trees in the Lauraceae family. The genus *Cinnamomum* contains 350 species around the world (Chakraborty and Das, 2010), and around 20 species originate in India. Lauraceae are comprised primarily of trees and tree-like bushes (Shah and Panchal, 2010). Numerous types of *Cinnamomum* have therapeutic and flavor value and are of great interest commercially. A few of the economically important species of *Cinnamomum* are *Cinnamomum verum* Presl. (True Cinnamom), *Cinnamomum Cassia* Presl. (Chinese Cinnamom), *Cinnamomum burmannii* Blume (Indonesia Cassia), *Cinnamomum camphora* (Camphor tree), and *Cinnamomum tamala* (Buch-Ham) Nees and Eberm. (Indian Cassia). *C. tamala* is known by various names in various languages: Tejpat (Manipuri), Tamalapatram (Malayalam), Tejpatta (Hindi and Bengali), Talisha (Telgu), Patraka (Kannada) and Tezpat (Urdu) (Hossain et al. 2012).

12.2 Taxonomy

Botanical name: *Cinnamomum tamala*

Authority: Nees & Eberm.

Family: Lauraceae

Synonym(s): *Laurus tamala* Buch.-Ham.

Common names: Tejpat, Kumaon

12.3 Distribution

C. tamala is found in Kashmir, Himachal Pradesh and Uttar Pradesh, Sikkim, Assam, Mizoram and Meghalya (Sharma and Nautiyal, 2011). The tree is dispersed from Indus to Bhutan. *C. tamala* is found in its native India, the Pacific Islands, Australia, South-East Asia, Bangladesh and Nepal (Mir et al. 2004). The natural habitat of *C. tamala* is in the tropical and subtropical rain forest at an altitude of 900–2500 m (Ahmed et al. 2000). The farming of *C. tamala* is very limited in Nainital (Uttar Pradesh) and the Kangra districts of Himachal Pradesh (Bradu and Sobti, 1988). The plantation of *C. tamala* occurs in Mikir Hills, Khasi and Jaintia Hills, Garo Hills, Manipur and Arunachal Pradesh (Pruthi et al. 1978).

12.4 Description

C. tamala is an aromatic evergreen tree, 8 m tall and 150 cm in circumference. The leaves are enormous, 12–20 cm long, 5–8 cm broad, glabrous, opposite, alternately placed, short stalked, shining, leathery, thick, acuminate, long pointed and 0.8–1.8 cm long. The leaves have a sea-green colour and contain a few brownish spots, but new leaves are slightly pinkish tinged (Jayaprakasha et al. 2003). The flowers of *C. tamala* are small, whitish, in axillary cymes, and are bisexual but on the same plant (monoecious). The flowers are produced at the end of March or the beginning of April. The stem of *C. tamala* is rough and greyish-brown with soft wrinkled bark, which produces mucilage. The seeds need 1 year to attain maturity. The fruit of *C. tamala* is an ellipsoidal drupe; the mature fruits have a dark purple colour and enclose a single seed. The seeds are dispersed mainly by birds, strong winds, hailstorms and arboreal mammals, and secondarily by rats and other small mammals (Sharma et al. 2009).

12.5 Medicinal Properties

Numerous plants have been used for long time as solutions for human illnesses. Plants have extraordinary potential for delivering new medications for human benefit. The restorative importance of plants lies in chemical substances that cause specific physiological activity in the human body. The World Health Organization noticed that most of the total population relies upon traditional medication for essential medical services. Plants contain an immense range of substances that can be utilized to treat resistant infections (Baquero, 1997).

C. tamala is utilized in the Indian system of conventional medications and has different therapeutic properties. It has been noticed that the bark and leaves of this tree possess carminative, stimulant, astringent and sweet-smelling properties. Its bark is valuable for the healing of gonorrhoea (Kirtikar and Basu, 1981). The dried leaves and bark of *C. tamala* have been given for fever, body odour and anaemia. Seeds of *C. tamala* were squashed and blended with sugar and nectar and administered to youngsters for dysentery or cough (Edwards, 1993). Ayurveda illustrates the utilization of Tejpatra leaves in the healing of different illnesses, such as dehydration of mouth, looseness of the bowels, vomiting, anorexia and bladder problems. It is additionally utilized restoratively as a carminative, as a diuretic and in the treatment of heart issues (Showkat et al. 2004), and relieves pain during dental treatment due to the presence of eugenol. It has been utilized in Indian medication as a tonic for the cerebrum, as an anthelmintic and for treating diseases of the anus and rectum (Kirtikar and Basu, 1995). *C. tamala* is utilized in numerous Ayurvedic formulations, such as *Sudarshanchoorna*, *Chandraprabhavati*, *Talisadichurna*, *Sitopaladichurna*, and in many herbal weight loss capsules.

12.6 Uses

Spices are dried ingredients from aromatic plants utilized as seasoning agents in cooking. *C. tamala* is one of them. The leaves and bark of this tree are fragrant, and both can be used as a spice (Dhar et al. 2002). Its leaves are used in the food processing industries because of its exceptional smell, that is, a clove-like taste and a pepper-like odour (Chang and Cheng, 2002). Because of its smell, leaves are kept in garments and also chewed to disguise bad breath. In Kashmir, the leaves are utilized as an alternative to paan. *C. tamala* as an analgesic is used in dental preparations and insect repellent. The bark oil of cinnamon has a discernible fragrance of the spice, a sweet and sharp taste, and confers a distinctive odour and flavour. It is used by the flavouring industries in mutton and fast food seasonings, sauces, pickles, soft drinks and tobacco flavours.

C. tamala leaves are also used as a clarifier, with *Emblica officinalis* fruits, for tanning and dyeing leather (Baruah and Nath, 2000). The essential oil (Tejpat oil) is also used in the formulation of liquors (Ahmed et al. 2000). *C. tamala* is utilized as foodstuff, grain, wood and medication in Uttarakhand Himalayan area (Nautiyal and Kaechele, 2007). It is likewise utilized in industry as a fragrance component in soaps, detergents, aromas, toothpastes and beauty care products, refreshments and pharmaceuticals (Atal and Kapur, 1982; Chauhan et al., 2009).

12.7 Phytoconstituents of *Cinnamomum tamala*

Previous phytochemical studies have revealed that essential oils separated from the leaves of *C. tamala* contain limonene, camphene, methyl eugenol, eugenol acetate, myrcene, cinnamyl acetate, camphor, β-caryophyllene and camphene), polyphenols, monoterpenoids and sesquiterpenoids including phellandrene, eugenol, linalool, and traces of p-cymene, β-pinene, α-pinene and phenylpropanoids (Shah and Panchal, 2010). Eugenol is one of the primary constituents of cinnamon oil (Fischer and Dengler, 1990). The leaves of *C. tamala* contain eugenol and isoeugenol, whereas 70–80% cinnamic aldehydes are present in the bark. The leaves of *Cinnamomum* contain quercetin, kaempferol and quercetrin (flavonoids), which are responsible for its antioxidant activity (Prasad et al. 2009; Sultana et al. 2010). Various chemotypes of *C. tamala* have been reported in different parts of the country: linalool-rich types (Assam) (Nath et al. 1994), eugenol type (north-east India) (Gulati, 1979), cinnamaldehyde type (Uttarakhand) and cinnamaldehyde-linalool (Himachal Pradesh) (Sood et al. 1979). Presence of secondary metabolites in *Cinnamomum tamala* is presented in Table 12.1.

12.8 Pharmacological Properties

12.8.1 Antidiabetic Activity

Diabetes is one of the significant complex issues that the globe faces today. Indian traditional medicine has utilized plants and spices for the treatment of diabetes since the Vedic days. *C. tamala* has powerful antidiabetic properties. A methanol and water concentrate of *C. tamala* bark was evaluated for antidiabetic activity utilizing the α-amylase inhibition test. The inhibition values of the bark of *C. tamala* were observed to be 97.49% and 93.78% in methanol and water, respectively. Likewise, the IC_{50} values of the methanol and following water extract of *C. tamala* were 1.80 and 5.53, respectively. It has been inferred that the methanol extract showed a more powerful action than the following water extract of *C. tamala* (Kumanan et al. 2010).

Palanisamy et al. 2011 evaluated that 50% ethanolic concentrate of *C. tamala* leaves caused a huge reduction in the blood glucose level in streptozotocin-induced diabetes. The extracts of leaves of *C. tamala* possess excellent antioxidant and anti-hyperglycaemic activities in streptozotocin-induced diabetic circumstances, and are hence utilized for the therapy of diabetes-related complications. Chakraborty and Das (2010) evaluated the anti-hyperglycaemic property of aqueous extracts of leaves of

TABLE 12.1

Secondary Metabolites Present in *Cinnamomum tamala*

Name	Structure	Classification	Activity
Eugenol	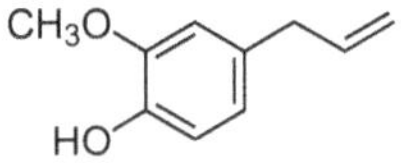	Phenylpropene	Antimicrobial activity (Bevilacqua et al. 2010), antifungal activity (Cheng et al. 2008), immunomodulatory activity (Farhath et al. 2013)
Eugenol acetate	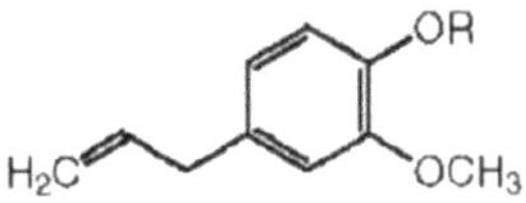	Phenylpropanoid	Anti-inflammatory property (Ozturk and Ozbek, 2005)
Cinnamaldehyde	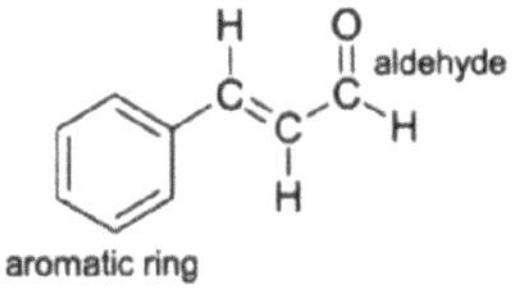	Aldehyde	Antioxidant, antifungal (Wang et al. 2005) and antimicrobial activity (Yang et al. 2011)
Cinnamyl acetate	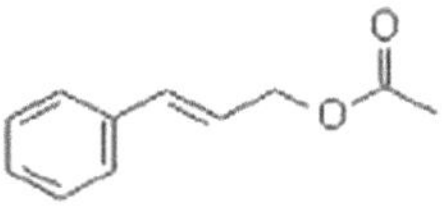	Organic compound	Antioxidant activity (Jayaprakasha et al. 2003), antimicrobial and fungicidal activity
Camphene	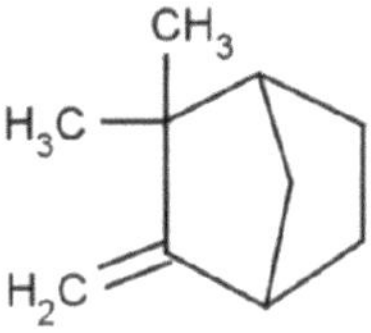	Monoterpene	Antimicrobial activity (Gerige and Ramjaneyulu 2007)
Camphor	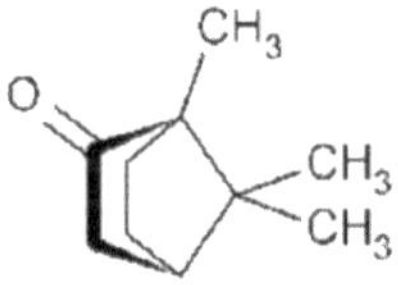	Terpenoid	Antibacterial activity (Imelouane et al. 2009), antioxidant (Hsu et al. 2012)
Pinene	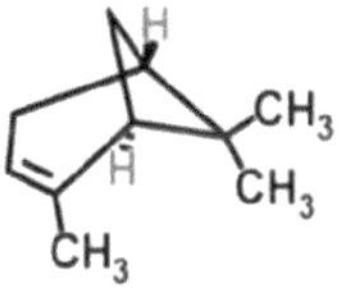	Monoterpene	Antibacterial activity (Imelouane et al. 2009)
p-Cymene	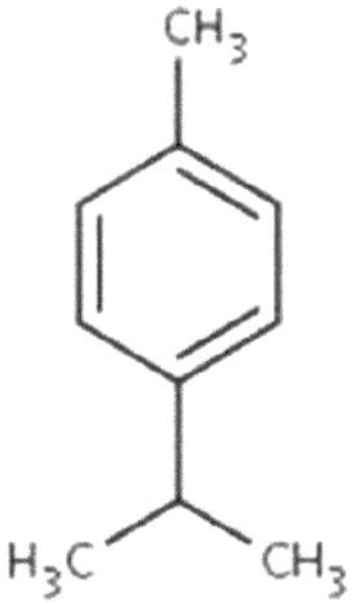	Alkyl benzene	Antifungal activity (Koba et al. 2009)

C. tamala on the blood glucose of albino rodents. It was found that administration of extract at a concentration of 250 mg/kg in streptozotocin-induced diabetic rats decreased the sugar level to normal.

12.8.2 Lipid-lowering Activity

The methanol extract of *C. tamala* was used to check lipid-lowering activity in rabbits. The methanol extract (500 mg/rabbit) was given to rabbits, and atorvastatin (0.005 mg) was used as a standard lipid-lowering agent. It was concluded that methanolic extract of *C. tamala* reduced the lipid profile by 1.0, 4.0, 14.0 and15 mg/dl for high-density lipoprotein (HDL-C), low-density lipoprotein (LDL-C), total cholesterol (TC) and triglycerides (TG), respectively.

It has been observed that the total cholesterol, triglycerides and LDL-C lowering activity of ethanol extract (400 mg/kg) of *C. tamala* leaves was more noteworthy as compared with aqueous extract. It has been revealed that the aqueous and ethanol extracts of leaves of *C. tamala* enhanced the serum lipid profile in rodents by lowering serum TC, TG and LDL-C while raising serum HDL-C, therefore improving the atherogenic index. As a result, *C. tamala* leaf extract is used as an antihyperlipidaemic agent and has a protective and remedial effect against hyperlipidaemia.

12.8.3 Antimicrobial Activity

Volatile oil of *C. tamala* possesses antimicrobial properties due to the presence of cinnamaldehyde and linalool oxide. Minimum inhibitory concentrations (MICs) of essential oil of *C. tamala* against bacterial and fungal pathogens were tested. It was shown that the strongest activity (MIC = 0.3–0.6 µl/ml) of *C. tamala* oil was responsible for inhibition of fungal strains, *Candida albicans*, *Candida parapsilosis*, *Aspergillus fumigates* and *Aspergillus niger*, whereas oil was found to be less effective against bacterial pathogens, *Staphylococcus aureus*, *Esherichia coli* and *Pseudomonas aeruginosa* (MIC = 2.5 µl/ml).

Antimicrobial activity of ethanol extract of *C. tamala* has been evaluated against several pathogens by disc diffusion assay. It was evaluated that methanolic extract (500 µg/disc) showed moderate antimicrobial activity against several pathogens, such as *Staphylococcus aureus* (8 mm), *Streptococcus pyogenes* (9 mm), *Streptococcus agalactiae* (9 mm), *Shigella sonnei* (9 mm), *Staphylococcus epidermidis* (10 mm), *Vibrio cholerae* (11 mm) and *Staphylococcus saprophyticus* (11 mm) (Hossain et al. 2012). Similarly, it has been found that butanol extract of *C. tamala* leaves adversely affects the growth of microorganisms and displays antibacterial activity against *Staphylococcus aureus*, *Escherichia coli*, *Klebsiella pneumoniae* and *Pseudomonas aeruginosa* (Jeyasree and Dasarathan, 2012).

The antibacterial effects of *C. tamala* leaf extracts (petroleum ether, acetone and aqueous) against Gram-positive and Gram-negative bacteria have been evaluated. It has been evaluated that petroleum ether, acetone and aqueous extract of leaves of *C. tamala* showed antibacterial activity against *Escherichia coli, Klebsiella pneumoniae* (showing zone of inhibition 12–23 mm), *Proteus vulgaris* and *Pseudomonas aeruginosa* (showing zone of inhibition 14–26 mm). Amongst all the leaf extracts of *C. tamala*, acetone and aqueous extracts at mg/disc concentrations exhibited maximum antibacterial efficacy against *Staphylococcus aureus*, *Streptococcus pneumoniae*, *Escherichia coli*, *Klebsiella pneumoniae*, *Pseudomonas aeruginosa and Proteus vulgaris* (Mishra et al. 2010).

It was found that aqueous extract of stem-bark of *C. tamala* was effective against *Bacillus cereus* and *Staphylococcus aureus* with zone of inhibition 11 mm. Methanol, ethanol and ethyl acetate extracts were effective against *Salmonella typhi* and *Streptococcus pyogenes*, with zone of inhibition ranging between 11 and 14 mm. Ethyl acetate extract was effective against *Staphylococcus aureus* with 15 mm zone of inhibition. Therefore, it has been clearly demonstrated that extracts of stem-bark of *C. tamala* revealed good antibacterial activity and can be utilized for the treatment of several contagious infections caused by microbes (Goyal et al. 2009).

Antibacterial activity of some other species of *Cinnamomum* has also been reported. It was found that the bark of *C. zeylanicum* showed no activity against *E. coli* but was moderately active against *Pseudomonas aeruginosa* and *Staphylococcus aureus* (Agaoglu et al. 2007). The alcoholic extract of *C. cassia* bark showed antibacterial activity (zone of inhibition 7–29 mm) against several microorganisms. *C. verum* bark showed mild activity against several *Enterobacteria*. Essential oils of *Cinnamomum cassia*, *Cinnamomum camphora*, *Cinnamomum iners*, *Cinnamomum osmophloe*, *Cinnamomum zeylanicum* and *Cinnamomum porrectum* also possess antimicrobial activity (Mishra et al. 1991; Chang et al. 2001; Phongpaichit et al. 2007).

12.8.4 Antidiarrhoeal Activity

The antidiarrhoeal action of ethanolic extract of dried leaves of *C. tamala* has been assessed on castor oil-induced diarrhoea in mice. It was evaluated that leaf extract of *C. tamala* at concentrations of 250 and 500 mg/kg postponed the onset of diarrhoea and reduced the mean numbers of defecations and stools by 24.9% and 40.82%, respectively (Hossain et al. 2012).

12.8.5 Antioxidant Activity

Plants are potential sources of natural antioxidants. The antioxidant effect is mostly due to the presence of phenolic acid, tannins, flavonoids and diterpenes (Duh et al. 1999). Reactive oxygen species play a crucial function in the progression of different diseases, such as inflammatory injury, atherosclerosis, malignancy and cardiovascular illness. Free radicals damage the cell and lead to pathological changes associated with ageing. Medicinal plants have the therapeutic potential to act as antioxidants to reduce the tissue injury caused by free radicals.

The anti-peroxidative impact of an alcoholic concentrate of *C. tamala* was examined in rodent liver homogenate. Ferrous sulphate was utilized as an inducer for lipid peroxidation. It was found that TBARS (thiobarbituric acid reactive oxygen species) production is very fast in rats treated with ferrous sulphate. TBARS are formed as a byproduct of lipid peroxidation. When alcoholic extract of *C. tamala* was added to ferrous sulphate-treated rat liver homogenate, the production of TBARS decreased from 400 to 250 nmole/100 mg protein. This decline in TBARS clearly showed the antioxidant property of *C. tamala* (Gupta and Sharma, 2010). Therefore, it has been inferred that addition of cinnamon compounds to the routine diet could reduce the dangerous factors related to the generation of free radicals and act as a superior antioxidant.

12.9 Micropropagation

As a growing innovation, plant tissue culture (PTC) greatly affects horticulture as well as industry by means of providing plants that are expected to meet the increasing demand of the world. Like other technologies, PTC is also a well-established technology of the day, and at the same time, it has experienced various phases of development, research tools, novel applications and mass exploitation. The technology has produced numerous employment opportunities and unfolded several entrepreneurial fields. The utilization of *in vitro* raised plantlets has enhanced the productivity per unit area, mainly in horticultural vegetation. This industry has made accessible, on a substantial scale, diverse exceptional commercial plant species that were no longer being produced by conventional strategies. Tissue culture has been one of the fundamental key sources that have contributed to the second green revolution and the gene revolution. The world is viewing India as the main source of technology for the production of economical plant varieties. With additional innovative work and intensive exploitation of our flora, the tissue culture technique will assist us in consolidating our leadership at the worldwide level (Singh and Shetty, 2011).

Micropropagation is the procedure of vegetative development and multiplication from plant tissues or seeds. It is carried out under sterile conditions on growth medium, utilizing different plant tissue culture procedures (Bhojwani and Razdan, 1996; Zhou and Wu, 2006). The micropropagation method has the ability to produce plants of better quality with improved disease and strain resistance capacities (Brown and Thorpe, 1995). Micropropagation holds significant promise for true-to-type, quick and mass multiplication that can lead to the production of disease-free plantlets (Gonzales et al. 2010; Thangavel et al. 2014). Micropropagation guarantees a regular supply of therapeutic plants, utilizing less space and time (Prakash and Staden, 2007).

In a large portion of tree species, the populace raised through seed proliferation does not guarantee hereditary fidelity, and there is a chance of losing their 'first class' characters. However, the vegetative proliferation strategy by cuttings has been rehearsed in numerous woody species; the frequency of established plants is very low, particularly when mature cuttings are used. To overcome these issues, different *in vitro* strategies like organogenesis, axillary shoot proliferation and somatic embryogenesis have been utilized for the micropropagation of forest plants and other significant tree species. Of the various strategies utilized for *in vitro* proliferation, the utilization of parallel buds or potentially stem sections with axillary buds has been an effective plant propagation technique to maintain the genetic stability of propagated plants (Mangal et al. 2008).

Sharma and Nautiyal (2009) observed 4.0 ± 0.0 shoots/explant with 100% rooting on MS medium containing BAP and IBA when petioles with nodal segments were used as explants. Further, these explants showed 90% survival when shifted to natural environmental conditions. It has been reported that microshoots were established by culturing explant on MS medium sustained with sucrose (3%) and α-naphthalene acetic acid (NAA) (3 μM), and it was found that up to 10 roots were formed per microshoot. The rooted plantlets were moved to pots containing soil, sand and rotted wood powder in 1:1:1 proportions and kept within the polyhouse at 75% shading of light for some time. Then, the plantlets were moved to the natural climatic conditions, and 70% of explants survived. It was also reported that MS medium containing 6 μM BAP produced five to six shoots when cotyledon segments were used as explants. and when shoots were cultured on rooting medium containing NAA. the maximum number of roots (five to six) was observed, with 65% survival when explants were shifted to natural climatic conditions.

12.10 Molecular Studies

Hereditary consistency is the maintenance of the genetic design make-up of a particular copy throughout its life expectancy (Chaterjee and Prakash, 1996). This is a vital pre-necessity in the multiplication of plant species and is confirmed through molecular investigation (Alizadeh and Singh, 2009). It is necessary to build up the best micropropagation procedure for the establishment of hereditarily identical plants before it is prepared for productive purposes. Besides, there is a need to regularly check the clonal fidelity of micropropagated plantlets to affirm their true-to-type nature so as to avoid variations, which, whenever they appear, can multiply quickly and cause harm to the advantageous characters of the parental genotypes (Alizadeh and Singh, 2009). Numerous factors may possibly influence the stability of the *in vitro* raised plantlets, such as genotype, time of culture period and nature of explants (Premvaranon et al. 2011). Various valuable tools, for example gas chromatographic profiling, molecular markers and flow cytometry, have been broadly utilized to confirm the biochemical stability of tissue culture-raised plantlets (Prasad et al. 2015). To examine the hereditary consistency and instability of *in vitro* culture-derived plantlets, random amplified polymorphic DNA (RAPD) and inter-simple sequence repeats (ISSR) are commonly utilized because they are easy, quick to execute, and utilize a minute quantity of DNA in the absence of any prior information regarding the genome (Williams et al. 1990). The utilization of more than one marker has been significant for the assessment of the hereditary strength of plants, as they target different regions of the genome (Lakshmanan et al. 2007).

RAPD (Williams et al. 1990) uses single, short, random oligonucleotide primers to reveal variations in nucleotide sequence by amplifying the unknown DNA sequences. RAPD has been used in the construction of linkage maps (Grattapaglia and Sedroff, 1994), resistance gene localization (Dweikat et al. 1997), identification of hybrid origin (Friesen et al. 1997) and estimation of genetic variation (Nesbitt et al. 1995). RAPD has been extensively used in many woody species, for example eucalyptus (Keil and Griffin, 1994), guava (Prakash et al. 2002) and mango species (Ravishankar et al. 2000).

Soulange et al. (2007) carried out molecular work on two different species of *Cinnamomum*, *C. camphora* and *C. verum*, which provided the useful information that hereditary dissimilarity exists among *C. camphora* and *C. verum*. It has been concluded that RAPD analysis is a reliable technique for assessing genetic diversity between these two species, as a large degree of polymorphism has been obtained by utilizing 11 primers. ISSR markers were found to be the best for genetic variation in *Cinnamomum tamala*.

This review chapter gives the important information that a small amount of work has been done on the micropropagation of *Cinnamomum tamala*. Interest in *C. tamala* is expanding step by step in the pharmaceutical industry, but its excessive use is leading towards its extinction, which is undesirable. Therefore, there is a great need to enhance the micropropagation of *C. tamala* to fulfill the needs of the pharmaceutical industry and to improve the production of secondary bioactives. There is a need to carry out more work in this field so as to protect this endangered medicinal plant from extinction. Extinction of medicinal plants is a very serious problem. It not only deprives us of important medicinal plants, which are useful to us on account of their economic importance, but also causes imbalance in our environment. Therefore, it is important for us to maintain the natural habitat of endangered medicinal plants, which will be a step towards the safety of these plants.

Conclusion

The numerous benefits of *C. tamala* make it a real wonder of nature. *C. tamala* shows a wide range of antimicrobial and cancer prevention activities against several microbes because of the presence of phytoconstituents such as alkaloids, glycosides and tannins. Broader research is important to investigate the standards responsible for these activities. Tissue culture of remedial plants with enhanced bioactive principles and cell culture strategies for production of particular metabolites are observed to be profoundly helpful for the profitable production of therapeutically significant compounds. Interest in *C. tamala* is expanding because of its high therapeutic value. To satisfy the growing need for drugs, plant tissue culture is helpful for increasing species that are difficult to recover by traditional techniques and saving them from elimination. Further examination and preservation of plant species are proposed to save nature's medications.

REFERENCES

Agaoglu, S., Dostbil, N., Alemdar, S., 2007. Antimicrobial activity of some spices used in the meat industry. *Bull Vet Inst Pulawy*. 51, 53–57.

Ahmed, A., Choudhary, M.I., Farooq, A., Demirei, B., Demirei, F., Baser, K.H.C., 2000. Essential oil constituents of the spice *Cinnamomum tamala* (Ham.) Nees and Eberm. *Flavour and Fragrance Journal*. 15, 388–390.

Alizadeh, M., Singh, S.K., 2009. Molecular assessment of clonal fidelity in micropropagated grape (Vitis spp.) rootstock genotypes using RAPD and ISSR markers. *Iran J Biotechnol*. 7(1), 37–44.

Atal, C.K., Kapur, B.M., 1982. *Cultivation and Utilization of Aromatic Plants*. CSIR, New Delhi.

Baquero, F., 1997. Gram positive resistance: Challenge for the development of new antibiotics. *Journal of Antimicrobial Chemotherapy*. 39, 1–6.

Baruah, A., Nath, S.C., 2000. Foliar epidermal characters in twelve species of *Cinnamomum Schaeffer* (Lauraceae) from Northeastern India. *Int. J. Plant Morphol*. 47, 127–134.

Bevilacqua, A., Corbo, M.R., Sinigaglia, M., 2010. *In vitro* evaluation of the antimicrobial activity of eugenol, limonene, and citrus extract against bacteria and yeasts, representative of the spoiling microflora of fruit juices. *J Food Prot*. 73(5), 888–894.

Bhojwani, S.S., Razdan, M.K., 1996. *Plant Tissue Culture: Theory and Practice*. Developments in crop science, Vol. 5. Elsevier, Amsterdam.

Bradu, B.L., Sobti, S.N., 1988. Cinnamomum tamala in North West Himalayas; evaluation of various chemical types for perfumery value. *Indian Perfumer*. 32, 334–340.

Brown, D.C.W., Thorpe, T.A., 1995. Crop improvement through tissue culture. *World J. Microbiol. Biotechnol.* 11, 409–415.

Chakraborty, U., Das, H., 2010. Antidiabetic and antioxidant activities of *Cinnamomum tamala* leaf extracts in Stz-treated diabetic rats. *GJBB*. 5(1), 12–18.

Chang, S.T., Chen, P.F., Chang, S.C., 2001. Antimicrobial activity of leaf essential oils and their constituents from *Cinnamomum osmophloeum*. *J. Ethnopharmacol.* 77, 123–127.

Chang, S.T., Cheng, S.S., 2002. Antitermitic activity of leaf essential oils and components from *Cinnamomum osmophloem*. *J. Agric. Food Chem.* 50(6), 1389–1392.

Chaterjee, G., Prakash, J., 1996. Genetic stability in commercial tiss cult. In Prakash, J. and Pierik, R.I.M. (eds), *Plant Biotechnology: Commercial Prospects and Problems*, 11–121, Oxford IBH Publishing Co., New Delhi, India.

Chauhan, N.K., Haider, S.Z., Lohani, H., Sah, S., Yadav, R.K., 2009. Quality evaluation of *Cinnamomum tamala* Nees. from different locations of Uttarakhand. *J Non-Tim Forest Prod.*16, 191–194.

Cheng, S.S., Liu, J.Y., Chang, H., Chang, S.T., 2008. Antifungal activity of cinnamaldehyde and eugenol congeners against wood-rot fungi. *Bioresour Technol.* 99, 5145–5149.

Dhar, U., Manjkhola, S., Joshi, M., Bhatt, A., Bisht, A.K., Joshi, M., 2002. Current status and future strategy for developmental of medicinal plant sector in Uttaranchal, India. *Curr Sci.* 83(8), 956–964.

Duh, P.D., Tu, Y.Y., Yen, G.C., 1999. Antioxidant activity of aqueous extract of Harnjyur (*Chrysanthemum morifolium* Ramat). *Lebensm Wiss Technol.* 32, 269–277.

Dweikat, I., Ohm, H., Patterson, F., Cambron S., 1997. Identification of RAPD markers for 11 Hesian fly resistance genes in wheat. *Theor Appl Genet* 94, 419–423.

Edwards, D.M., 1993. The marketing of non-timber forest products from the Himalayas: The trade between East Nepal and India. In Rural Development Forestry Network, Paper 15, 1–21, Overseas Development Institute, London.

Farhath, S., Vijaya, P.P., Manivannan, V., 2013. Immunomodulatory activity of geranial, geranial acetate, gingerol, and eugenol essential oils: Evidence for humoral and cell-mediated responses. *Avicenna J Phytomed.* 3(3), 224–230.

Fischer, I.U., Dengler, I.J., 1990. Sensitive high performance liquid chromatographic assay for the determination of eugenol in body fluids. *J Chromatogr A.* 525, 369–377.

Friesen, N., Fritsch, R., Bachmann, K., 1997. Hybrid origin of some ornamentals of *Allium* subgenus Melanocrommyum verified with GISH and RAPD. *Theor Appl Genet.* 95, 1229–1238.

Gerige, S.J., Ramjaneyulu, 2007. Antimicrobial activity of *Melia dubia* leaf volatile oil and camphene compound against skin pathogens. *Int J Plant Sci.* 2(2), 166–168.

Gonzales, G.R., Quiroz, K., Carrasco, B., Caligari, P., 2010. Plant tissue culture: Current status, opportunities and challenges. *Ciencia Investigacion Agraria.* 37(3), 5–30.

Goyal, P., Chauhan, A., Kaushik, P., 2009. Laboratory evaluation of crude extracts of *Cinnamomum tamala* for potential antibacterial activity. *Electron J Biol.* 5(4), 75–79.

Grattapaglia, D., Sedroff. R.,1994. Genetic linkage maps of *Eucalyptus grandis* and *Eucalyptus urophylla* using a Pseudo-Testcross mapping strategy and RAPD markers. *Genetics.* 137, 1121–1137.

Gulati, B.C., 1979. Production and utilization of forest products. *RRL Jammu.* 1, 130–139.

Gupta, V., Sharma, M., 2010. Protective effect of *Cinnamomum tejpata* on lipid peroxide formation in isolated rat liver homogenate. *Curr Res J Biol Sci.* 2(4), 246–249.

Hossain, M.H., Islam, M.S., Howlader, M.S.I., Dey, S.K., Hira, A., Ahmed, A., Jahan, F., Sarkar, R.P., 2012. Evaluation of antidiarrhoeal, antimicrobial and cytotoxic activities of *Cinnamomum tamala* leaves from Bangladesh. *Int J Pharm.* 2(4), 731–736.

Hsu, F.L., Li, W.H., Yu, C.W., Hsieh, Y.C., Yang, Y.F., Liu, J.T., Shih, J., Chu, Y.J., Yen, P.L., Chang, S.T., Liao, V.H., 2012. *In vivo* antioxidant activities of essential oils and their constituents from leaves of the Taiwanese *Cinnamomum osmophloeum. J Agric Food Chem.* 60(12), 3092–3097.

Imelouane, B., Amhamdi, H., Wathelet, J.P., Ankit, M., Khedid, K., Bachiri, A.E.L., 2009. Chemical composition and antimicrobial activity of essential oil of thyme (*Thymus vulgaris*) from Eastern Morocco. *Int J Agric Biol.* 205–208.

Jayaprakasha, G.K., Rao, L.J., Sakariah, K.K., 2003. Volatile constituents from *Cinnamomum zeylanicum* fruit stalks and their antioxidant activities. *J Agric Food Chem.* 51(15), 4344–4348.

Jeyasree, P., Dasarathan, P., 2012. Screening of phytochemicals and immunomodulatory potential of a medicinal plant, Cinnamomum tamala. *IJPSR.* 3(4), 1049–1052.

Keil, M., Griffin, R.A., 1994. Use of random amplified polymorphic DNA (RAPD) markers in the discrimination and verification of genotypes in Eucalyptus. *Theor Applied Genet.* 89, 442–450.

Kirtikar, K.R., Basu, B.D., 1995. *Cinnamomum tamala* (Plate 826). In Blatter, E., Caius, J.F., Mhaskar, K.S. (eds), *Indian Medicinal 261 Plants*, 2nd Edition, Vol. 3, 262, 2146–2147. L.M.B. Publishers, Allahabad, India.

Koba, K., Poutouli, P.W., Raynaud, C., Sanda, K., 2009. Antifungal activity of the essential oils from *Ocimum gratissimum* L. grown in Togo. *J Sci Res.* 1(1), 164–171.

Kumanan, R., Manimaran, S., Khan, S., Dhanabal, S.P., Nanjan, M.J., 2010. Screening of bark of *Cinnamomum tamala* (Lauraceae) by using α-amylase inhibition assay for anti-diabetic activity. *Int J Pharm Biomed Res.* 1(2), 69–72.

Lakshmanan, V., Venkataramareddy, S.R., Neelwa, R.N.E.B., 2007. Molecular analysis of genetic stability in long-term micropropagated shoots of banana using RAPD and ISSR markers. *Electron J Biotechnol.* 10, 106–113.

Mangal, M., Sharma, D., Sharma, M., Kher, R., Singh, A.K., 2008. *In vitro* plantlet regeneration in guava from nodal segments. *Phytomorphol.* 58, 103–108.

Mir, S.R., Ali, M., Kapoor, R., 2004. Chemical composition of essential oil of *Cinnamomum tamala* Neesand Eberm leaves. *Flavour Fragr J.* 19(2), 112–114.

Mishra, A.K, Singh, B.K., Pandey, A.K., 2010. *In vitro* antibacterial activity and phytochemical profiles of *Cinnamomum tamala* (Tejpat) leaf extracts and oil. *RIF.* 1(3), 134–139.

Mishra, A.K., Dwivedi, S.K., Kishoe, N., Dubey, N.K., 1991. Fungistatic properties of essential oil of *Cinnamomum camphora. Int J Pharmacog.* 29, 259–262.

Nath, S.C., Baruah, A., Hazariak, A.K., Singh, R.S., 1994. Essential oil of leaves of *Cinnamomum tamala* Nees and Eberm from north east India. *J Spices Aromatic Crops.* 3, 33–35.

Nautiyal, S., Kaechele., H., 2007. Adverse impacts of pasture abandonment in Himalayan protected areas: Testing the efficiency of a natural resource management plan (NRMP). *Environmental Impact Assessment Review.* 27(2), 109–125.

Nesbitt, K.A., Potts, B.M., Vaillancourt, R.E., West, A.K., Reid, J.B., 1995. Partitioning and distribution of RAPD variation in a forest tree species, *Eucalyptus globulus* (Myrtaceae). *Heredity.* 74, 628–637.

Ozturk, A., Ozbek, H., 2005. The anti-inflammatory activity of *Eugenia caryophyllata* essential oil: An animal model of anti-inflammatory activity. *Eur J Med.* 2(4), 159–163.

Palanisamy, P., Srinath, K.R., Kumar, D.Y., Chowdary, P., 2011. Evaluation of antioxidant and antidiabetic activities of *Cinnamomum tamala* Linn leaves in streptozotocin-induced diabetic rats. *Int Res J Pharm.* 2(12), 157–162.

Phongpaichit, S., Kummee, S., Nilrat, L., Itarat, A., 2007. Antimicrobial activity of oil from the *Cinnamomum porrectum* Songklanakarin. *J Sci Technol.* 29, 11–16.

Prakash, D.P., Narayanasswamy, P., Sondur, S.N., 2002. Analysis of molecular diversity in guava using RAPD markers. *J Hort Sci Biotechnol.* 77, 287–293.

Prakash, S., Staden, V.J., 2007. Micropropagation of *Hoslundia opposita* Vahl--a valuable medicinal plant. *S Afr J Bot.* 73, 60–63.

Prasad, A., Shukla, S.P., Mathur, A., Chanotiya, C.S., Mathur, A.K., 2015 Genetic fidelity of long-term micropropagated *Lavandula officinalis* Chaix: An important aromatic medicinal plant. *Plant Cell, Tiss Org Cult.* 120, 803–811.

Prasad, N.K., Yang, B., Dong, X., Jiang, G., Zhang, H., Xie, H., Jiang, Y., 2009. Flavonoid contents and antioxidant activities from Cinnamomum species. *Innov Food Sci Emerg Tech.* 10, 627–632.

Premvaranon, P., Vearasilp, S., Thanapornpoonpong, S., Karladee, D., Gorinstein, S., 2011 *In vitro* studies to produce double haploid in Indica hybrid rice. *Biologia.* 66(6), 1074–1081.

Pruthi, J.S., Varkey, A.G., Bhat, A.V., 1978. Post harvest technology of cinnamon and utilization of cinnamon waste (leaves and twigs). *Indian Cocoa, Arecanut and Spices, J.* 1(4), 87–93.

Ravishankar, K.V., Anand, L., Dinesh, M.R., 2000. Assessment of genetic relatedness among mango cultivars of India using RAPD markers. *J Hort Sci Biotechnol.* 75, 198–201.

Shah, M., Panchal, M., 2010. Ethnopharmacological properties of *Cinnamomum tamala*—a review. *International Journal of Pharmaceutical Sciences Review and Research.* 5(3), 141–144.

Sharma, G., Nautiyal, A.R., 2009. Influence of explants type and plant growth regulators on *in vitro* multiple shoots regeneration of a Laurel from Himalaya. *Nat Sci.* 7(9), 1–7.

Sharma, G., Nautiyal, A.R., 2011. *Cinnamomum tamala*: A valuable tree from Himalayas. *Int J Med Arom Plants.* 18, 2249–4340.

Sharma, G., Nautiyal, B.P., Nautiyal, A.R., 2009. Seedlings emergence and survival in *Cinnamomum tamala* under varying micro-habitat conditions: Conservation implications. *Tropical Ecology.* 50(1), 201–209.

Showkat, R.M., Ali, M., Kapoor, R., 2004. Chemical composition of essential oil of *Cinnamomum tamala* Nees et Eberm leaves. *Flavour Fragr J.* 19, 112–114.

Singh, G., Shetty, S., 2011. Impact of tissue culture on agriculture in India. *Biotechnology, Bioinformatics and Bioengineering.* 1(3), 279–288.

Sood, R.P., Padha, C.D., Talwar, Y.P., Jamwal, R.K., Rao, P.R., 1979. Essential oil from the leaves of *Cinnamomum tamala* Nees and Eberm. growing in Himachal Pradesh. *Indian Perfumer.* 23, 75–789.

Soulange, J.G., Ranghoo, V.M.S., Seeburrun, S.D., 2007. Tissue culture and RAPD analysis of *Cinnamonum camphora* and *Cinnamomum verum*. *Biotechnol.* 6(2), 239–244.

Sultana, S., Ripa, F.A., Hamid, K., 2010. Comparative antioxidant activity study of some commonly used spices in Bangladesh. *Pak J Biol Sci.* 13, 340–343.

Thangavel, K., Ebbie, M.G., Ravichandran, P., 2014. Biotechnology and *in vitro* conservation of medicinal plants. *Ann Plant Sci.* 3(6), 734–744.

Wang, S.Y., Chen, P.F., Chang, S.T., 2005. Antifungal activities of essential oils and their constituents from indigenous cinnamon (*Cinnamomum osmophloeum*) leaves against wood decay fungi. *Bioresource Technol.* 96, 813–818.

Williams, J.G.K., Kubelik, A.R., Livak, K.J., Rafalski, J.A., Tingey, S.V., 1990. DNA polymorphisms amplified by arbitrary primers are useful as genetic markers. *Nucleic Acids Res.* 18, 6531–6535.

Yang, C.H., Yang, C.S., Hwang, M.L., Chang, C.C., Chuang, R.X., 2011. Antimicrobial activity of various parts of *Cinnamomum cassia* extracted with different extraction methods. *J Food Biochem.* 36(6), 690–698.

Zhou, L.G., Wu, J.Y., 2006. Development and application of medicinal plant tissue cultures for production of drugs and herbal medicinals in China. *Nat Prod Rep.* 23, 789–810.

13

Quantitative Trait Locus (QTL) Mapping in Crop Improvement

Sharanabasappa B. Yeri
Zonal Agricultural Research Station
Kalaburgi
University of Agricultural Sciences
Raichur
Karnataka, India

Varsha Kumari
Sri Karan Narendra Agriculture University
Jobner-Jaipur
Rajasthan, India

Radheshyam Sharma
Jawaharlal Nehru Krishi Vishwa Vidhalya
Jabalpur
Madhya Pradesh, India

Sumer Singh Punia
Sri Karan Narendra Agriculture University
Jobner-Jaipur
Rajasthan, India

CONTENTS

DOI: 10.1201/9781003239932-13

13.1 Quantitative Traits

The phenotypic characteristics that distinguish different individuals into discrete categories led to Mendelian genetics of inheritance, revealing differences among simple traits like height, color, size, shape and many others, exhibiting clearly dichotomous phenotypes known as qualitative traits. These traits can be easily studied using simple statistical tools. On the contrary, the majority of the directly observable individual traits that build up the architecture of the plant are quantitative in nature. **Quantitative traits** are those that are governed by many genes collectively. The phenotypes of such traits are determined by many genes having a small effect under the influence of environmental factors, thus showing polygenic inheritance (polygenic or multifactorial or complex traits). They cannot easily classify specific individuals in the population into a limited set of discrete categories.

The quantitative traits show continuous variation due to the effects of genetic and non-genetic factors (environmental) contributing to the traits. It is obvious that under constant environmental factors, the observed variation in a trait is due to a genetic component. However, the estimate of the genetic component is limited by the fact that the traits are governed by polygenes.

Quantitative traits are very interesting traits from the evolutionary standpoint. Generally, one phenotype tends to be slightly masked by others, but this is unnoticeable. However, when many individuals in a population are examined, significant differences can be found among them. Often, this type of variation can be quantified by measuring the trait under question in a population by sampling the individuals (Figure 13.1).

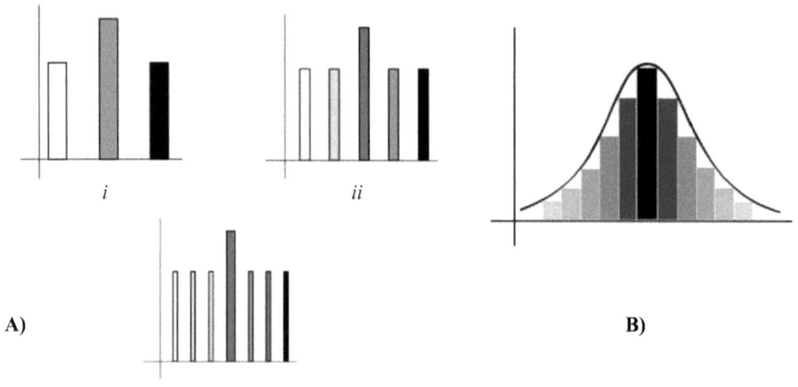

FIGURE 13.1 **A)** Qualitative traits (discrete distinction): *i*) one locus, *ii*) two loci, *iii*) three loci: Qualitative traits show the absolute distinction between two individuals with discrete values. **B)** Quantitative traits (continuous variation) controlled by several genes having a small effect under the influence of the environment tend to show a bell curve.

In nature, the majority of traits, such as grain color, bean size, kernel numbers and many more, show continuous variation rather than discrete phenotypic classes. These traits are quantitative, and segregation ratios are difficult or impossible to discriminate as a result of the large number of phenotypes and one phenotype showing the tendency to blend indiscernibly into the next. Thus, such traits do not segregate as simple Mendelian factors; rather, they show complex inheritance. A growing body of evidence suggests that the complex inheritance of quantitative traits is attributed to the influence of many genes and many factors in the environment. These traits can often be shown to have a component that is heritable, and that must therefore involve one or more genes.

The very understanding of quantitative traits or the nature and behavior of the polygenes comes from the classic studies of Fisher's "Infinitesimal model", which showed linearity between the Mendelian traits and the quantitative traits. Fisher successfully fused micro mutations contributing to a phenotype with Mendelism, showing the observed correlations of a large number of discrete inherited differences, each inherited according to Mendel's laws and contributing a small effect to the quantitative trait under question. Thus, the cumulative effect of such variants and many loci could tend towards a normal distribution assuming the bell curve. Therefore, the analysis of quantitative traits was studied with a higher degree of statistics using variance and covariance. Consequently, the "infinitesimal model" is considered to be the founding principle of quantitative genetics.

13.2 Quantitative Trait Loci (QTLs)

It is widely accepted that the chromosomal theory of inheritance is a plausible explanation for how traits are physically transmitted from parents to offspring. The theory also substantiated the evidence of genes governing a trait being located on the chromosomes and spatially separated from each other. Further, the majority of traits are governed by more than one gene (polygenic), as advocated by the "infinitesimal model". The quantitative trait is controlled by an infinite number of loci, and each locus contributes with an infinitely small effect. Since the effect of each locus is unrecognizable, these loci must be collectively studied. Thus, it is possible to assume that such traits are linearly arranged in a genome that confirms the phenotype of a quantitative trait. The segment of the chromosome or genomic region that contains the genes associated with a particular quantitative trait is known as the QTL.

A QTL is a genetic locus that affects the variation in a trait by its alleles. This is basically multifactorial, influenced by several genes and environmental conditions. Thus, a solitary QTL or many QTLs can influence a trait/phenotype. Occasionally, the phenotypic variation can be solely influenced by the environment or through gene–environment interactions. Sometimes, closely linked genes are responsible for the quantitative variation of the trait. Thus, such traits are to be counted as single QTLs. Overall, it is difficult to define a QTL precisely, as the exact number of genes for a given QTL governing the trait in question is not well known, as well as the fact that for any given gene on its own, there are several allelic forms segregating in Mendelian fashion in a population, and their effects are roughly additive.

13.3 Quantitative Trait Loci (QTL) Analysis

Exploring the QTLs and dissecting them to discover their actions, their interactions and their precise location in the genome is of crucial importance in agriculture and medicine. However, identifying and understanding the QTLs with conventional phenotypic analysis is difficult due to their complex inheritance. Sax (1923), in his experiment on beans, demonstrated that the effect of an individual locus on a quantitative trait could be identified through a series of crosses resulting in randomization of the genetic background with respect to all genes not linked to the genetic markers under observation. Even though all the markers used by Sax were morphological seed markers with complete dominance, he was able to show a significant effect on seed weight associated with some of his markers. On the premise that many genes and the environment act and interact to determine the phenotype of the trait, it would be difficult, if not impossible, to determine the action of individual trait genes. Though statistical methods such as analysis of variance and path coefficients, respectively, can be used to partition the variation and describe the resemblance between relatives, the effect still remains unanswered. A major breakthrough in the characterization of quantitative traits that created opportunities to select for QTLs was initiated by the development of DNA (or molecular) markers in the 1980s.

13.4 Molecular Markers and Linkage Mapping

Molecular markers are the DNA segments that announce their presence in the genome/chromosomes. They are characterized by signature sequences that are readily amenable to detection. They may be the scaffold, motifs, repetitive DNA, restriction sites or housekeeping genes. They are a site of heterozygosity for some types of silent DNA variation not associated with any measurable phenotypic variation. Such a "DNA locus", when heterozygous, can be used in mapping analysis just like a conventional heterozygous allele pair. These molecular markers can be easily detected and abundantly found in a genome when they are mapped by linkage analysis; they fill the void between genes of known phenotypes. The significance of the DNA markers in mapping is with reference to the heterozygous site, which can be a convenient reference point, useful in finding one's way around the chromosomes. Thus, the resolution with the marker system can be used to map the QTL on to the chromosomes/linkage groups, enabling the development of linkage maps and QTL maps. Molecular markers are preferred for genotyping, because these markers are unlikely to affect the trait of interest. Hence, they have enabled an array of QTL mapping studies in most crop plants for diverse traits like yield, quality, disease and insect pest resistance, abiotic stress tolerance and environmental adaptation.

13.5 Principles of QTL Mapping

The important application of molecular markers in agricultural research has been utilized in constructing linkage maps in diverse crop species and cultivated species. These linkage maps aid in identifying the chromosomal segments containing the genes controlling simple (Mendelian) traits and quantitative traits. Such linkage maps indicate the position and relative distance of the markers/genes along the length of the chromosomes, which can be taken as the signs or milestones on the highway. The process of constructing linkage maps and performing QTL analysis to identify genomic regions associated with traits of interest is known as QTL mapping (Collard et al. 2005). Identifying any specific gene or QTL within a plant genome is like finding a needle in a haystack. QTL analysis is based on the principle of detecting an association between the phenotype and the genotype of markers, whereas "QTL mapping" is based on the fact that genes and markers segregate during the chromosomal recombination (called crossing-over) during meiosis (i.e. sexual reproduction), consequently allowing their analysis in the progeny. Genes or markers that are close together or tightly linked will be transmitted together from parent to progeny more frequently than genes or markers that are located further apart. Basically, the markers are used to partition a population (mapping the population) into different genotypic classes based on genotypes at the marker locus followed by correlative statistics to determine whether individuals of one genotype differ significantly from individuals of another genotype with respect to the trait under study.

A significant difference between the phenotypic means of two or more groups, depending on the marker system and the type of population, indicates that the marker locus is linked to a QTL. If linked, the recombinants show a significant P value between the marker and the QTL. The smaller the distance between the marker and a QTL, the lower the chance of occurrence of recombination between the marker and the QTL. Thus they both tend to be inherited together in the progeny, and the mean of the group with the tightly linked marker will be significantly different ($P < 0.05$) from the mean of the group without the marker. On the contrary, when a marker is loosely linked or unlinked to a QTL, they tend to segregate independently. Thus, there will be no significant difference between their means in the genotype.

13.6 Steps in QTL Mapping

Developing a QTL map involves four major steps, which will be discussed under the following subheadings (Figure 13.2): 1) Identification of diverse parents and development of the mapping population, 2) identification of parental polymorphism for the markers and generating the saturated linkage map, 3) phenotyping and genotyping the mapping population, and 4) performing the QTL analysis using the statistical tools.

13.6.1 Developing the Mapping Population

A suitable mapping population generated using diverse parents (e.g. highly resistant and susceptible lines) will enable the possibility of identifying a large set of polymorphic markers that are well distributed across the genome. Several different populations are also used for QTL mapping, as shown in Figure 13.2. The mapping population could vary based on the objective of study, the time frame and the resources available for undertaking QTL mapping. However, the ability to detect QTLs in F_2 or F_2-derived populations and recombinant inbred lines (RILs) is relatively higher than in other mapping populations.

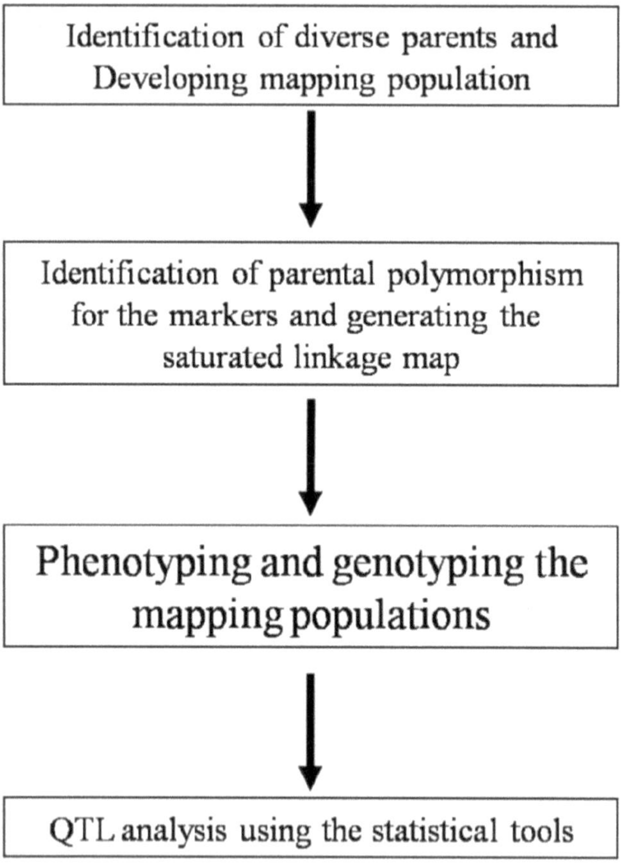

FIGURE 13.2 Steps involved in QTL mapping.

The $F_{2:3}$ families have the advantage that it is possible to measure the effects of additive and dominant gene actions at specific loci. The RILs are essentially homozygous, and only additive gene action can be measured; the advantage of RILs is that the experiments can be performed at several locations in multiple years. The size of the mapping population for QTL analysis depends on the type of mapping population, the genetic nature of the target trait, the objective of the study, and the resources available for handling a sizable mapping population in terms of phenotyping and genotyping. From the practical point of view, the purpose of QTL mapping is to detect the QTLs with major effects, and this is possible only when a large number of individuals, say 500 or more, are used for QTL analysis. So, in general, the size of the mapping population is around 200–300 individuals. Even backcross and near isogenic lines (NILs) are used for mapping the QTLs. Recently, the standing genetic variation encompassing the availability of diverse germplasm pools has been used in combination with single nucleotide polymorphism (SNP) markers to identify the haplotypes and for fine mapping.

13.6.2 Generating Saturated Linkage Map

Before generating the linkage map, the population is investigated for parental polymorphism using the appropriate marker system. Once the polymorphic markers are made available, they are screened across the population to analyze the segregation patterns for each of the markers, generally known as genotyping. Linkage is a process of assigning the markers in order, indicating the relative genetic distance between them, and assigning them to their linkage groups, based on the recombination values generated from all pair-wise combinations between the markers. The linkage map indicates the position and relative genetic distance between markers along chromosomes. Various molecular markers, RFLPs, RAPD, SSRs, AFLP, SNPs, etc., have been used to identify individual QTLs and to find the effects and position of these QTLs. The genetic segregation ratio at the marker locus is jointly determined by the nature of markers, i.e. dominant or codominant, and the type of mapping populations. Advances in molecular biology and biotechnology have led to the development of genome-wide genetic association studies (GWAS), which has enabled the identification of single polymorphic variants with identifiable functional effects on complex traits.

13.6.3 Phenotyping of Mapping Population

The target quantitative traits in question have to be measured precisely. As far as possible, there be no missing data, but to a certain extent, the missing data can be managed by built-in algorithms of the software used for analysis. The missing data can influence the effect of the sample size and in turn affect the power of QTL mapping. The experiments may be conducted in different locations and seasons if possible. The data generated across the locations is pooled to obtain a single quantitative value for the line. Such multi-location trials would provide a better understanding of the QTL × Environment interaction.

13.6.4 QTL Detection

The statistical parameters used basically postulate the null hypothesis as the occurrence or non-occurrence of association between a marker and the quantitative trait under question. The basic purpose of QTL mapping is to detect QTLs while minimizing the occurrence of false positives (Type I error), i.e. declaring an association between a marker and QTL when in fact it does not exist. The tests for QTL or trait association are often performed by the following approaches:

13.6.4.1 Single Marker Analysis (SMA)

Also referred to as single point analysis, this is the simplest method for detecting QTLs associated with single markers. The statistical method used for the single point analyses includes t-test, analysis of variance (ANOVA) and linear regression. SMA is performed with each marker locus independently of other loci. This method does not require a complete linkage map and can be executed with basic statistical software programs. However, the major disadvantage is that the further the QTL is from a marker, the less likely it is to be detected. This is because recombination may occur between the marker and the QTL, the linkage between them can be broken down, and they tend to segregate independently. The effect of QTLs is likely to be underestimated because these are baffled recombination frequencies. The use of a large number of polymorphic DNA markers covering the entire genome may minimize these problems (Tanksley, 1993).

13.6.4.2 Simple Interval Mapping (SIM)

Simple interval mapping was first proposed by Lander and Botstein in 1989. The method makes use of linkage maps followed by analysis of intervals between the adjacent pairs of linked markers along the chromosomes, simultaneously, as against the analysis of single markers. The presence of a putative QTL is estimated if the logarithm of odds ratios (LOD) exceeds a critical threshold, which is most often fixed as ≥3. The use of linked markers for analysis is considered statistically more powerful than SMA as it compensates for the recombination between the marker and the QTL. Mapmaker/QTL (Lincoln et al., 1993) and QGene (Nelson, 1997) are the choices of software for conducting the SIM.

13.6.4.3 Composite Interval Mapping (CIM)

Composite internal mapping is one of the popular methods used to detect QTLs. CIM was developed by Zeng (1993, 1994; Manly et al., 2001) and MQM (multiple QTL model or marker–QTL marker analysis) by Jansen and Stam (1994). This method combines internal mapping with linear regression. It gives scope for the inclusion of additional markers in the statistical model along with an adjacent pair of linked markers for interval mapping. The main advantage of CIM is that it is more precise and effective at mapping QTLs compared with single point analysis and simple interval mapping, particularly when linked QTLs are involved. Many researchers have used QTL Cartographer (Basten et al., 1994, 2001) and Map manager QTL to perform CIM.

13.6.4.4 Multiple Interval Mapping (MIM)

Most recently, MIM has gained popularity for mapping QTLs. It is the extension of internal mapping to multiple QTLs, just as multiple regression extends analysis of variance. It uses multiple marker intervals simultaneously to fit multiple putative QTLs directly for mapping QTLs. MIM allows one to infer the location of QTLs to a position between markers, makes proper allowance for missing genotype data, and can allow interaction between QTLs. The MIM approach is more precise and powerful to map QTLs. Further, the epistasis between QTLs and genotypic values of individuals can be easily estimated. The different software used for QTL analysis, along with its salient features, is presented in Table 13.1.

Recently, the advances in genomics to sequence the whole genome have enabled a more robust and powerful tool known as GWAS, which is used to identify single polymorphic variants with identifiable functional effects on complex traits.

TABLE 13.1

Software for QTL Mapping

SOFTWARE	FEATURES
MAPMAKER/QTL	Interval mapping (IM)
QGene	Single marker analysis (SMA), IM and multiple-trait analysis
MapQTL	IM, Composite interval mapping (CIM), non-parametric mapping with the Kruskal–Wallis rank sum test per marker (for non-normally distributed data), permutation tests, etc.
PLABQTL	SIM, CIM, also analysis for QTL × Environment (QE) interactions
MQTL	SIM, CIM, analysis for main effect, QE interactions, and can perform permutation tests
MapManager	QTXSMA, SIM, CIM, searches for interacting QTLs, etc.
QTL Cartographer	SMA, SIM, CIM, Bayesian interval mapping (BIM), Multiple interval mapping (MIM), multiple trait analysis, permutation tests, etc.
QTLMapper	Mapping QTL with epistatic effects, QE interaction effects, etc.
QTLNetwork	Mapping QTL with epistatic effects, QE interaction effects, etc.

13.7 Application of QTL Mapping

The important goal of QTL mapping in plant breeding is to understand the nature of inheritance and the genetic architecture of quantitative/complex traits. The knowledge of their inheritance within and across the species and identification of the markers linked to the important QTLs in various agriculturally important crops have enabled breeders to perform indirect selection of complex traits. Markers that are linked to agronomically important traits can be directly deployed in MAS and marker-assisted backcrossing (MAB). The introgression of QTLs into elite lines/germplasm, and MAS for QTLs in crop improvement, has been undertaken in some crops, such as maize (Li et al., 2008), tomato (Stevens et al., 2007) and wheat (Naz et al., 2008). In maize, the QTLs with major effects conferring resistance to downy mildew have been identified and transferred into CM139, an elite but downy mildew-susceptible inbred line (George et al., 2003). QTLs identified for diverse traits in different crops have been used in crop improvement, especially to enhance the yield and to develop disease-resistant elite lines.

REFERENCES

Basten, C.J., Weir B.S. and Z.-B. Zeng, 1994. Zmap-a QTL cartographer. In: J.S.G.C. Smith, B.J. Benkel, W.F. Chesnais, J.P. Gibson, B.W. Kennedy and E.B. Burnside (Eds.), *Proceedings of the 5th World Congress on Genetics Applied to Livestock Production: Computing Strategies and Software, Guelph, Ontario, Canada.* Published by the Organizing Committee, 5th World Congress on Genetics Applied to Livestock Production.

Basten, C., Weir, B. and Zeng, Z.-B., 2001. *QTL Cartographer.* Department of Statistics, North Carolina State University, Raleigh, NC.

Collard, B.C.Y., Jahufer, M.Z.Z., Brouwer, J.B. and Pang, E.C.K., 2005. An introduction to markers, quantitative trait loci (QTL) mapping and marker-assisted selection for crop improvement: The basic concepts. *Euphytica,* 142: 169–196.

George, M.L., Prasanna, C., Rathore, B.M., Setty, R.S., Kasim, T.A.S., Azrai, F., Vasal, M., Balla, S.O., Hautea, D., Canama, A., Regalado, E., Vargas, M., Khairallah, M., Jeffers, D. and Hoisington, D., 2003. Identification of QTLs conferring resistance to downy mildews of maize in Asia. *Theor. Appl. Genet.,* 107: 544–551.

Jansen, R. and Stam, P., 1994. High resolution of quantitative traits into multiple loci via interval mapping. *Genetics,* 136: 1447–1455.

Li, Y.L., Dong, Y., Niu, S., Cui, D., Wang, Y., Liu, Y., Wei, M. and Li, X., 2008. Identification of agronomically favorable quantitative trait loci alleles from a dent corn inbred Dan232 using advanced backcross QTL analysis and comparison with the F2:3 populations in popcorn. *Mol. Breed,* 21: 1–14.

Lincoln, S., Daly, M. and Lander, E., 1993. Mapping genes controlling quantitative traits using MAPMAKER/QTL. Version 1.1. Whitehead Institute for Biomedical Research Technical Report, 2nd Edn.

Manly, K.F., Cudmore, H., Robert, Jr. and Meer, J.M., 2001. Map Manager QTX, cross-platform software for genetic mapping. *Mamm. Genome,* 12: 930–932.

Naz, A.A., Kunert, A., Lind, V., Pillen, K. and Léon, J., 2008. AB-QTL analysis in winter wheat: II. genetic analysis of seedling and field resistance against leaf rust in a wheat advanced backcross population. Theor. Appl. Genet, 116: 1095–1104. doi: 10.1007/s00122-008-0738-y

Nelson, J.C. 1997. Qgene—software for marker-based genomic analysis and breeding. *Mol. Breed.* 3: 239–245.

Sax, K., 1923. The association of size differences with seed coat pattern and pigmentation in Phaseolus vulgaris. *Genetics,* 8: 552–560.

Stevens, R., Buret, M., Duffé, P., Garchery, C., Baldet, P., Rothan, C. and Causse, M., 2007. Candidate genes and quantitative trait loci affecting fruit ascorbic acid content in three tomato populations (2007). *Plant Physiol.,* 143: 1943–1953.

Tanksley, S.D., 1993. Mapping polygenes. *Annu. Rev. Genet.,* 27: 205–233.

Zeng, Z.-B., 1993. Theoretical basis for separation of multiple linked gene effects in mapping quantitative trait loci. *Proc. Natl. Acad. Sci. USA,* 90: 10972–10976.

Zeng, Z.-B., 1994. Precision mapping of quantitative trait loci. *Genetics* 136: 1457–1468.

14

*Progress in Genetic Engineering of Pigeonpea [*Cajanus cajan *(L.) Millsp.]: A Review*

Gourab Ghosh
National Institute for Plant Biotechnology
Pusa Campus
New Delhi, India

Jasdeep Chatrath Padaria
National Institute for Plant Biotechnology
Pusa Campus
New Delhi, India

CONTENTS

14.1 Introduction

The global human population is expected to reach approximately 10 billion by 2050. A large population, coupled with changes in dietary habits towards high-quality food, has created tremendous pressure on the existing agricultural system (Fróna et al. 2019). Dependence on animal-based diets has taken a heavy toll on the environment, and people throughout the world are reconsidering various vegetarian dietary options. In India, almost 40% of the population are vegetarian, which is the highest percentage in the world, followed by Africa, the Middle East and Latin America (Ruby, 2012). India and Africa house nearly 900 million people who live in extreme poverty, accounting for around 70% of the worldwide total. Vast tracts of these areas fall into the semi-arid zones, where agriculture is predominantly dependent on sporadic rainfall, poor

DOI: 10.1201/9781003239932-14

irrigation facilities and various climatic stresses. These problems pose a huge threat to the resource-starved farmers of the region, who are already affected by limited economic resources as well as poverty. For these reasons, the majority of these populations are dependent on vegetarian diets, as they cannot afford animal-based proteins.

One of the major sources of proteins are pulses, which are extensively grown in the semi-arid tropical zones of these developing regions. From the nutritional point of view, soybean, mungbean, pigeonpea, chickpea, urdbean, cowpea, lentil and peanut are important grain legumes for millions of people worldwide. Globally, pulses are grown in about 171 countries, of which 52 countries grow chickpea, contributing about 16.77%. This is followed by 96 countries growing peas, contributing around 8.50%, and 21 pigeonpea-producing countries, contributing about 7.70% to the worldwide production (Parankusam et al. 2018). Farmers in these regions readily grow pulses, as they are an important source of proteinaceous food and fodder. Pulse crops play a pivotal role in global agriculture, since they have a shorter growing duration, a tendency to adapt to different cropping schemes, and tolerance to abiotic stresses, particularly drought and heat. Pulses or legumes have great potential for mitigating protein hunger and malnutrition among resource-constrained peoples in the developing countries. Additionally, grain legumes have symbiotic nitrogen-fixing bacteria in root nodules, which fix their own nitrogen, thereby reducing the cost of nitrogen inputs by farmers.

14.2 The Crop Pigeonpea

Pigeonpea [*Cajanus cajan* (L.) Millsp.] occupies an important place among grain legumes due to its capability to flourish under varied cropping systems and environments and to recuperate from losses caused by various biotic and abiotic stresses. The estimated global area of pigeonpea is more than 5.6 mha, and the major pigeonpea-growing countries are India, Myanmar, Malawi, Tanzania, Kenya and Uganda (FAO, 2019). Pigeonpea is considered as the sixth most important edible grain legume after *Phaseolus* beans, peas, chickpeas, broad beans and lentils. In the last decade, pigeonpea production has increased globally, but the yield per hectare has declined. Pigeonpea plays a significant role in supplying valuable nutrition to the downtrodden population due to the high protein content in seeds along with essential amino acids, complementing the nutritional profile mainly based on cereals and tubers. India, where pigeonpea is one of the most consumed pulses after chickpea, contributes approximately 90% of the global production. Despite being the highest producer, India has to import 200,000 tonnes of grains annually to cope with the enormous market demand for dry and split pigeonpea seeds used for edible purposes (Shiferaw et al. 2008). Adding to the problems are low-input and rain-fed conditions, which result in lower yield and poor nutritional quality.

14.3 Yield Constraints in Pigeonpea

Pigeonpea is subject to a number of biotic (bacteria, fungi, insects, viruses and nematodes) and abiotic (drought, heat, salinity, waterlogging and cold) stresses, which severely affect the yield and quality of this crop, especially in the rain-fed agricultural zones (Ghosh et al. 2017). Fusarium wilt, phytophthora blight and sterility mosaic are some of the most important diseases of pigeonpea. Fusarium wilt, caused by *Fusarium udum* Butler, is the most devastating soilborne fungal disease of pigeonpea. Phytophthora blight is another fungal disease, caused by *Phytophthora drechsleri* Tucker f. sp. cajani, while sterility mosaic is the most damaging viral disease of pigeonpea, caused by pigeonpea sterility mosaic virus transmitted by the eriophyid mite, *Aceria cajani* (Kumar et al. 2005). It is responsible for losses to the tune of US$300 million per annum. As well as these, insect pests are a major threat to pigeonpea production. The major insect pests of pigeonpea can be categorized into three groups: (1) the flower and pod-munching Lepidopteran larvae (mainly *Helicoverpa armigera* Hübner, *Maruca vitrata*, *Etiella zinkenella*), (2) the pod-sucking Hemipterans (*Clavigralla* spp.), and (3) the seed-feeding Dipteran group (*Melanagromyza* sp.) and Hymenoptera. Among these, the most destructive insect is *Helicoverpa armigera*, followed by the pigeonpea pod fly, *Melanagromyza obtuse* Malloch. These insects are responsible for the loss of approximately US$317 and US$256 million annually, respectively (Shanower et al. 1999).

14.4 Biotechnology to the Rescue

Many approaches have been applied to reduce the insect damage by developing resistant pigeonpea varieties, but the success rate is extremely low compared with other model crops. Although conventional breeding has been successful in producing a large number of improved varieties, suitable remedies for the various stresses have not yet been provided. One of the major reasons behind this is the absence of desirable characteristics in the primary gene pool. The usage of wild species to access secondary and tertiary gene pools has not been successful due to sterility, cross incompatibility and restricted recombination. Insect pest resistance could not be developed due to the unavailability of complete resistance in the germplasm. Also, its recalcitrant nature, with poor regeneration response in tissue culture, has been considered as one of the major bottlenecks in handling pigeonpea (Ghosh et al. 2014).

To evade pathogenic and insect-related losses, farmers have resorted to rampant usage of pesticides and fungicides. The application and acceptance of pesticides to improve agricultural yields have now become an integral part of our modern life. However, careless and continual use of pesticides has led to harmful consequences for the farming system and wreaked havoc in our ecosystem, allowing the entry of lethal residues into our food chain (Abhilash and Singh, 2009). Additionally, Lepidopteran pests have a tendency to develop resistance against pesticides, making them ineffective. Therefore, the development of high-yielding, disease-resistant crop varieties in a sustainable manner with reduced use of harmful chemicals is a major challenge in modern agriculture.

One of the solutions to the constraints imposed by sexual incompatibility and pest resistance is genetic engineering, which can be considered as a complementary tool in breeding strategies. Following the success of modern biotechnology, the establishment of multi-disciplinary research methodologies, along with the allocation of sufficient resources, has enabled rapid development in legume biotechnology for overcoming the severe bottlenecks associated with the improvement of important crops like soybean, chickpea and pigeonpea (Christou, 1997). Along with plant transformation technology, fast-evolving genomics data for pigeonpea (Varshney et al. 2012) may unravel various molecular genetics approaches, generating knowledge that can be applied for ingenious breeding strategies. Furthermore, extensive research has been conducted to isolate and characterize genes and molecular mechanisms controlling abiotic stress responses in both model plants and crops that cope with drought stress conditions. Recent statistics reveal that the adoption of transgenic technology has improved agricultural productivity by 22%, along with a significant expansion of the transgenic crop production area from 1.7 million hectares to 191.7 million hectares, a 113-fold increase in the last 22 years (ISAAA, 2018).

14.5 Source of Explants for Pigeonpea Transformation

In pigeonpea genetic transformation, organogenesis-mediated plant regeneration has been the most important choice due to its high regeneration frequency. Somatic embryogenesis has not been preferred due to its lower rate of embryo germination (Krishna et al. 2010). On the other hand, cotyledonary node is a much better and more regenerative explant compared with other tissues (Krishna et al. 2010, Ghosh et al. 2014). As well as this, particle bombardment has been deployed in leaf explants (Dayal et al. 2003) and in cotyledonary node (Thu et al. 2003). Stable regenerants in pigeonpea have been obtained through organogenesis from apical meristem (Cheema and Bawa, 1991), undifferentiated callus (Kumar et al. 1983; George and Eapen, 1994), differentiated non-meristematic tissues like leaf (Eapen and George, 1993; Eapen et al. 1998; Geetha et al. 1998; Singh et al. 2002; Dayal et al. 2003; Villiers et al. 2008), and various seedling explants such as hypocotyls (Geetha et al. 1998), cotyledons (George and Eapen, 1994; Geetha et al. 1998), cotyledonary nodes (Mehta and Mohan Ram, 1980; Kumar et al. 1983, 1984; Shiva Prakash et al. 1994; Naidu et al. 1995; Geetha et al. 1998), epicotyls (Kumar et al. 1984; Naidu et al. 1995; Geetha et al. 1998) and embryonal axes (Franklin et al. 2000).

Pigeonpea is a highly recalcitrant legume. Thus, in tissue culture-based transformation systems, there are chances of obtaining chimeric shoots with poor rooting response. Lack of root formation was found to be responsible for loss of regenerated shoots even after prolonged antibiotic selection (Krishna et al. 2010). To avoid the dilemma of root formation in transformants, a novel shoot-grafting (transgenic scion and wild-type stock) strategy after pigeonpea transformation was applied for the first time to obtain insect-resistant transgenic pigeonpea lines (Ghosh et al. 2014, 2017). But grafting has been a tedious and technique-oriented process, where the chances of losing transformants are high if compatibility between stock and scion fails. Under these circumstances, a tissue culture-independent in planta transformation strategy was first introduced in pigeonpea, which also provided a broad avenue in transformation technology for the biotechnological improvement of this recalcitrant crop (Rao et al. 2008). In this strategy, in vitro co-cultivation and selection steps were completely bypassed to generate a huge number of pigeonpea transformants. This method was developed by Ramu et al. (2012) and Kaur et al. (2016) to express *cry1AcF* and *cry1Ac* genes, respectively, in transgenic pigeonpea. A separate method for transformation by using the meristematic region of the plumular axis has also been reported for the first time to obtain primary pigeonpea transformants (Ganguly et al. 2018)

14.6 Genetic Transformation

Through genetic transformation technology, more than 15 genotypes of pigeonpea have been used by researchers for the development of biotic and abiotic stress-resilient varieties and also to upgrade the nutritional quality (Ghosh et al. 2017). Among them, ICPL87 was found to be the most frequently used genotype, with the achievement of 80% transformation frequency (Krishna et al. 2010). Over the last few decades, several attempts have been made in pigeonpea to introduce different foreign genes through *Agrobacterium*-mediated transformation strategies, but the success rate was immensely constrained by its poor tissue culture response (Ghosh et al. 2014). Various genes, like *Bt*-encoded *cry1A(b)*, *cry1Ab*, *cry1Aabc*, *cry1AcF*, *cry1Ac* and *cry2Aa* (Verma and Chand, 2005; Sharma et al. 2006; Ramu et al. 2012; Das et al. 2016; Kaur et al. 2016; Ghosh et al. 2017; Singh et al. 2018) and cowpea *protease inhibitor* (*CPI*) (Lawrence and Koundal, 2001), were used in pigeonpea transformation with a higher level of toxicity against Lepidopteran insects, whereas *cry1 E-C* was found to be effective against *Spodoptera litura* (Surekha et al. 2005). Among the *Bt* group of genes, *cry1Ac* has been found to be advantageous, with a significant response, to the transgenic establishment of insect-resistant varieties (Sanahuja et al. 2011). Recently, synthetic *cry1Ab* was expressed in pigeonpea under the influence of a tissue-specific promoter of the RuBP carboxylase/oxygenase small subunit (*rbcS*) gene (Sarkar et al. 2021).

Additionally, the *rice chitinase* (*Rchit*) gene has also been incorporated in pigeonpea for enhancing the resistance level to fungal pathogens (Kumar et al. 2004a). Apart from improving stress tolerance, pigeonpea transformation has also been done with genes like *hemagglutinin* gene of rinderpest virus (*RPVH*) and *hemagglutinin neuraminidase* gene of peste des petits ruminants virus (*PPRV-HN*) to improve the goat and sheep immune response against rinderpest virus (Satyavathi et al. 2003) and peste des petits ruminants virus, respectively (Prasad et al. 2004). In comparison to other grain legumes, pigeonpea is nutritionally deficient due to the presence of low amounts of sulfur-containing amino acids. Isolated *dihydrodipicolinate synthase* (*dhdps-r1*) gene from *Nicotiana sylvestris* was expressed in pigeonpea to enhance the lysine levels in seed (Thu et al. 2003, 2007).

14.7 Tools of Gene Transfer

Over the last three decades, *Agrobacterium tumefaciens* has been established as a versatile tool for plant transformation. Detailed knowledge from *A. tumefaciens*–plant cell interaction and T-DNA transfer and integration is being used for transformation of nearly every plant species of interest (Tzfira and Citovsky, 2006). Various types of *Agrobacterium* strains, such as LBA4404, EHA101, EHA105, AGL1 and C58, have been reported to infect a wide range of legume species. EHA101, EHA105 and AGL1 contain *vir* genes from the oncogenic strain A281 (Hellens et al. 2000). Surekha et al. (2007) focused on comparison of the transformation efficiency between two *Agrobacterium* strains, LBA 4404 and GV2260. Among all the strains used for pigeonpea transformation, EHA105 and AGL1 have been reported to have significantly better transformation efficiency (Satyavathi et al. 2003; Ramu et al. 2012; Ghosh et al. 2017; Ganguly et al. 2018). Details of the transformation efficiency achieved by using different strains of *A. tumefaciens* are mentioned in Table 14.1. It can be noted that no

TABLE 14.1
Progress in Pigeonpea Transformation

Cultivar	Explants used	*Agrobacterium* strain	Plasmid	Promoter	Transgene
Hyderabad C	SA and EA	LBA4404	pBI121	*CaMV35S*	*uid A*
Pusa 855	Precultured EA	GV2260	pCPI	*CaMV35S*	*CPI*
ICPL 88039	LE	Particle bombardment	pRT99GUS	*CaMV35S*	*uid A*
T 15-15	EA	LBA4404	pBIN19	*CaMV35S*	*uid A, gfp*
Hyderabad	EA and CN	EHA 105	pBI121	*CaMV35S*	*RVPH*
ICPL 87	CN	LBA 4404 and particle bombardment	pdhdps-GUS, pphas-dhdps-r1	*dhdps, Phas*	*uid A, dhdps-r1*
Hyderabad	CN	GV3101	pBI 121	*CaMV35S*	*PPRV-HN*
LRG 30	CN	C58	pCAMBIA1301	*CaMV35S*	*uid A*
LRG 30	CN	C58	pCAMBIA1302	*CaMV35S*	*gfp, Rchit*
ICPL 87	EA	GV2260	pBI121 pAD288	*CaMV35S*	*cry 1E-C*
UPAS 120, Bahar	CN, EA and LE	LBA4404	pBI121	–	*uid A, cry1Ab*
ICPL 87	Seedlings, auxillary bud region	C58	pHS723	*CaMV35S CaMV35SDE*	*uid A*
ICPL 87, ICPL85063, LRG30	Plumule, EA and CN	LBA4404,GV 2260	pBI121, pAD288	*CaMV35S*	*uid A, GS-TAPI*
ICPL 87	CN	LBA4404	pphas-dhdps-r1 p2S2-dhdps-r1	*Phas, 2S2*	*dhdps-r1*
TTB 7	CN, in planta	LBA4404	pKIWI1105	*CaMV35S*	*uid A*
cv. JKVL	EA	GV3101	pPZP211	*CaMV35S*	*cry1Ac*
TTB7	CN, in planta	EHA105	pBinBt8	*CaMV35S*	*cry1AcF*
Asha (ICPL87119), ICPL87	CN, in-vitro shoot grafting	EHA105	pBI121	*CaMV35S*	*uid A*
Asha (ICPL87119)	AME	EHA105	pBinAR	*CaMV35S*	*cry1Aabc*

Marker gene	Transformation efficiency (%)	Transgenic analysis	References
nptII	45.0–62.0	ND	Geetha et al. 1999
nptII	30.0–59.0	T_0 Northern	Lawrence and Koundal, 2001
nptII	50	T_0 PCR, Southern; T_1 RT PCR; T_1 segregation analysis	Dayal et al. 2003
nptII	–	T_0 Southern	Mohan and Krishnamurthy 2003
nptII	51.0–67.0	T_0 PCR, Southern, western, ELISA; T_1 segregation analysis	Satyavathi et al. 2003
nptII	6.5	T_0 PCR, Southern; T_1 segregation analysis	Thu et al. 2003
nptII	60.0–65.0	T_0 PCR, western; T_1 PCR, bioassay	Prasad et al. 2004
hpt	45	T_0 PCR, Southern, RT- PCR; T_1 Southern	Kumar et al. 2004a
hpt	2.8	T_0 PCR, Southern; T_1 segregation, RT-PCR	Kumar et al. 2004b
nptII	15	T_0 PCR	Surekha et al. 2005
nptII	0.20–0.33	NF	Verma and Chand 2005
nptII	60	T_2 PCR, Southern; T_2 segregation; T_3 ELISA and PCR	Sharma et al. 2006
nptII	50.0–80.0	T_0 PCR, Southern, western; T_1 PCR, Southern; $T_1$and T_2 detached leaf bioassay	Surekha et al. 2007
nptII	–	T_1 PCR, dhdps assay	Thu et al. 2007
nptII	13.7–60	T_1 and T_2 Southern	Rao et al. 2008
nptII	11.53–44.61	T_0 Dot blot, Southern, bioassay	Krishna et al. 2011
nptII	44	T_1 and T_2 Southern, nested PCR, RT-PCR, ELISA, western; T_3 PCR, bioassay	Ramu et al. 2012
nptII	8.75; 9.39	T_0 gus and PCR	Ghosh et al. 2014
nptII	0.06	T_0 and T_1 PCR, T_2 bioassay, T_3 and T_4 PCR, ELISA, Western, Southern	Das et al. 2016

(*continued*)

TABLE 14.1 (Continued)
Progress in Pigeonpea Transformation

Cultivar	Explants used	*Agrobacterium* strain	Plasmid	Promoter	Transgene
PAU 881	Cotyledonary embryonic axes; in planta	EHA105	pBin19	*CaMV35S* and *NOS*	*cry1Ac*
UPAS 120	EA	AGL-1	pBinAR	*CaMV35S*	*cry1Ac* and *cry2Aa*
ICPL 87119 (ASHA); ICPL 87	CN	AGL-1	pBI121, pCAMBIA3300	*CaMV35S*	*uid A*
ICPL87119 (ASHA)	Plumular meristem	EHA105	pBI121	*CaMV35S*	*uid A*
PUSA 992	In planta	EHA105	pBinAR	*CaMV35S*	*cry2Aa*
Asha (ICPL87119)	Lateral branching	EHA105	pCAMBIA1300	*CaMV35S, rbcs*	*loxP-syn bar-loxP;* synthetic *cry1ab; cre recombinase*

conclusions can be drawn regarding which *Agrobacterium* strain is most efficient in transforming pigeonpea, and that careful comparison of different strains is advisable for separate cultivars of pigeonpea.

Physical injuries to explants before *Agrobacterium* infection have always been recommended in all transformation protocols. Wounding provides an entry point for *Agrobacterium* and also activates the secretion of phenolic substances necessary for *Agrobacterium vir* gene induction (Zupan et al. 2000). Normally, plant tissues are injured by scalpels, but additional wounding has been inflicted by the use of hypodermic needles in the cotyledonary nodes and decapitated plumular regions (Rao et al. 2008; Ramu et al. 2012; Ganguly et al. 2018). In various publications concerning pigeonpea transformation, acetosyringone has been used as a supplement for inducing *vir* gene to enhance the gene transfer process. It is added to the bacterial re-suspension medium as well as in co-cultivation medium to boost transformation efficiency, but transgenic pigeon pea has also been obtained without using acetosyringone (Kumar

Marker gene	Transformation efficiency (%)	Transgenic analysis	References
nptII	NF	T_1 and T_2 PCR, T_2, RT-PCR, bioassay; T_3 segregation	Kaur et al. 2016
nptII	1.8 for *cry1Ac*; 1.3 for *cry2Aa*	T_0, T_1, T_2 and T_3 PCR, T_0 Strip assay, T_0, T_1 and T_2 western, T_1 segregation, T_1 and T_2 Southern, ELISA, bioassay T_3; Immunohistofluorescence localization	Ghosh et al. 2017
nptII, bar	NF	T_0 and T_1 PCR	Ganguly et al. 2018
nptII	41–72	T_0 gus and PCR, T_1, Southern	Ganguly et al. 2018
nptII	0.8	T_1 and T_2 selection, PCR, T_1-T_3 in vitro bioassay, T_2 Z distribution analysis, T_3 PCR, Dot blot, Southern, qRT-PCR, Western, Pod bioassay	Singh et al. 2018
bar; *hpt*	2–5	T_1 dot blot, PCR, Western, Southern, bioassay	Sarkar et al. 2021

2S2: *Arabidopsis thaliana 2S2* promoter; AME: axillary meristem explants; *bar*: bialaphos gene; *cry1Ab*: gene for crystal protein 1Ab; *CaMV35S*: cauliflower mosaic virus 35S promoter; *CaMV35SDE:* cauliflower mosaic virus 35S double-enhanced promoter; CN: cotyledonary node; *CPI*: *cowpea protease inhibitor*; *cre*: *cry1Ab*: gene for crystal protein 1Ab; *cry 1Ac*: gene for crystal protein 1Ac; *cry1E-C*: gene for crystal protein 1E-C; *cry1AcF*: fusion of the N-terminal along with domain II from *cry1Ac* and the C-terminal domain from *cry1F*; *dhdps*: *dihydrodipicolinate synthase* promoter; EA: embryonic axis; *gfp*: *green fluorescent protein*; *GS-TAP1*: *glutamine synthetase* and tobacco *anodic peroxidase*; *hpt*: *hygromycin phosphotransferase*; LE: leaf disc explants; ND: not done; NF: not found; *nptII*: *neomycin phosphotransferase II*; *phas*: *phaseolin* promoter; *PPRV-HN*: *hemagglutinin neuraminidase* gene of peste des petits ruminants virus; *rbcs*: *rubisco small subunit promoter*; *RChit*: rice *chitinase* gene; *RVPH*: *hemagglutinin* gene of rinderpest virus; SA: shoot apices; *uid A*: *β-glucuronidase* gene.

et al. 2004a; Surekha et al. 2005). So, the efficacy of acetosyringone as an important additive is debatable in the absence of comparative studies.

The choice of specific promoters has been an integral part of genetic transformation programs. It regulates the activity of foreign genes in host plant tissue and also influences the expression level in a spatio-temporal manner (Kummari et al. 2020). The promoter drives the expression of a gene and may be a key factor in determining the success of a particular transformation experiment. Mainly constitutive promoters have been used for pigeonpea transformation. Constitutive promoters direct the expression of genes in almost all tissues and are independent of any environmental or developmental conditions. The majority of studies conducted in pigeonpea thus far have largely relied on the *Cauliflower mosaic virus 35S* (*CaMV35S*) constitutive promoter, as is evident from Table 14.1. Beside this, reports are also available on the usage of tissue-specific promoters like flower and leaf specific double enhanced *CaMV35S* promoter (*CaMV35S*DE) and seed-specific *phaseolin* and *Arabidopsis thaliana 2S2 albumin* promoters to drive the tissue-specific transgene expression in pigeonpea (Sharma et al. 2006; Thu et al. 2007). With the usage of *CaMV35SDE*, expression levels of the insecticidal cry1Ab were 0.10% and 0.025% of total soluble protein in flowers and leaves, respectively (Sharma et al. 2006), whereas the expression level of *hemagglutinin* gene of rinderpest virus (RVPH) under *CaMV35S* was relatively higher (0.12–0.49%) (Satyavathi et al. 2003). Seed-specific promoters like bean *phaseoline* and *2S2* showed up to 400–600-fold expression of dhdps-r1 at late stages of seed development in comparison to its wild-type counterparts (Thu et al. 2007).

14.8 Selection of Transformants

Apart from the gene transfer techniques used, the number of transformants that stably integrate and express the alien gene is very limited. A robust selection method is imperative in distinguishing these transformants among the large number of non-transformants. Selectable marker genes (SMGs) are mostly based on a negative selection method, which encodes proteins to confer resistance against a selection agent that is lethal for the non-transgenic tissues (Goodwin et al. 2005). Conventionally, antibiotic and herbicide resistance genes have been extensively used in transgenic research (Miki and McHugh, 2004). The majority of the selection methods deployed in pigeonpea transformation were conducted with kanamycin, while a few reports exist of hygromycin (Kumar et al. 2004a and b; Krishna et al. 2010). A wide range of kanamycin concentration (25.0–125.0 mgl^{-1}) have been utilized for making effective selections of transgenic shoots (Sharma et al. 2006; Lawrence and Koundal, 2001; Thu et al. 2003; Satyavathi et al. 2003; Prasad et al. 2004; Geetha et al. 1999). Root initiation from regenerated shoots has been reported to be inhibited on root-inducing medium containing kanamycin at high selection pressure (>50.0 mgl^{-1}), but it was successful at low selection pressure (<50.0 mgl^{-1}) (Sharma et al. 2006; Lawrence and Koundal, 2001; Thu et al. 2003; Satyavathi et al. 2003; Prasad et al. 2004; Geetha et al. 1999). In the case of hygromycin, a concentration as low as 5.0 mgl^{-1} was used for the selection of shoots generated from co-cultivated explants (Kumar et al. 2004b).

A comparative analysis of the minimum inhibitory concentration to determine the dosage of three important selection agents, kanamycin, hygromycin and glufosinate, was conducted for the first time by Ganguly et al. (2018). They concluded that kanamycin has been the most popular choice among legume researchers, while hygromycin has been proven to be too lethal even at very low doses. A technique of leaf painting with glufosinate ammonium was reported for the first time to select primary progenies of pigeonpea transformants (Ganguly et al. 2018). Production of chimeric transgenics and escapes of non-transformants have always been a serious problem in legume transformation (Krishna et al. 2010). But antibiotics equally impact the quality of root formation even in stably transformed pigeonpeas. To bypass the effect of antibiotics, the tissue culture-independent in planta method resorted to screening some of the primary transformants with *uid A* gene or gus gene with considerable success, but also faced the problems of chimeric T_0 plants (Rao et al. 2008). This technique was later applied by other groups to develop transgenic pigeonpea (Ramu et al. 2012; Kaur et al. 2016). However, to avoid the chimeric T_0 condition, the plumular meristem pricking is a considerably better technique than the in planta method, because in the latter method, too many primary transformants are obtained, and it becomes strenuous to screen them in the T_1 stage by polymerase chain reaction (PCR). Although SMGs are fundamental to all plant transformation protocols, they are of little use once the transgenic plants have been released. On the contrary, their retention has no practical utility, raising a major concern for consumer acceptance (Jaiwal et al. 2002). Hence, it is desirable to remove the marker genes. Over the last few decades, several approaches have been utilized for removing the SMGs from transgenic crops (Yau and Stewart, 2013). There have been very few reports on research in marker-free legumes (Li et al. 2007) until recently, when a marker gene-free transgenic pigeonpea plant expressing *cry1Ab* gene has been reported through cre-lox-mediated recombination (Sarkar et al. 2021).

14.9 Transgenic Analysis

Transgenic research should not be limited to the confines of the laboratory. It must have an ample scope of application in addressing the concerns regarding crop yield and food security. Initiatives for establishing transgenic crop species should be attempted, keeping in mind a longstanding goal rather than just the analysis of initial transformants. In the case of pigeonpea, initial reports of transgenic research had no prolonged repercussions. Initially, pigeonpea transformation with *uid A* gene was reported to have 45–62% transformation efficiency, but this not backed by any molecular analysis (Geetha et al. 1999). Dayal and co-workers (2003) performed PCR and segregation studies on the T_1 transformants. Green fluorescent protein was first used in pigeonpea for probing transformed cells (Mohan and Krishnamurthy, 2003). Gus analysis has been a primary choice of reporter gene in most research groups (Surekha et al. 2005; Sharma et al. 2006; Thu et al. 2003; Mohan and Krishnamurthy, 2003; Ganguly et al. 2018). Various molecular screenings were conducted by most of the researchers in the T_0 and T_1 stages to confirm the integration pattern and copy number of the transgene (Krishna et al. 2010). The first stable transgene integration in the T_3 generation was reported by Sharma and co-workers (2006). Rao et al. (2008) conducted Southern hybridization of transgenic plants in the T_1 and T_2 generations, as the T_0 transformants were chimeric due to the in planta technique of transformation.

Amidst several reports of recovery of transgenic pigeonpea with trait genes, Surekha et al. (2005), Sharma et al. (2006) and Ramu et al. (2012) reported the expression and inheritance of the fused *Bt* proteins Cry1 E-C, Cry1Ab and Cry1AcF beyond the T_0 generation. Krishna et al. (2011) reported an expression study of Cry1Ac protein through enzyme-linked immunosorbent assay (ELISA) analysis from different parts of established transgenic plants in the T_0 generation. A tissue culture-independent transformation study of pigeonpea demonstrated 3–15 $\mu g g^{-1}$ fresh weight of Cry1AcF protein accumulation until the T_2 generation (Ramu et al. 2012). Transgenic pigeonpea plants accumulated Cry2Aa protein in the range of 25–80 $\mu g g^{-1}$ fresh weight (Singh et al. 2018). Levels of Cry1Aabc were expressed up to15.82–44.89 ng mg^{-1} of total soluble protein (Das et al. 2016). The highest levels of expression of Cry1Ac and Cry2Aa proteins in T_1 progeny lines were reported to be 0.172% and 0.170% of total soluble protein, respectively (Ghosh et al. 2017). Immunohistofluorescence analysis on T_3 transgenic events has also detected the constitutive nature of protein expression in different tissue layers (Ghosh et al. 2017).

Bioassay is an important parameter to check mortality levels of the insect in which resistance is being developed. Various groups have developed transgenic pigeonpea resistant against various insects. Constitutive expression of Cry1Ac protein conferred insect resistance and only 55% mortality was recorded in T_0 transgenic lines (Krishna et al. 2011). Ramu and his co-workers performed insect bioassay in T_1 and the successive two generations of pigeonpea, where Cry1AcF protein showed 80–100% mortality of *H. armigera* (Ramu et al. 2012). Transgenic T_1 and T_2 generations of pigeonpea expressing Cry1Ac and Cry2Aa proteins showed a high level of mortality against *H. armigera* (80–100%) (Ghosh et al. 2017). In another study, Cry2Aa protein expressed in the T_1 to T_3 generations conferred around 80–100% mortality of pod borer larvae (Singh et al. 2018). Other reports of transgenic pigeonpea expressing Cry1Ac and Cry1Aabc proteins have presented 90–100% mortality against pod borer larvae (Kaur et al. 2016; Das et al. 2016). Transgenic pigeonpea expressing Cry1 E-C proteins has been found to express 60–80% mortality against first and second instar larvae of *Spodoptera litura* (Surekha et al. 2005).

Besides insecticidal bioassay, accumulation of lysine in transgenic pigeonpea seeds has also been assessed. Activity of the lysine degradative enzyme lysine ketoglutarate reductase (LKR) was probed to determine lysine accumulation in transgenic pigeonpeas (Thu et al. 2007). It has been observed that, during late stages of seed development, LKR had enhanced activity in some of the transgenic lines, indicating an increase in free lysine and free threonine. The biological activity of *PPRV-HN* gene of peste des petits ruminants virus has also been studied for the possible development of an edible vaccine for consumption by cattle (Prasad et al. 2004). Neuraminidase activity was seen in extracts from transgenic plants when fetuin had been used as substrate.

14.10 Future Prospects

The future of pigeonpea genetic engineering looks promising, as various new protocols of transformation are being reported to enhance the transformational efficiency. The in planta method of tissue culture-independent transformation has opened up newer avenues, not only in pigeonpea research but also in other leguminous crops. Similarly, the plumular meristem is another option for optimizing transformation efficiency. The efficacy of *CaMV35S* and other tissue-specific promoters should be explored on a comparative basis. Probing for pigeonpea-specific endogenous promoters to enforce high levels of constitutive transgene expression will become imperative in the near future. Due to its importance as an edible pulse, endeavors should be undertaken to inhibit expression of transgene in the seed tissues. This will enhance the chances of commercialization and acceptability of genetically modified food. Thus, research on seed-specific promoters of pigeonpea might become an important field in times to come. One of the major issues of transgenic research to develop tolerance against insects is the development of insect resistance against the genetically modified crop. To tackle the issue and delay the evolution of toxin resistance in insects, stacking multiple insecticidal genes can be explored, which has already been proved to be successful in soybean. With transgenic research come biosafety concerns. To prevent the antibiotic resistance gene from escaping into nature, marker-free, clean gene techniques have recently been introduced in pigeonpea. This will open further avenues in generating marker-free pigeonpeas possessing important agronomic qualities. Finally, due to their high protein content, transgenic pigeonpeas can be alluring tools for innovative applications in the arena of molecular farming, such as manufacturing vaccines or antibodies.

REFERENCES

Abhilash, P.C., Singh, N., 2009. Pesticide use and application: An Indian scenario. *J. Hazard. Mater.* 165(1–3),1–2.

Cheema, H.K., Bawa, J., 1991. Clonal multiplication via multiple shoots in some legumes (*Vigna unguiculata* and *Cajanus cajan*). *Acta. Hortic.* 289,93–94.

Christou, P., 1997. Biotechnology applied to grain legumes. *Field Crops Res.* 53(1–3),83–97.

Das, A., Datta, S., Sujayanand, G.K., Kumar, M., Singh, A.K., Shukla, A., Ansari, J., Faruqui, L., Thakur, S., Kumar, P.A., Singh, N.P., 2016. Expression of chimeric Bt gene, *Cry1Aabc* in transgenic pigeonpea (cv. Asha) confers resistance to gram pod borer (*Helicoverpa armigera* Hubner.). *Plant Cell Tiss. Org. Cult.* 127(3),705–715.

Dayal, S., Lavanya, M., Devi, P., Sharma, K.K., 2003. An efficient protocol for shoot regeneration and genetic transformation of pigeonpea (*Cajanus cajan* (L.) Millsp.) by using leaf explants. *Plant Cell. Rep.* 21, 1072–1079.

Eapen, S., George, L., 1993. Plant regeneration from leaf discs of peanut and pigeonpea: Influence of benzyladenine, indole acetic acid and indoleacetic acid-amino acid conjugates. *Plant Cell Tissue Organ. Cult.* 35,223–227.

Eapen, S., Tivarekar, S., George, L., 1998. Thidiazuron-induced shoot regeneration in pigeonpea (*Cajanus cajan* L.). *Plant Cell Tissue Organ Cult.* 53,217–220.

FAO, 2019. Food and Agricultural Organization of the United Nation, FAO Statistical Database. http://faostat.fao.org. Accessed 29 September 2021.

Franklin, G., Jeyachandran, R., Ignacimuthu, S., 2000. Factors affecting regeneration of pigeon pea (*Cajanus cajan* L. Millsp) from mature embryonal axes. *Plant Growth Regul.* 30,31–36.

Fróna, D., Szenderák, J., Harangi-Rákos, M., 2019. The challenge of feeding the world. *Sustainability.* 11(20),5816.

Ganguly, S., Ghosh, G., Purohit, A. et al., 2018. Effective screening of transgenic pigeonpea in presence of negative selection agents. *Proc. Natl. Acad. Sci., India, Sect. B Biol. Sci.* 88,1565–1571. https://doi.org/10.1007/s40011-017-0895-3

Geetha, N., Venkatachalam, P., Prakash, V., Lakshmi Sita, G., 1998. High frequency induction of multiple shoots and plant regeneration from seedling explants of pigeon pea (*Cajanus cajan* L.). *Curr. Sci.* 75,1036–1041.

Geetha. N., Venkatachalam, P., Lakshmi Sita, G., 1999. *Agrobacterium* mediated genetic transformation of pigeonpea (*Cajanus cajan* [L.]) and development of transgenic plants via direct organogenesis. *Plant Biotechnol.* 16,213–218.

George, L., Eapen, S., 1994. Organogenesis and embryogenesis from diverse explants in pigeon pea (*Cajanus cajan* L.). *Plant Cell Rep.* 13,417–420.

Ghosh, G., Purohit, A, Ganguly, S., Chaudhuri, R.K., Chakraborti, D., 2014. In vitro shoot grafting on rootstock: An effective tool for *Agrobacterium*-mediated transformation of pigeonpea (*Cajanus cajan* (L.) Millsp.). *Plant Biotechnol.* 31,301–308.

Ghosh, G., Ganguly, S., Purohit, A., Chaudhuri, R.K., Das, S., Chakraborti, D., 2017. Transgenic pigeonpea events expressing Cry1Ac and Cry2Aa exhibit resistance to *Helicoverpa armigera*. *Plant Cell Rep.* 36(7),1037–1051.

Goodwin, J.L., Pastori, G.M., Davey, M.R., Jones, H.D., 2005. Selectable markers: Antibiotic and herbicide resistance. In: Pena, L. (ed) *Methods in Molecular Biology, Vol 286. Transgenic Plants: Methods and Protocols*. Humana Press Inc., Totowa, pp. 191–201.

Hellens, R., Mullineaux, P., Klee, H., 2000. A guide to *Agrobacterium* binary Ti vectors. *Trends in Plant Science* 5,446–451.

ISAAA. 2018. *Global Status of Commercialized Biotech/GM Crops: 2018. ISAAA Brief No. 54*. ISAAA, Ithaca, NY.

Jaiwal, P.K., Sahoo, L., Singh, N.D., Singh, R.P., 2002. Strategies to deal with the concern about marker genes in transgenic plants: Some environment-friendly approaches. *Curr. Sci.* 83, 128–136.

Kaur, A., Sharma, M., Sharma, C., Kaur, H., Kaur, N., Sharma, S., Arora, R., Singh, I., Sandhu, J.S., 2016. Pod borer resistant transgenic pigeon pea (*Cajanus cajan* L.) expressing *cry1Ac* transgene generated through simplified *Agrobacterium* transformation of pricked embryo axes. *Plant Cell Tiss. Org. Cult.* 127(3),717–727.

Krishna, G., Reddy, P.S., Ramteke, P.W., Bhattacharya, P.S., 2010. Progress of tissue culture and genetic transformation research in pigeonpea pea [*Cajanus cajan*, L. Millsp.]. *Plant Cell Rep.* 29,1079–1095.

Krishna, G., Reddy, P.S., Ramteke, P.W., Rambabu, P., Tawar, K.B., Bhattacharya, P., 2011 *Agrobacterium*-mediated genetic transformation of pigeon pea [*Cajanus cajan* (L.) Millsp.] for resistance to legume pod borer *Helicoverpa armigera*. *J. Crop Sci. Biotechnol.* 14(3),197–204.

Kumar, A.S., Reddy, T.P., Reddy, G.M., 1983. Plantlet regeneration from different callus cultures of pigeon pea (*Cajanus cajan* L.). *Plant Sci. Lett.* 32,271–278.

Kumar, A.S., Reddy, T.P., Reddy, G.M., 1984. Multiple shoots from cultured explants of pigeon pea and *Atylasia* species. *SABRAO J.* 16,101–105.

Kumar, S.M., Kumar, B.K., Sharma, K.K., Devi, P., 2004a. Genetic transformation of pigeon pea with rice chitinase gene. *Plant Breed.* 123,485–489.

Kumar, S.M., Syamala, D., Sharma, K.K., Devi, P., 2004b. *Agrobacterium tumefaciens* mediated genetic transformation of pigeon pea [*Cajanus cajan* (L.) Millsp.]. *J. Plant Biotechnol.* 6,69–75.

Kumar, P.L., Latha, T.K.S., Kulkarni, N.K., Raghavendra, N., Saxena, K.B., Waliyar, F., Rangaswamy, K.T., Muniyappa, V., Doriswamy, S., Jones, A.T., 2005. Broad-based resistance to pigeonpea sterility mosaic disease in wild relatives of pigeonpea (*Cajanus*: Phaseoleae). *Ann. Appl. Biol.* 146(3),371–379.

Kummari, D., Palakolanu, S.R., Kishor, P.K., Bhatnagar-Mathur, P., Singam, P., Vadez, V., Sharma, K.K., 2020. An update and perspectives on the use of promoters in plant genetic engineering. *J. Biosci.* 45(1),1–24.

Lawrence, P.K., Koundal, K.R., 2001. *Agrobacterium tumefaciens* mediated transformation of pigeon pea (*Cajanus cajan* L. Millsp.) and molecular analysis of regenerated plants. *Curr. Sci.* 80,1428–1432.

Li, Z., Xing, A., Moon, B.P., Burgoyne, S.A., Guida, A.D., Liang, H. et al. 2007. A Cre/loxP-mediated self-activating gene excision system to produce marker gene free transgenic soybean plants. *Plant Molecular Biology* 65,329–341.

Mehta, U., Mohan Ram, H.Y., 1980. Regeneration of plantlets from the cotyledons of *Cajanus cajan* L. *Ind. J. Exp. Biol.* 18,800–802.

Miki, B., McHugh, S., 2004. Selectable marker genes in transgenic plants: Applications, alternatives and biosafety. *J. Biotechnol.* 107,193–232.

Mohan, M.L., Krishnamurthy, K.V., 2003. Plant regeneration from decapitated mature embryonic axis and *Agrobacterium*-mediated genetic transformation of pigeonpea. *Biol. Plant* 49,519–527.

Naidu, R.B., Kulkarni, D.D., Krishnamurthy, K.V., 1995. Genotype dependent morphogenic potentiality of various explants of a food legume, the pigeon pea (*Cajanus cajan* L.). *In Vitro Cell Dev. Biol. Plant* 31,26–30.

Parankusam, S., Katamreddy, S., Bommineni, P.R., Bhatnagar-Mathur, P., Sharma, K.K., 2018. Insights into insect resistance in pulse crops: Problems and preventions. In: *Pulse Improvement*, Springer, Cham, pp. 137–173.

Prasad, V., Satyavathi, V.V., Sanjaya, Valli, K.M., Khandelwal, A., Shaila, M.S., Lakshmi Sita, G., 2004. Expression of biologically active hemagglutinin-neuraminidase protein of Peste des petits ruminants virus in transgenic pigeonpea [*Cajanus cajan*, L., Millsp.]. *Plant Sci.* 166,199–205.

Ramu, S.V., Rohini, S., Keshavareddy, G., Neelima, M.G., Shanmugam, N.B., Kumar, A.R.V., Sarangi, S.K., Ananda Kumar, P., Udayakumar, M., 2012. Expression of a synthetic *cry1AcF* gene in transgenic Pigeonpea confers resistance to *Helicoverpa armigera*. *J. Appl. Entomol.* 136,675–687.

Rao, K.S., Sreevathsa, R., Sharma, P.D., Keshamma, E., Udaya, K.M., 2008. In planta transformation of pigeon pea: A method to overcome recalcitrancy of the crop to regeneration in vitro. *Physiol. Mol. Biol. Plant* 14,321–328.

Ruby, M.B., 2012. Vegetarianism: A blossoming field of study. *Appetite.* 58,141–150. doi: 10.1016/j.appet.2011.09.019.

Sanahuja, G., Banakar, R., Twyman, R.M., Capell, T., Christou, P., 2011. *Bacillus thuringiensis*: A century of research, development and commercial applications. *Plant Biotechnology Journal* 9,283–300.

Sarkar, S., Roy, S., Ghosh, S.K., 2021. Development of marker-free transgenic pigeon pea (*Cajanus cajan*) expressing a pod borer insecticidal protein. *Sci. Rep.* 11,10543. https://doi.org/10.1038/s41598-021-90050-8

Satyavathi, V.V., Prasad, V., Khandelwal, A., Shaila, M.S., Lakshmi Sita, G., 2003. Expression of hemagglutinin protein of Rinder pest virus in transgenic pigeon pea (*Cajanus cajan* (L.) Millsp.) plants. *Plant Cell Rep.* 21,651–658.

Shanower, T.G., Romeis, J., Minja, E.M., 1999. Insect pests of pigeon pea and their management. *Ann. Rev. Entomol.* 44,77–96.

Sharma, K.K., Lavanya, K., Anjaiah, A., 2006. *Agrobacterium tumefaciens*-mediated production of transgenic pigeonpea (*Cajanus cajan* [L.] Millsp.) expressing the synthetic bt *cry1ab* gene. *In Vitro Cell Dev. Biol-Plant.* 42,165–173.

Shiferaw, B., Okello, J., Muricho, G., Omiti, J., Silim, S.N., Jones, R., 2008. Unlocking the potential of high value legumes in the semi-arid regions: Analyses of the pigeon pea value chains in Kenya. International Crops Research Institute for the Semi Arid Tropics.

Shiva Prakash, N., Pental, D., Bhalla Sarin, N., 1994. Regeneration of pigeonpea (*Cajanus cajan*) from cotyledonary node via multiple shoot formation. *Plant Cell Rep.* 13,623–627.

Singh, N.D., Sahoo, L., Sonia, Jaiwal, P.K., 2002. In vitro shoot organogenesis and plant regeneration from cotyledonary node and leaf explants of pigeonpea (*Cajanus cajan* L. Millsp.). *Physiol. Mol. Biol. Plant* 8,133–140.

Singh, S., Kumar, N.R., Maniraj, R., Lakshmikanth, R., Rao, K.Y., Muralimohan, N., Arulprakash, T., Karthik, K., Shashibhushan, N.B., Vinutha, T., Pattanayak, D., 2018. Expression of Cry2Aa, a *Bacillus thuringiensis* insecticidal protein in transgenic pigeon pea confers resistance to gram pod borer, *Helicoverpa armigera. Sci. Rep.* 8(1),1–2.

Surekha, Ch., Beena, M.R., Arundhati, A., Singh, P.K., Tuli, R., Dutta-Gupta, A., Kirti, P.B., 2005. *Agrobacterium*-mediated genetic transformation of pigeon pea (*Cajanus cajan* (L.) Millsp.) using embryonal segments and development of transgenic plants for resistance against Spodoptera. *Plant Sci.* 169(6),1074–1080.

Surekha, Ch., Arundhati, A., Rao, G.S., 2007. Differential response of *Cajanus cajan* varieties to transformation with different strains of *Agrobacterium. J. Biol. Sci.* 7(1),176–181.

Thu, T.T., Mai, T.T.X., Dewaele, E., Farsi, S., Tadesse, Y., Angenon, G., Jacobs, M., 2003. In vitro regeneration and transformation of pigeonpea (*Cajanus cajan* (L.) Millsp.). *Mol. Breed.* 11,159–168.

Thu, T.T., Dewaele, E., Trung, L.Q., Claeys, M., Jacobs, M., Angenon, G., 2007. Increasing lysine levels in pigeonpea (*Cajanus cajan* (L.) Millsp.) seeds through genetic engineering. *Plant Cell Tiss. Org. Cult.* 91,35–143.

Tzfira, T., Citovsky, V., 2006. *Agrobacterium*-mediated genetic transformation of plants: Biology and biotechnology. *Curr. Opin. Biotechnol.* 17,147–154.

Varshney, R.K., Chen, W., Li, Y., Bharti, A.K., Saxena, R.K., Schlueter, J.A., et al., 2012. Draft genome sequence of pigeonpea (*Cajanus cajan*), an orphan legume crop of resource-poor farmers. *Nat. Biotechnol.* 30,83–89.

Verma, A.K., Chand, L., 2005. *Agrobacterium*-mediated transformation of pigeon pea (*Cajanus cajan* L.) with *uidA* and *CryIA(b)* genes. *Physiol. Mol. Biol. Plants* 11,99–109.

Villiers, S.D., Emongor, Q., Njeri, R., Gwata, E., Hoisington, D., Njagi, I., Silim, S., Sharma, K., 2008. Evaluation of the shoot regeneration response in tissue culture of pigeon pea (*Cajanus cajan* (L.) Millsp.) varieties adapted to eastern and southern Africa. *Afr. J. Biotechnol.* 7,587–559.

Yau, Y.Y., Stewart, C.N., 2013. Less is more: Strategies to remove marker genes from transgenic plants. *BMC Biotechnol.* 13(1),36.

Zupan, J., Muth, T.R., Draper, O., Zambryski, P., 2000. The transfer of DNA from *Agrobacterium tumefaciens* into plants: A feast of fundamental insights. *The Plant Journal* 23,11–28.